CMP

Acknowledgement is made to C. Haack, P. Gravert and V. Schlegel for the use of Figure 7 on p.101, which appears on the front cover of this book.

Computational Modelling of Free and Moving Boundary Problems

Vol. 1 Fluid Flow

Proceedings of the First International Conference, held 2-4 July 1991, Southampton, U.K.

Editors: L.C. Wrobel
C.A. Brebbia

Computational Mechanics Publications
Southampton Boston

Co-published with

Walter de Gruyter
Berlin New York

L.C. Wrobel
Wessex Institute of Technology
Computational Mechanics Institute
Ashurst Lodge, Ashurst
Southampton SO4 2AA

C.A. Brebbia
Wessex Institute of Technology
Computational Mechanics Institute
Ashurst Lodge, Ashurst
Southampton S04 2AA

A Catalogue record for this book is available from the British Library

ISBN 1-85312-157-6 Computational Mechanics Publications, Southampton
ISBN 1-56252-084-9 Computational Mechanics Publications, Boston
ISBN 3-11-013172-2 de Gruyter, Berlin
ISBN 0-89925-908-1 de Gruyter, New York

set

ISBN 1-85312-142-8 Computational Mechanics Publications, Southampton
ISBN 1-56252-070-9 Computational Mechanics Publications, Boston
ISBN 3-11-013174-9 de Gruyter, Berlin
ISBN 0-89925-910-3 de Gruyter, New York

Library of Congress Catalog Card Number 91 72243

Die Deutsche Bibliothek - Cataloging-in-Publication Data

Computational modelling of free and moving boundary problems:
proceedings of the ... international conference ... -Southampton ; Boston :
Computational Mechanics Publ. ; Berlin ; New York : de Gruyter

1. Held 2-4 July 1991, Southampton, U.K.
   Vol.1. Fluid flow.-1991
   ISBN 3-11-013172-2 (de Gruyter, Berlin)
   ISBN 3-11-013174-9 (de Gruyter, Berlin, 1, Vol. 1 und 2)
   ISBN 0-89925-908-1 (de Gruyter, New York)
   ISBN 0-89925-910-3 (de Gruyter, New York, 1, Vol. 1 und 2)
   ISBN 1-85312-157-6 (Computational Mechanics Publ., Southampton)
   ISBN 1-56252-084-9 (Computational Mechanics Publ., Boston)
   ISBN 1-85312-142-8 (Computational Mechanics Publ., Southampton, 1, Vol. 1 und 2)
   ISBN 1-56252-070-9 (Computational Mechanics Publ., Boston, 1, Vol. 1 und 2)

Printed in Great Britain by The Alden Press, Ltd, Oxford

PREFACE

The classification "free boundary problem" or "moving boundary problem" can be applied to a large variety of physical situations. The common feature is the presence, in their mathematical modelling, of an initially unknown (free) boundary or a boundary which moves throughout the analysis, the determination of which is an important part of the solution procedure. Practical examples are ingot solidification and scrap melting in metallurgy, the re-entry problem of hypersonic missiles in aeronautical sciences, the cooling, freezing and cold storage of foodstuffs in the industry, to name but a few. Although completely different in nature, these problems have several common features regarding their mathematical modelling and computer methods of solution.

The present book and its companion volume contain edited versions of the papers presented at the First International Conference on Computational Modelling of Free and Moving Boundary Problems, held in Southampton, England, in July 1991. The purpose of the conference was to promote the interaction between engineers, applied mathematicians and numerical analysts involved in the creation, development and application of computational methods to free and moving boundary problems. In particular, the conference intended to provide a useful connection between scientists working in different areas of application but using similar numerical techniques.

Although the papers have been typographically edited they have been reproduced directly from material submitted by the authors, and their content is a reflection of the authors' research and opinion.

The editors are grateful for the assistance of the staff of Computational Mechanics Publications in the preparation of the volumes.

Southampton, July 1991
L.C. Wrobel and C.A. Brebbia

CONTENTS

SECTION 3: CAVITATIONAL FLOW

SECTION 4: FREE SURFACE FLOW

FIRST INTERNATIONAL CONFERENCE
ON
COMPUTATIONAL MODELLING OF FREE AND MOVING BOUNDARY PROBLEMS 91

SCIENTIFIC COMMITTEE

SECTION 1: FLOW THROUGH POROUS MEDIA

Multisplitting and Domain Decomposition Techniques Applied to Free Surface Flow Through Porous Media

J.C. Bruch, Jr.

Department of Mechanical and Environmental Engineering, University of California, Santa Barbara, CA 93106, USA

ABSTRACT

Free surface flow in porous media is a common occurrence in nature. Such problems are difficult to solve numerically since the problem boundary locations are not all known a priori. A solution approach which can capably handle this difficulty is one developed by Baiocchi in the early 1970's which uses an integral transformation in a fixed domain method. This method will be applied to the steady, two-dimensional seepage through a variable shaped dam with a toe drain. In the numerical solutions given for the fixed domain formulated problem several efficient multisplitting and domain decomposition techniques are used and comparison numerical results given. These techniques are not only very useful from the theoretical point of view but also from the numerical, especially in regards to their use in conjunction with parallel computation.

INTRODUCTION

Since the early 70's the Baiocchi method and transformation have been used on a large number of seepage problems. (See Baiocchi and Capelo [1], Kinderlehrer and Stampacchia [2], Oden and Kikuchi [3], Bruch [4] and Crank [5]). In this approach the a priori unknown solution region is extended across the free surface into a known region. The dependent variable is also continuously, similarly extended. Then a new dependent variable is defined using Baiocchi's transformation within these regions. The resulting problem formulation leads to a "complementarity system" associated with its respective variational or quasi-variational inequality formulation. This method has proven effective not only from the purely theoretical point of view, but also from the point of view of yielding new, simple, and efficient numerical solution schemes. The numerical schemes are the focus of this paper.

The physical problem on which the method and numerical schemes will be applied is the steady, two-dimensional seepage through a variable shaped dam with a toe drain. Initially as a reference the Baiocchi method and transformation will be applied to the problem along with the usual numerical solution approach over the extended solution domain [6]. Next a multisplitting technique will be applied to the extended solution domain problem with the numerical solution being obtained in each subdomain after application of the numerical schemes and interchange of information across the subdomain overlap [7]. The next two, and more recent, numerical approaches will be domain decomposition schemes in which Dirichlet data and Neumann data are interchanged along the boundary separating the domains [8].

These techniques are not only very useful from the theoretical point of view but also from the numerical, especially in regards to their use in conjunction with parallel computation. The multisplitting technique has been used to solve free surface seepage problems by White [9] and Wang and Bruch [10]. A domain decomposition approach different from the ones used herein was applied by White [9] using a parallel computer to the problem of free surface seepage through a two-dimensional rectangular dam with reservoirs on both sides.

DIFFERENTIAL EQUATION FORMULATION

Considered here is the steady, two-dimensional seepage through a variable shaped dam, i.e., $y = \alpha(x)$, with a toe drain (see Fig. 1). In Baiocchi [11] the problem of a dam with arbitrary shape on an impermeable, horizontal base separating two reservoirs is formulated using quasi-variational inequalities and efficient and rigorous algorithms are given for obtaining numerical results. In order to simplify the problem, several common assumptions are made. First, the soil in the flowfield is taken to be homogeneous and isotropic. Second, capillary and evaporation effects are neglected. Third, the flow follows Darcy's Law.

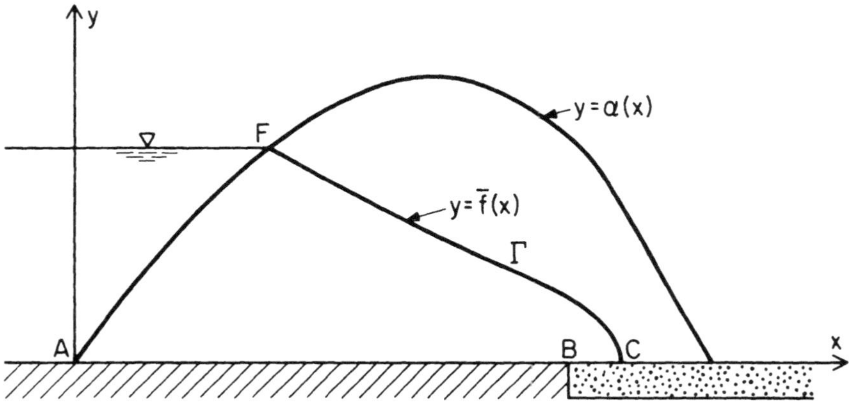

Figure 1. Flow problem.

Since a free surface problem is being considered, the free surface location is not known a priori. However, on the free surface two boundary conditions hold: (1) $\phi = ky$; and (2) ψ = constant; or equivalently for (2) $\phi_\eta = 0$, where η is the outward normal direction, ϕ is the velocity potential, and ψ is the stream function. On the remainder of the boundaries, either Neumann or Dirichlet data are given. Let $\Gamma = \{(x,y) \mid x_F < x < x_C, y = \bar{f}(x)\}$ be the free surface which is the boundary line between wet and dry soil. It should be noted that x's and y's having letter subscripts on them designate the x-distance and y-distance to that lettered point, respectively. Furthermore, k (the permeability) will be set equal to 1 without loss of generality.

Thus the following quantities need to be found: $y = \bar{f}(x)$, (the function describing the shape of the free surface Γ); Ω, (the seepage domain); $\phi(x,y) = \dfrac{p}{\rho g} + y$, where p(x,y) is the pressure in the flowfield, ρ is the density of the fluid and g is the acceleration constant; ψ(x,y); and q(the flowrate through the flowfield). The function $\bar{f}(x)$ is to be defined and smooth on its interval of existence. Furthermore, it is strictly decreasing on this interval and must satisfy

$$\bar{f}(x_F) = y_F, \quad \bar{f}(x_C) = 0, \quad \frac{d\bar{f}(x_F)}{dx} = -\frac{1}{\alpha'(x_F)} \text{ and } \frac{d\bar{f}(x_C)}{dx} = -\infty .$$

The functions ϕ(x,y) and ψ(x,y) are defined on $\bar{\Omega}$ and are to be in $H^1(\Omega) \cap C^0(\bar{\Omega})$. Further, for the problem shown in the Fig. 1:

$$\Omega = \{(x,y) \mid 0 < x \le x_F, 0 < y < \alpha(x); x_F < x < x_C, 0 < y < \bar{f}(x)\},$$

$$\phi_x - \psi_y = 0 \text{ and } \phi_y + \psi_x = 0 \text{ in } \Omega, \tag{1a}$$

$$\begin{aligned} \phi &= y_F && \text{on } \overset{\frown}{AF}, \\ \phi &= 0 && \text{on } [BC], \\ \psi &= q && \text{on } [AB], \\ \psi &= 0 && \text{on } \Gamma \text{ [i.e., curve } (x, \bar{f}(x))], \\ \text{and} \quad \phi &= y && \text{on } \Gamma. \end{aligned} \tag{1b}$$

FIXED DOMAIN FORMULATION

Extended Solution Domain

Let the solution domain Ω be extended to the known region $D = \{(x,y) \mid 0 < x \le x_F, 0 < y < \alpha(x); x_F < x < x_{C'}, 0 < y < y_F\}$ in Fig. 2. Then extend ϕ and ψ continuously to

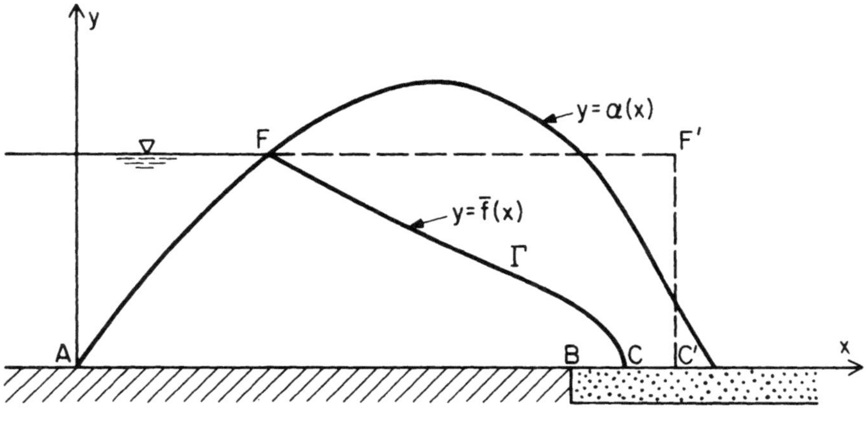

Figure 2. Flow problem and fixed solution domain.

be defined on $\overline{D}$ by setting

$$\tilde{\phi}(x,y) = \phi(x,y) \text{ in } \overline{\Omega}, \quad \text{and} \quad \tilde{\psi}(x,y) = \psi(x,y) \text{ in } \overline{\Omega}$$
$$= y \text{ in } \overline{D} - \overline{\Omega} \qquad\qquad = 0 \text{ in } \overline{D} - \overline{\Omega} \quad . \tag{2}$$

This yields

$$\tilde{\phi}_x - \tilde{\psi}_y = 0 \text{ and } \tilde{\phi}_y + \tilde{\psi}_x = \chi_{D-\overline{\Omega}} \text{ in D,} \tag{3}$$

in the sense of distributions where $\chi_{D-\overline{\Omega}} = 1$ in $D - \overline{\Omega}$ and $\chi_{D-\overline{\Omega}} = 0$ in Ω.

Formulation Using an Integral Transformation

Next define a new dependent variable w using the Baiocchi transformation

$$w(P) = \int_{\widehat{FP}} -\tilde{\psi}\, dx + (y - \tilde{\phi})\, dy, \tag{4}$$

where $\widehat{FP}$ is a smooth path in D joining F to P in D in Fig. 2. The integration is independent of the path due to Equations (3). If Equations (1) hold, then for all w in $H^2(D) \cap C^1(\overline{D})$:

$$\Delta w = \chi_\Omega \qquad \text{in D,} \tag{4a}$$

$$\begin{aligned} w_y &= y - y_F && \text{on } \widehat{AF}, \\ w &= (q^2/6) + q(x_B - x) && \text{on } [AB], \\ w_y &= 0 && \text{on }]BC[, \\ w &= 0 && \text{in } \overline{D} - \overline{\Omega} \text{ (in particular on } \Gamma), \\ \text{and} \quad w &> 0 && \text{in } \Omega \text{ } (w \geq 0 \text{ in D}). \end{aligned} \tag{4b}$$

Hence the complementarity system in D is,

$$w(x,y) \geq 0, \; 1 - \Delta w(x,y) \geq 0, \; w(1 - \Delta w) = 0 \; . \tag{5}$$

If a w is found satisfying Equation (4a) subject to the conditions (4b), then the following quantities can be obtained:

$$\begin{aligned} &\Omega = \{(x,y) \mid (x,y) \text{ in D, } w > 0\}; \\ &\text{graph } \bar{f} = \partial\Omega - \partial D \equiv \text{ points of } \partial\Omega \text{ not in } \partial D; \\ &\phi = y - w_y \text{ in } \Omega; \; \psi = -w_x \text{ in } \Omega; \text{ and } q = \psi(x,0) \text{ on [AB].} \end{aligned} \tag{6}$$

Numerical Scheme

The flowrate through the flowfield, Ω, is unknown a priori. Assuming q known, the numerical scheme uses a finite difference successive over-relaxation method with projection and is given by:

$$w_{q,i,j}^{(n+1/2)} = \frac{\Delta x^2 \Delta y^2}{2(\Delta x^2 + \Delta y^2)} \left[\frac{1}{\Delta x^2} \left(w_{q,i-1,j}^{(n+1)} + w_{q,i+1,j}^{(n)} \right) + \frac{1}{\Delta y^2} \left(w_{q,i,j-1}^{(n+1)} + w_{q,i,j+1}^{(n)} \right) - 1 \right] \tag{7a}$$

$$w^{(n+1)}_{q,i,j} = \max\left\{0,\, w^{(n)}_{q,i,j} + \overline{\alpha}\left(w^{(n+1/2)}_{q,i,j} - w^{(n)}_{q,i,j}\right)\right\} , \tag{7b}$$

where $\overline{\alpha}$ is the relaxation parameter, Δx and Δy are the mesh spacings in the respective directions, n is the iteration number, and the i and j are the column and row mesh point numbers, respectively.

Thus, in addition to the inner iterations to solve for w for a given q, there is also an outer iteration on the q to determine the flowrate. A compatibility condition is necessary for this outer iteration. The condition used herein is similar to that given by Sloss and Bruch [6], i.e.,

$$f_h\left(q^{(r)}\right) = w_{q^{(r)}}(x_F\,,\, y_F - \Delta y) - \frac{\Delta y^2}{2} \;, \qquad r = 0,1,2,\ldots \tag{8}$$

Then $f_h(q^{(r)}) = 0$ represents a compatibility condition, which if imposed on the solutions $w_{q,h}$ (one for every q), permits the determination of an unique $\overline{q}$ such that $w_{\overline{q},h}$ will be a solution of Equations (4a) and (4b), respectively, as well as satisfy Equation (8). Due to the monotonicity of $f_h(q^{(r)})$ in Equation (8), the secant method is used to approximate the root.

The initial guess for $w_{q^{(0)},i,j}$ in D is taken equal to zero; while for each following iteration on $q^{(r)}$ the initial guess for $w_{q^{(r)},i,j}$ in D is $w_{q^{(r-1)},i,j}$. To begin the problem two initial guesses of the discharge, q, are also needed. The computational procedure using the finite-difference SOR method described previously using the stopping criteria

$$\max_{i,j}\left|w^{(n+1)}_{q^{(r)},i,j} - w^{(n)}_{q^{(r)},i,j}\right| < \varepsilon_1 \tag{9}$$

and

$$\left|f_h(q^{(r)})\right| < \varepsilon_2 \;, \qquad r = 0,1,2,\ldots\,, \tag{10}$$

where ε_1 and ε_2 are small, given positive constants, is:

(1) Using the first-guessed value of q, $q^{(0)}$, and the initial guess $w^{(0)}_{q^{(0)},i,j}$ solve the problem using Equations (7a) and (7b) for $w^{(1)}_{q^{(0)},i,j}$, then $w^{(2)}_{q^{(0)},i,j}$,... until the error criterion on $w_{q^{(0)},i,j}$, Equation (9), is satisfied. Then evaluate $f_h(q^{(0)})$ from Equation (8) and check its error criterion, Equation (10).

(2) Using the second-guessed value of q, $q^{(1)}$, which is larger than the first and using $w_{q^{(0)},i,j}$ as the initial guess for $w^{(0)}_{q^{(1)},i,j}$, solve for $w_{q^{(1)},i,j}$ and evaluate $f_h(q^{(1)})$ and check Equation (10). The value of $q^{(2)}$ is found by inserting $q^{(0)}$ and $q^{(1)}$ along with $f_h(q^{(0)})$ and $f_h(q^{(1)})$ into the secant method

$$q^{(r+1)} = q^{(r)} - \frac{q^{(r)} - q^{(r-1)}}{f_h(q^{(r)}) - f_h(q^{(r-1)})} f_h(q^{(r)}) \quad , \qquad r \geq 1. \tag{11}$$

(3) Using $q^{(2)}$ and $w_{q^{(1)},i,j}$ as input, the scheme is repeated to obtain $w_{q^{(2)},i,j}$ and $f_h(q^{(2)})$.

(4) This procedure then continues until the stopping error criteria are both satisfied.

FORMULATION USING A MULTISPLITTING TECHNIQUE

Modified Problem

Now instead of using the problem formulation given by Equations (4a) and (4b), a modified form of the problem will be considered. Domain D will be split into two regions having a common overlap. Let (see Fig. 3)

$$D_\phi = \{(x,y) \mid 0 < x \leq x_F, 0 < y < \alpha(x); x_F < x < \bar{g}(y), 0 < y < y_F\},$$
$$D_w = \{(x,y) \mid x_F < x < x_{C'}, 0 < y < y_F\},$$
$$\Omega_{f.s.} = \{(x,y) \mid x_F < x < x_C, 0 < y < \bar{f}(x)\}, \text{ and } \gamma'_1 = \{(x,y) \mid x = x_F, 0 < y < y_F\},$$

where $x = \bar{g}(y)$ is the equation of the curve γ_1, i.e.

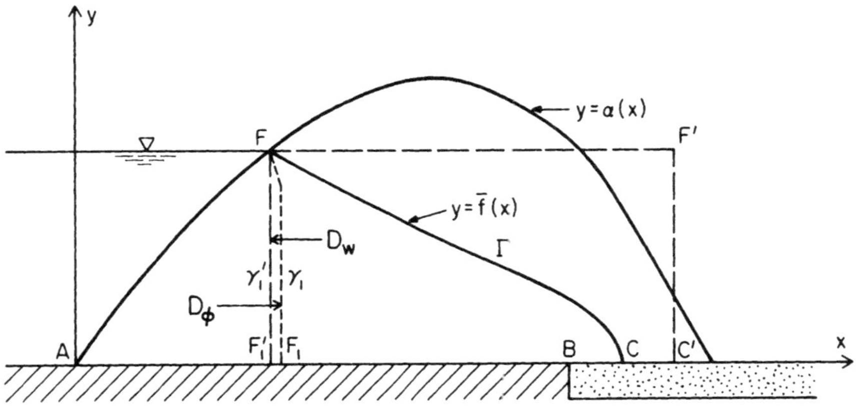

Figure 3. Solution domains in multisplitting approach.

$$\gamma_1 = \{(x,y) | x = \bar{g}(y),\ 0 < y < y_F\}.$$

In domain D_ϕ the following boundary value problem is solved:

$$\begin{aligned} &\Delta\phi = 0 && \text{in } D_\phi, && (12a)\\ &\phi = y_F && \text{on } \widehat{AF}, &&\\ &\phi_y = 0 && \text{on }]AF_1[, && (12b)\\ &\phi = y - w_y && \text{on } \gamma_1. && \end{aligned}$$

The boundary value problem to be solved in D_w is:

$$\begin{aligned} &\Delta w = \chi_{\Omega_{f.s.}} && \text{in } D_w\left(\chi_{\Omega_{f.s.}} = 1 \text{ in } \Omega_{f.s.}, \chi_{\Omega_{f.s.}} = 0 \text{ in } D - \overline{\Omega}_{f.s.}\right), && (13a)\\ &w = (q^2/6) + q(x_B - x) && \text{on } [F_1'B], &&\\ &w_y = 0 && \text{on }]BC[, &&\\ &w = \int_{y_F}^{y} (\bar{y} - \phi)d\bar{y} && \text{on } \gamma'_1, && (13b)\\ &w = 0 && \text{in } \overline{D}_w - \overline{\Omega}_{f.s.} \text{ (in particular on } \Gamma), &&\\ &w \ge 0 && \text{in } D_w \text{ and } w > 0 \text{ in } \Omega_{f.s.}. && \end{aligned}$$

Hence the complementarity system in D_w is,

$$w(x,y) \ge 0,\ \ 1 - \Delta w(x,y) \ge 0,\ \text{ and } w(1 - \Delta w) = 0. \tag{14}$$

It can be seen that information is interchanged between the two domains along the line γ'_1 and the curve γ_1. The approach used to solve this coupled problem is similar to that presented by Bruch and Sloss [12]. It is an alternating iteration scheme which for the complementarity system given in Equations (14) yields for each iteration the associated variational inequality for $w \in K_q$

$$\iint_{D_w} \nabla(v - w)\cdot\nabla w \, dxdy \ge -\iint_{D_w} (v - w)\, dxdy, \quad \forall\, v \in K_q, \tag{15}$$

with $K_q = \{v | v \in H^1(D_w),\ v = (q^2/6) + q(x_B - x) \text{ on } [F_1'B],\ v = 0 \text{ on } [FF']$, $[F'C']$ and $[C'C]$, $v = \int_{y_F}^{y} (\bar{y} - \phi)d\bar{y}$ on γ'_1, $v \ge 0$ a.e. on $D_w\}$.

In this form uniqueness and existence of the solution w can be shown (Bruch and Sloss [12]).

Numerical Scheme

A modified alternating iteration scheme similar to that given in Remar et al. [13] will be used to solve the free surface problem as modified. The D_ϕ region will be discretized and a finite difference successive over-relaxation scheme will be used there in the solution method. A finite difference successive over-relaxation scheme with projection will be used for the D_w region.

The schemes used are:

$$w^{(n+1/2)}_{q,i,j} = \frac{\Delta x^2 \Delta y^2}{2(\Delta x^2 + \Delta y^2)} \left[\frac{1}{\Delta x^2}\left(w^{(n+1)}_{q,i+1,j} + w^{(n)}_{q,i-1,j} \right) + \frac{1}{\Delta y^2}\left(w^{(n+1)}_{q,i,j+1} + w^{(n)}_{q,i,j-1} \right) - 1 \right] \tag{16a}$$

and $$w^{(n+1)}_{q,i,j} = \max\left\{ 0, w^{(n)}_{q,i,j} + \bar{\alpha}\left(w^{(n+1/2)}_{q,i,j} - w^{(n)}_{q,i,j} \right) \right\} \tag{16b}$$

in D_w, and

$$\phi^{(n+1)}_{q,i,j} = (1-\bar{\alpha})\phi^{(n)}_{q,i,j} + \frac{\bar{\alpha}\Delta x^2 \Delta y^2}{2(\Delta x^2 + \Delta y^2)} \bullet$$

$$\bullet \left[\frac{1}{\Delta x^2}\left(\phi^{(n+1)}_{q,i+1,j} + \phi^{(n)}_{q,i-1,j} \right) + \frac{1}{\Delta y^2}\left(\phi^{(n+1)}_{q,i,j+1} + \phi^{(n)}_{q,i,j-1} \right) \right] \tag{17}$$

in D_ϕ. The same procedure in regards to the a priori unknown flowrate as discussed previously is used, i.e., two guessed flowrates and Equation (11) are used to find a new flowrate.

The idea of the approach for this problem is a modified version of the alternating iteration scheme and is the following using the stopping criteria:

$$\max_{i,j} \left| \phi^{(n+1)}_{q^{(r)},i,j} - \phi^{(n)}_{q^{(r)},i,j} \right| < \varepsilon_1 \qquad \text{in } D_\phi \tag{18}$$

Equation (9) in D_w and Equation (10):

1. Choose $\phi^{(0)}_{q^{(0)},i,j}$ in D_ϕ equal to zero and $w^{(0)}_{q^{(0)},i,j}$ in D_w equal to zero.

2. Setting $w^{(1)}_{q^{(0)},i,j} = \int_{y_F}^{y} \left(\bar{y} - \phi^{(0)}_{q^{(0)},i,j} \right) d\bar{y}$ on γ'_1, sweep once through the mesh in D_w to obtain $w^{(1)}_{q^{(0)},i,j}$. Sweep means one pass through the nodal points using the discretized equations with relaxation and projection (projection is used for w but not for ϕ).

3. Setting $\phi^{(1)}_{q^{(0)},i,j} = y - \left(w^{(1)}_y\right)_{q^{(0)},i,j}$ on γ_1, sweep once through the mesh in D_ϕ to obtain $\phi^{(1)}_{q^{(0)},i,j}$.

4. Setting $w^{(2)}_{q^{(0)},i,j} = \int_{y_F}^{y} \left(\bar{y} - \phi^{(1)}_{q^{(0)},i,j}\right) d\bar{y}$ on γ'_1, sweep once through the mesh in D_w to obtain $w^{(2)}_{q^{(0)},i,j}$.

5. Setting $\phi^{(2)}_{q^{(0)},i,j} = y - \left(w^{(2)}_y\right)_{q^{(0)},i,j}$ on γ_1, sweep once through the mesh in D_ϕ to obtain $\phi^{(2)}_{q^{(0)},i,j}$.

6. Continue this process until Equations (9) and (18) are satisfied with $r = 0$.

7. Evaluate $f_h(q^{(0)})$ using Equation (8) and check whether Equation (10) is satisfied. If not, proceed.

8. Using the second-guessed value of q, $q^{(1)}$, which is larger than $q^{(0)}$, and using $w_{q^{(0)},i,j}$ and $\phi_{q^{(0)},i,j}$ as the initial guess for $w^{(0)}_{q^{(1)},i,j}$ and $\phi^{(0)}_{q^{(1)},i,j}$, respectively, repeat steps 2 to 6 until the stopping criteria are satisfied with $r = 1$ and then evaluate $f_h(q^{(1)})$ and check whether Equation (10) is satisfied. If not, proceed. ($w_{q^{(0)},i,j}$ and $\phi_{q^{(0)},i,j}$ are the (n+1) iteration values that satisfied the criteria in step 6.)

9. Using Equation (11) evaluate $q^{(2)}$.

10. Using $q^{(2)}$ with $w_{q^{(1)},i,j}$ and $\phi_{q^{(1)},i,j}$ as the initial guess for $w^{(0)}_{q^{(2)},i,j}$ and $\phi^{(0)}_{q^{(2)},i,j}$ repeat steps 2 to 6 until the stopping criteria are satisfied with $r = 2$ and then evaluate $f_h(q^{(2)})$ and check whether Equation (10) is satisfied. If not, use Equation (11) and obtain $q^{(3)}$.

11. Repeat step 10 updating the indices appropriately until Equation (10) is satisfied, $r \geq 3$.

FIRST FORMULATION USING DOMAIN DECOMPOSITION

Decomposed Problem

The region D is decomposed into the subsets $D_1 = \{(x,y) \mid 0 < x < x_{F'_1},\ 0 < y < \alpha(x)\}$ and $D_2 = \{(x,y) \mid x_{F'_1} < x < x_{C'},\ 0 < y < y_F\}$ with the boundary of D given by ∂D and the interface between D_1 and D_2 given by γ'_1 (see Fig. 4). See Funaro et al. [14]. (Also see

Marini and Quarteroni [15] and [16]). If w_1 denotes the restriction of w to D_1 and w_2 the one to D_2, then Equations (4a) and (4b) can be equivalently written in the following split form:

$$\Delta w_1 = 1 \quad \text{in } D_1 , \tag{19a}$$

$$(w_1)_y = y - y_F \quad \text{on } \widehat{AF} ,$$

$$w_1 = \left(q^2/6\right) + q(x_B - x) \quad \text{on } [AF_1'] , \tag{19b}$$

$$\frac{\partial w_1}{\partial x} = \frac{\partial w_2}{\partial x} \quad \text{on } \gamma'_1 , \tag{19c}$$

and

$$\Delta w_2 = \chi_{\Omega_{f.s.}} \quad \text{in } D_2, \tag{20a}$$

$$w_2 = \left(q^2/6\right) + q(x_B - x) \quad \text{on } [F_1'B] ,$$
$$(w_2)_y = 0 \quad \text{on }]BC[, \tag{20b}$$
$$w_2 = 0 \quad \text{in } \overline{D}_2 - \Omega_{f.s.} \text{ (in particular on } \Gamma),$$

$$w_2 = w_1 \quad \text{on } \gamma'_1. \tag{20c}$$

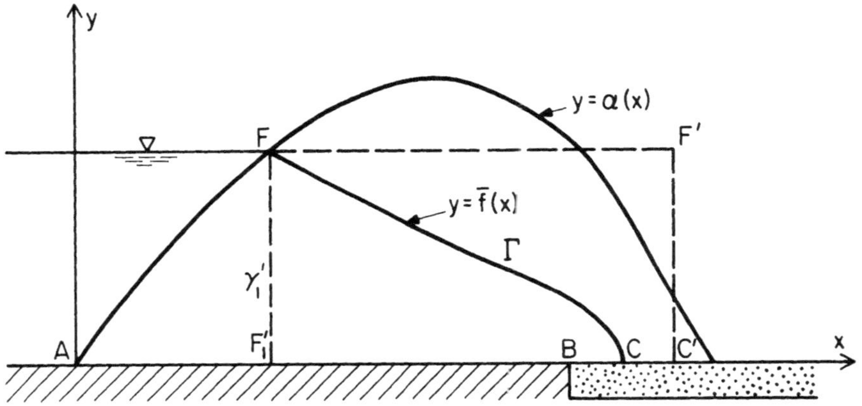

Figure 4. Solution domains used in the domain decomposition approaches.

The following iterative procedure similar to that stated in Papadopoulos et al. [8] is now given for this problem. Let $w_2^1 = g^1$ be given on γ_1. We consider the two sequences of functions w_2^n and w_1^n, $n \geq 1$ satisfying, respectively, the problems:

$$\Delta w_2^n = \chi_{\Omega f.s.}, \quad \text{in } D_2 , \tag{21a}$$

$$w_2^n = \left(q^2/6\right) + q(x_B - x), \quad \text{on } [F_1'B],$$
$$\left(w_2^n\right)_y = 0\,, \quad \text{on }]BC[, \qquad (21b)$$
$$w_2^n = 0 \quad \text{in } \overline{D}_2 - \Omega_{f.s.},$$
$$w_2^n = g^n \quad \text{on } \gamma'_1 \qquad (21c)$$

and

$$\Delta w_1^n = 1 \quad \text{in } D_1, \qquad (22a)$$
$$\left(w_1^n\right)_y = y - y_F \quad \text{on } \widehat{AF},$$
$$(22b)$$
$$w_1^n = (q^2/6) + q(x_B - x) \quad \text{on } [AF_1'],$$
$$\frac{\partial w_1^n}{\partial x} = \frac{\partial w_2^n}{\partial x} \quad \text{on } \gamma'_1\,, \qquad (22c)$$

where

$$g^{n+1} = \theta w_1^n + (1-\theta)g^n \quad \text{on } \gamma'_1 \text{ for } n \geq 1\,. \qquad (23)$$

Numerical Scheme

An iteration scheme to solve the problem given by Equations (21a)-(23) will now be presented. It is a modified version of that stated in these equations and will be designated in particular by using the superscript iteration numbers in parentheses. The D_1 and D_2 regions will be discretized and the problem solved using a finite difference successive over-relaxation scheme with projection.

The flowrate through the flowfield, Ω, is also an unknown a priori. Thus, in addition to the inner iterations to solve for w for a given q, there is also an outer iteration on the q to determine the flowrate. The same procedure in regards to determining the flowrate as discussed previously is used.

The stopping error estimates for the numerical scheme are given by:

$$\max_{i,j}\left|\left(w_1\right)_{q^{(r)},i,j}^{(n+1)} - \left(w_1\right)_{q^{(r)},i,j}^{(n)}\right| < \varepsilon_1 \quad \text{in } D_1, \qquad (24a)$$

$$\max_{i,j}\left|\left(w_2\right)_{q^{(r)},i,j}^{(n+1)} - \left(w_2\right)_{q^{(r)},i,j}^{(n)}\right| < \varepsilon_1 \quad \text{in } D_2, \qquad (24b)$$

$$\max_{j}\left|\left(w_1\right)_{q^{(r)},NC1,j}^{(n+1)} - \left(w_2\right)_{q^{(r)},1,j}^{(n+1)}\right| < \varepsilon_1 \quad \text{on } \gamma'_1 \qquad (24c)$$

and Equation (10), where NC1 is the number of columns in D_1.

The basic numerical scheme consists in choosing initial guessed values for $\left(w_1\right)^{(1)}_{q^{(0)},i,j}$ and $\left(w_2\right)^{(1)}_{q^{(0)},i,j}$ in D_1 and D_2 equal to zero, respectively, as well as equal to zero on γ_1. The first iteration pass (which consists in evaluating the algorithm given in Equations (26a) and (26b) once at each mesh point) is performed in D_2 using the Dirichlet data boundary condition Equation (21c) with $(w_2)^{(1)} = g^{(1)} = 0$ and $(w_1)^{(1)} = 0$ in Equation (23) which can be written as

$$w_2^{(n+1)} = \theta w_1^{(n)} + (1-\theta) w_2^{(n)} \text{ on } \gamma'_1 \tag{25a}$$

or in discretized notation as

$$\left(w_2\right)^{(n+1)}_{q^{(r)},1,j} = \theta\left(w_1\right)^{(n)}_{q^{(r)},NC1,j} + (1-\theta)\left(w_2\right)^{(n)}_{q^{(r)},1,j}, \text{ on } \gamma'_1,\ 1 \le j \le NR, \tag{25b}$$

where NR is the number of rows. The basic numerical algorithm used in D_2 is:

$$\left(w_2\right)^{(n+1/2)}_{q^{(r)},i,j} = \frac{\Delta x^2 \Delta y^2}{2[\Delta x^2 + \Delta y^2]} \left\{ \frac{1}{\Delta x^2}\left[\left(w_2\right)^{(n)}_{q^{(r)},i-1,j} + \left(w_2\right)^{(n+1)}_{q^{(r)},i+1,j}\right] + \frac{1}{\Delta y^2}\left[\left(w_2\right)^{(n)}_{q^{(r)},i,j-1} + \left(w_2\right)^{(n+1)}_{q^{(r)},i,j+1}\right] - 1 \right\}, \tag{26a}$$

$$\left(w_2\right)^{(n+1)}_{q^{(r)},i,j} = \max\left\{0, \left(w_2\right)^{(n)}_{q^{(r)},i,j} + \overline{\alpha}\left[\left(w_2\right)^{(n+1/2)}_{q^{(r)},i,j} - \left(w_2\right)^{(n)}_{q^{(r)},i,j}\right]\right\}. \tag{26b}$$

For the first iteration pass (which consists in evaluating the basic algorithm once at each mesh point) in D_1 the Neumann boundary condition (22c) is used along γ_1. This shows up in the basic numerical scheme in D_1 (which is similar to Equations (26a) and (26b) with w_1 replacing w_2) using

$$\left(w_1\right)^{(n+1)}_{q^{(r)},NC1+1,j} = \left(w_1\right)^{(n)}_{q^{(r)},NC1-1,j} - \left\{3\left(w_2\right)^{(n+1)}_{q^{(r)},1,j} - 4\left(w_2\right)^{(n+1)}_{q^{(r)},2,j} + \left(w_2\right)^{(n+1)}_{q^{(r)},3,j}\right\}. \tag{27}$$

This iteration scheme which consists in evaluating the basic algorithm once at each mesh point with interchange of data on γ_1 after each pass proceeds until Equations (24a)-(24c) are satisfied for the inner iteration. $f_h(q^{(0)})$ is then evaluated and checked against ε_2. If condition (10) is not satisfied, $q^{(1)}$ is used along with the values of $\left(w_1\right)_{q^{(0)},i,j}$ and $\left(w_2\right)_{q^{(0)},i,j}$ just obtained as starting values to make another set of inner and outer iterations. If $f_h(q^{(1)})$ does not satisfy condition (10), $q^{(0)}$, $q^{(1)}$, $f_h(q^{(0)})$, and $f_h(q^{(1)})$ are used to obtain $q^{(2)}$ from the secant method, Equation (11). This procedure is then repeated until all error criteria, (24a)-(24c) and (10), are satisfied.

The iteration approach presented (a modified version of that given in Equations (21a)-(23) and designated by the superscript iteration numbers in parentheses) passes from one region, D_2, to the other, D_1, interchanging data along γ_1 after each iteration (i.e., one pass through the mesh points with the algorithm). The approach given in Equations (21a)-(23) designated by the superscript notation without parentheses was also tried on the example problem to be discussed later in which the problem in region D_2 was solved first to convergence. Data was then interchanged on γ_1 and the problem in region D_1 was solved. The error criterion (10) was then checked and was found not to be satisfied. Data was again transferred through γ'_1 and the right hand region, D_2, solved again. This pattern was repeated until all the error criteria were met. There were roughly two and a half times as many iterations needed for this latter approach over the former in addition to a change in the θ-value.

SECOND FORMULATION USING DOMAIN DECOMPOSITION

The same subdomains, D_1 and D_2, as described previously, will also be used here. The new aspect of this formulation is that in region D_1 the velocity potential ϕ will be the dependent variable while in D_2 the dependent variable is w_2 as defined by Equation (4). The problem is written in the split form again:

$$\Delta\phi = 0 \qquad \text{in } D_1\,, \tag{28a}$$

$$\begin{aligned} \phi &= y_F && \text{on } \widehat{AF}\,, \\ \phi_y &= 0 && \text{on }]AF_1[\,, \end{aligned} \tag{28b}$$

$$\phi = y - (w_2)_y \qquad \text{on } \gamma'_1 \tag{28c}$$

and

$$\Delta w_2 = \chi_{\Omega_{f.s.}} \qquad \text{in } D_2\,, \tag{29a}$$

$$\begin{aligned} w_2 &= (q^2/6) + q(x_B - x) && \text{on } [F'_1B]\,, \\ (w_2)_y &= 0 && \text{on }]BC[\,, \\ w_2 &= 0 && \text{in } \overline{D}_2 - \Omega_{f.s.} \text{ (in particular on } \Gamma)\,, \end{aligned} \tag{29b}$$

$$\left(w_2\right)_x = -q - \int_0^y \phi_x d\overline{y} \qquad \text{on } \gamma_1\,. \tag{29c}$$

The following iterative procedure somewhat similar to that already stated is now given for the formulation. Let $\phi^0 = g^0 = 0$ be given on γ'_1. We consider the two sequences of functions w_2^n and ϕ^n, $n \geq 1$, satisfying, respectively, the problems:

$$\Delta w_2^n = \chi_{\Omega_{f.s.}} \qquad \text{in } D_2\,, \tag{30a}$$

$$\begin{aligned} w_2^n &= (q^2/6) + q(x_B - x) && \text{on } [F'_1B]\,, \\ \left(w_2^n\right)_y &= 0 && \text{on }]BC[\,, \\ w_2^n &= 0 && \text{on } \overline{D}_2 - \Omega_{f.s.} \text{ (in particular on } \Gamma)\,, \end{aligned} \tag{30b}$$

$$\left(w_2^n\right)_x = -q - \int_0^y \phi_x^{n-1} d\bar{y} \qquad \text{on } \gamma'_1 \tag{30c}$$

and

$$\Delta\phi^n = 0 \qquad \text{in } D_1 \,, \tag{31a}$$

$$\phi^n = y_F \qquad \text{on } \widehat{AF} \,,$$

$$\phi_y^n = 0 \qquad \text{on }]AF_1[\,, \tag{31b}$$

$$\phi^n = g^n \qquad \text{on } \gamma'_1 \,, \tag{31c}$$

where

$$g^{n+1} = \theta\left[y - \left(w_2^{n+1}\right)_y\right] + (1-\theta)g^n \qquad \text{on } \gamma'_1 \text{ for } n \geq 0 \,. \tag{32}$$

Numerical Scheme

The numerical approach for this formulation is similar to that given in the previous section with the following exceptions: Equation (17) is used in D_1 with the stopping error criterion

$$\max_{i,j}\left|\phi^{(n+1)}_{q^{(r)},i,j} - \phi^{(n)}_{q^{(r)},i,j}\right| < \varepsilon_1 \qquad \text{in } D_1 \,. \tag{33a}$$

along with

$$\max_j\left|\phi^{(n+1)}_{q^{(r)},NC1,j} - \left\{j\Delta y - \left[\left(w_2\right)^{(n+1)}_{q^{(r)},1,j+1} - \left(w_2\right)^{(n+1)}_{q^{(r)},i,j-1}\right] / (2\Delta y)\right\}\right| \tag{33b}$$

$$\text{on } \gamma'_1,\ 2 < j < NR \;.$$

For j = 1 the term in the { } in Equation (33b) is

$$-\left[-3\left(w_2\right)^{(n+1)}_{q^{(r)},1,1} + 4\left(w_2\right)^{(n+1)}_{q^{(r)},1,2} - \left(w_2\right)^{(n+1)}_{q^{(r)},1,3}\right] / (2\Delta y) \;.$$

Equation (32) is written as

$$\phi^{(n+1)} = \theta\left[y - \left(w_2\right)_y^{(n+1)}\right] + (1-\theta)\phi^{(n)} \text{ on } \gamma_1 \text{ for } n \geq 0 \tag{34}$$

or in discretized notation as

$$\phi^{(n+1)}_{q^{(r)},NC1,j} = \theta\left\{j\Delta y - \left[\left(w_2\right)^{(n+1)}_{q^{(r)},i,j+1} - \left(w_2\right)^{(n+1)}_{q^{(r)},i,j-1}\right] / (2\Delta y)\right\}$$

$$+(1-\theta)\phi^{(n)}_{q^{(r)},NC1,j}, \qquad \text{on } \gamma'_1,\ 2 \le j \le NR\ . \tag{34a}$$

For j = 1 the term in the { } brackets in Equation (34a) is

$$-\left[-3\left(w_2\right)^{(n+1)}_{q^{(r)},1,1}+4\left(w_2\right)^{(n+1)}_{q^{(r)},1,2}-\left(w_2\right)^{(n+1)}_{q^{(r)},1,3}\right]/\,(2\Delta y).$$

The discretized form of Equation (30c) is

$$\left(w_2\right)^{(n)}_{q^{(r)},-1,j}=\left(w_2\right)^{(n)}_{q^{(r)},2,j}+2\Delta x\, q^{(r)}$$

$$+\frac{\Delta y}{2}\sum_{k=2}^{j}\left[\left(\phi^{(n-1)}_{q^{(r)},NC1-2,k-1}-4\phi^{(n-1)}_{q^{(r)},NC1-1,k-1}+3\phi^{(n-1)}_{q^{(r)},NC1,k-1}\right)\right.$$

$$\left.+\left(\phi^{(n-1)}_{q^{(r)},NC1-2,k}-4\phi^{(n-1)}_{q^{(r)},NC1-1,k}+3\phi^{(n-1)}_{q^{(r)},NC1,k}\right)\right],\quad 2\le j\le NR-2 \tag{35}$$

with a modification (a two point approximation for ϕ_x) in the second term in the summation for j = NR-1.

RESULTS

The example to be investigated (one solved by Sloss and Bruch [6]) using the previously given approaches has the following data: y_F = 30 ft; x_F = 30 ft; x_B = 60 ft; $x_{C'}$, which should be greater than $(x_B + \frac{q}{2})$, (see Harr [17] and Sloss and Bruch [6]), $q^{(0)}$ = 15.0 ft^3/sec per foot depth normal to the plane of flow; $q^{(1)}$ = 16.0 ft^3/sec per foot depth normal to the plane of the flow; stopping error estimates: ε_1 = 0.01 , ε_2 = 0.01; a relaxation factor equal to 1.85; $\Delta x = \Delta y$ = 2.5 ft; NR = 13 and NC1 = 13.

Table 1 lists the iteration information for the example problem. The first column gives the outer iteration number. The next four columns list the values of the flowrate for the various outer iterations and show they converge after the use of the two guessed values. The next four columns list the compatibility condition values showing how they converge after the first two outer iterations. The last four columns list the number of iterations (the number of passes through the mesh points) to obtain convergence for each outer iteration. The legend for the column designations within these groupings of four is presented on the Table.

Since columns three and four in each grouping were obtained using the domain decomposition techniques, they are functions of the relaxation parameter θ. The number of iterations depends upon this parameter and numerical experiments were performed to determine the best value. Tables 2 and 3 present these results for the first and second domain decomposition techniques, respectively. The word 'Iterations' in the Tables stands for the total number of iterations to obtain overall convergence (the sum of the outer iteration convergence results). These best values were used to obtain the results shown in Table 1 for the respective schemes.

Table 1. Iteration information for example.

	$q^{(r)}$				$f_h(q^{(r)})$				n			
r	FD*	MT*	FDD* $\theta=0.06$	SDD* $\theta=0.115$	FD	MT	FDD $\theta=0.06$	SDD $\theta=0.115$	FD	MT	FDD $\theta=0.06$	SDD $\theta=0.115$
0	15.00000	15.00000	15.00000	15.00000	47.13327	73.08655	51.50678	64.56488	134	79	123	92
1	16.00000	16.00000	16.00000	16.00000	59.45287	91.45297	64.87255	81.33057	95	67	97	59
2	11.17412	11.02064	11.14637	11.14899	2.23226	3.33601	2.25730	3.04526	114	80	109	70
3	10.98586	10.83213	10.97139	10.96028	0.27792	0.42894	0.23439	0.42973	74	60	73	49
4	10.95909	10.80431	10.95112	10.92928	-0.00152	0.00061	0.00326	0.00564	51	39	47	34
									468	325	449	304

*

FD - Fixed Domain

MT - Multisplitting Technique

FDD - First Domain Decomposition

SDD - Second Domain Decomposition

Table 2. Iteration information obtained by varying θ in the first domain decomposition technique.

θ	0.025	0.050	0.055	0.060	0.065	0.075	0.085	0.100
Iterations	976	481	450	449	453	458	467	479

Table 3. Iteration information obtained by varying θ in the second domain decomposition technique.

θ	0.070	0.095	0.105	0.115	0.125	0.150	0.200	0.250
Iterations	393	323	313	304	312	329	378	527

In comparing the results for the free surface location obtained by Sloss and Bruch [6] and those of the multisplitting and domain decomposition techniques, they all gave exactly the same results, i.e., the same first mesh points to have a value of zero as you proceed vertically for fixed x's.

ACKNOWLEDGEMENT

This material is based upon work partially supported by the National Science Foundation under award No. ECS-9006516.

REFERENCES

1. Baiocchi, C. and Capelo, A. Variational and Quasivariational Inequalities, John Wiley and Sons, New York, 1984.

2. Kinderlehrer, D. and Stampacchia, G. An Introduction to Variational Inequalities and Their Applications, Academic Press, New York, 1980.

3. Oden, J.T. and Kikuchi, N. Theory of Variational Inequalities with Applications to Problems of Flow Through Porous Media, Int. J. of Engng. Sci., Vol. 18, pp. 1173-1284, 1980.

4. Bruch, Jr., J.C. A Survey of Free Boundary Value Problems in the Theory of Fluid Flow Through Porous Media: Variational Inequality Approach, Advances in Water Resources, Part I, Vol. 3, pp. 65-80, Part II, Vol. 3, pp. 115-124, 1980.

5. Crank, J. Free and Moving Boundary Problems, Clarendon Press, Oxford, England, 1984.

6. Sloss, J.M. and Bruch, Jr., J.C. Free Surface Seepage Problems, J. Eng. Mech. Div. Proc., ASCE, Vol. 104, EM5, pp. 1099-1111, 1978.

7. Bruch, Jr., J.C., Sloss, J.M. and Remar, J., An Alternating Iteration Scheme for Certain Free Boundary Value Problems, Mathematics of Finite Elements and Applications, (Ed. Whiteman, J.R.), pp. 251-258, Academic Press, London, 1982.

8. Papadopoulos, C.A., Sloss, J.M. and Bruch, J.C. Jr., A Domain Decomposition Technique Applied to a Free Surface Seepage Problem, to appear in Proceedings of the Conference on Numerical Methods for Free Boundary Problems, (Ed. Neittaanmaki, P.), Jyvaskyla, Finland, 1990.

9. White, R.E., Multisplitting and Parallel Iterative Methods, Comp. Meth. Appl. Mech. Engr., Vol. 64, pp. 567-577, 1987.

10. Wang, K.P. and Bruch, J.C., Jr., Solutions of a Steady State Free Surface Seepage Problem on a Hypercube Concurrent Computer, Engineering Computations, Vol. 6(3), pp. 225-236, 1989.

11. Baiocchi, C., Free Boundary Problems in the Theory of Fluid Flow Through Porous Media, Proceeding of the Int. Congr. Math., Vol. II, pp. 237-243, Vancouver, 1975.

12. Bruch, Jr., J.C. and Sloss, J.M. Coupled Variational Inequalities for Flow from a Nonsymmetric Ditch, Contemporary Mathematics, Vol. 11, Plane Ellipticity and Related Problems, (Ed. Gilbert, R.P.), pp. 49-70, AMS, 1982.

13. Remar, J., Bruch, Jr., J.C. and Sloss, J.M. Axisymmetric Free Surface Seepage, Num. Math., Vol. 40, pp. 143-168, 1982.

14. Funaro, D., Quarteroni, A. and Zanolli, P. An Iterative Procedure with Interface Relaxation for Domain Decomposition Methods, SIAM J. Numer. Anal., Vol. 25, pp. 1213-1236, 1988.

15. Marini, L.D. and Quarteroni, A. An Iterative Procedure for Domain Decomposition Methods: A Finite Element Approach, Domain Decomposition Methods for Partial Differential Equations (Eds. Glowinski, R., Golub, G.H. and Periaux, J.), SIAM, Philadelphia, PA, pp. 129-143, 1988.

16. Marini, L.D. and Quarteroni, A. A Relaxation Procedure for Domain Decomposition Methods Using Finite Elements, Num. Math., Vol. 55, pp. 575-598, 1989.

17. Harr, M.E. Groundwater and Seepage, McGraw-Hill, New York, 1962.

Application of B-Spline Boundary Elements to Groundwater Flow Problems

J.J.S.P. Cabral (*), L.C. Wrobel

Computational Mechanics Institute, Wessex Institute of Technology, Ashurst Lodge, Ashurst, Southampton SO4 2AA, England

(*) On leave from Water Resources and Environmental Engineering Group, Federal University of Pernambuco, Recife, PE, Brazil

ABSTRACT

Continuity of the derivatives of the main variable is an important feature to obtain an accurate representation of free and moving boundaries with iterative or time-marching schemes. A recently developed BEM formulation using cubic B-splines provides up to C^2 continuity between adjacent elements. This formulation is applied to groundwater flow problems related to flow through earth dams and salt water intrusion.

INTRODUCTION

The Boundary Element Method (BEM) is now established as a suitable and efficient technique for the analysis of free and moving boundary problems. However, as in other discretization procedures, inaccuracies may be introduced due to the lack of derivative continuity between adjacent elements.

This occurs when using Lagrangian elements for free surface motion as reported by Chang [1], Bruch and Grilli [2], and may sometimes cause divergence either of iterative procedures or transient algorithms. The main reason is the appearance, during the motion, of points of geometric tangent discontinuity between adjacent elements, at which the velocity is not unique. Chang [1] has used a smoothing procedure by spacewise averaging the velocity value. However, an approach that intrinsically provides continuity of derivatives of the variable is much more desirable.

The most popular of such approaches involve the use of splines. In particular, the use of cubic B-spline elements [3], [4] preserves continuity of the geometric tangent and curvature (C^2 continuity) between adjacent elements and produces accurate results even with coarse discretizations.

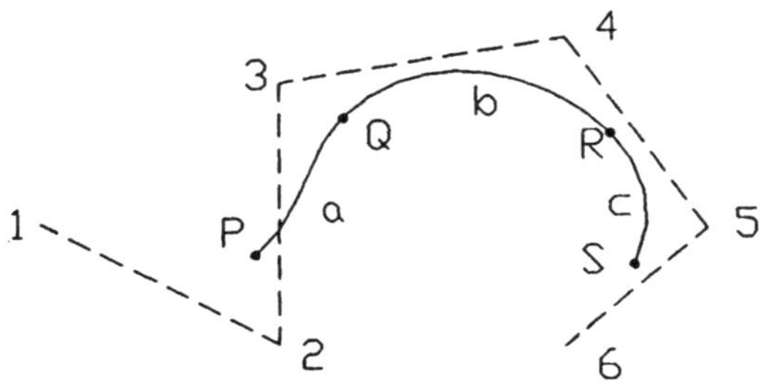

segments	knots	control points
a	P-Q	1,2,3,4
b	Q-R	2,3,4,5
c	R-S	3,4,5,6

Figure 1: Example of cubic B-spline

This work applies the BEM with cubic B-spline elements to groundwater flow problems including seepage through earth dams and salt water intrusion in confined aquifers. The applications included herein attest the excellent convergence and stability properties achieved with the present formulation.

B-SPLINES

A spline is a piecewise polynomial function. It can be viewed as a generalized polynomial that has certain points of derivative discontinuity.

In figure 1 points P, Q, R, S are called *knots* and the pieces of the curve in the interval between these points are called *span* or *segment* of the spline. Points 1 to 6 are the *control points* used to define the shape of the B-spline.

B-splines are a class of *approximating splines* because they approach but do not intersect their control points (fig. 1); they present continuity between the pieces of the curve and are infinitely differentiable at points which are not knots.

Although it is possible to use approximations of any order, most applications in engineering employ cubic B-splines and adopt uniform knot spacing. With uniform spacing the knots are assumed to be integers and a local parameter ω (varying from 0 to 1) takes the spline from one integer knot to the next. It is noted that uniform knot spacing does not imply that all segments necessarily have the same size.

Each segment is defined by four control points. Two neighbour segments share three control points and have a different one, so for m segments there are $m + 1$ knots and $m + 3$ control points. The three shared control points are responsible for the continuity between adjacent segments.

One problem that appears on applying B-splines in conjunction with the BEM to the analysis of a particular problem is that some specified points on the boundary of the problem region are known in advance (given the region geometry) but not the control points. Therefore, it is necessary to perform the so-called *inversion procedure* in order to find the position of the control points. Here, the algorithm proposed by Yamaguchi [5] was used.

When the region under study presents points of geometric discontinuity - *e.g.* corners - there is a need to introduce multiple knots (not to be confused with usual BEM double nodes) and consequently the knot spacing will no longer be uniform; thus, we must have different bases for the segments on the boundary [4].

In the present work only integer knots have been used; all possible cases of different bases are classified and, for each segment, it is verified which basis it corresponds to.

Cubic B-splines normally present C^2 continuity between segments of the curve. Using double knots we obtain C^1 continuity, *i.e.* continuity up to the first derivative, and using triple knots we obtain C^0 continuity, *i.e.* continuity of the function only. The use of quadruple knots imposes no continuity at all.

Bartels *et al.* [6] showed that the blending functions for segments near triple knots are the same as for segments near quadruple knots, the difference being that, at quadruple knots they are multiplied by different control points. As the blending functions for triple or quadruple knots are the same we henceforth refer to triple knot at the corners and to single knot elsewhere.

MATHEMATICAL FORMULATION

Combination of Darcy's law and the mass balance equation for two-dimensional flow through a homogeneous isotropic porous medium gives, for a vertical model,

$$K\left(\frac{\partial^2 u}{\partial x^2}+\frac{\partial^2 u}{\partial y^2}\right)=S\frac{\partial u}{\partial t} \tag{1}$$

where K is the hydraulic conductivity, S the specific storage and the potential u is the piezometric head

$$u=y+\frac{p}{\gamma}$$

in which p is pressure, γ is specific weight and y is elevation.

For practical purposes, transient flow can be treated as a sequence of successive steady states in a step-by-step procedure. So, the problem can be modelled using Laplace's equation

$$\frac{\partial^2 u}{\partial x^2}+\frac{\partial^2 u}{\partial y^2}=0 \tag{2}$$

The following integral equation, equivalent to (2), can be obtained after some transformations [7]

$$c^i u^i + \int_\Gamma u q^* d\Gamma = \int_\Gamma u^* q d\Gamma \tag{3}$$

where $c^i = \theta/2\pi$, θ is the internal angle at the point i on the boundary Γ, u^* is the fundamental solution of Laplace's equation, $q = \partial u/\partial n$ and $q^* = \partial u^*/\partial n$.

Discretizing the boundary Γ into N elements, we obtain

$$c^i u^i + \sum_{j=1}^{N} \int_{\Gamma_j} u q^* d\Gamma = \sum_{j=1}^{N} \int_{\Gamma_j} u^* q d\Gamma \tag{4}$$

The values of the potential and its normal derivative using cubic B-splines as interpolation functions are given at any point by

$$u(\omega) = E_0(\omega)a_{i-1} + E_1(\omega)a_i + E_2(\omega)a_{i+1} + E_3(\omega)a_{i+2} \tag{5}$$

$$q(\omega) = E_0(\omega)b_{i-1} + E_1(\omega)b_i + E_2(\omega)b_{i+1} + E_3(\omega)b_{i+2} \tag{6}$$

These expressions apply to the segment between nodes i and $i+1$; a_i are the coefficients (control points) of the potential representation and b_i the coefficients of the normal derivative; $E_k(\omega)$ are the blending functions for cubic B-splines given by Cabral *et al.* [3], [4] and ω is a local parameter which varies between 0 and 1 within each segment. To represent corners, multiple knots are used and non-uniform blending functions must be applied as reported in [4].

The present formulation employs isoparametric boundary elements for which the coordinates x and y of each point are also expressed by cubic B-spline functions

$$x = E_0 X_{i-1} + E_1 X_i + E_2 X_{i+1} + E_3 X_{i+2} \tag{7}$$

$$y = E_0 Y_{i-1} + E_1 Y_i + E_2 Y_{i+1} + E_3 Y_{i+2} \tag{8}$$

where X and Y are the coordinates of the control points.

Application of the BEM equation (4) to all nodal points, incorporating approximations (5) to (8) within each boundary element, generates a system of algebraic equations which, in the present case, will have the form

$$\mathbf{Ha} = \mathbf{Gb} \tag{9}$$

Evaluation of the coefficients of matrices **H** and **G** is discussed in detail in [3] and [4].

Boundary conditions

The boundary conditions of the problem specify the potential or flux on the boundary, but generally the coefficients a_i and b_i are not known. So one must modify the system and substitute the coefficients by their relation with the potential and flux.

Writing expression (5) for all collocation points and collecting the result in matrix form gives

$$\begin{Bmatrix} u_1 \\ u_2 \\ \cdot \\ \cdot \\ u_n \end{Bmatrix} = \begin{bmatrix} E_{11} & E_{12} & \dots & E_{1n} \\ E_{21} & \dots & \dots & E_{2n} \\ \cdot & \cdot & \cdot & \cdot \\ \cdot & \cdot & \cdot & \cdot \\ E_{n1} & E_{n2} & \dots & E_{nn} \end{bmatrix} \begin{Bmatrix} a_1 \\ a_2 \\ \cdot \\ \cdot \\ a_n \end{Bmatrix} \tag{10}$$

Each row of matrix $\mathbf{E}$ of nodal values of blending functions has three non-zero terms (four blending functions but one of them is null at each node), and all the other terms are zero.

Inverting equation (10) and substituting for vector $\mathbf{a}$ in (9), adopting a similar procedure for vector $\mathbf{b}$, produces

$$\mathbf{H}\mathbf{E}^{-1}\mathbf{u} = \mathbf{G}\mathbf{E}^{-1}\mathbf{q} \tag{11}$$

It can be noted that matrices $\mathbf{H}$ and $\mathbf{G}$ are post-multiplied by the inverse of the blending function matrix and after that the solution procedure follows the usual BEM way.

STEADY FLOW

The B-spline BEM formulation was initially applied to study the steady flow through an earth dam. Considering, for simplicity, a rectangular dam of isotropic material, the mathematical description of the problem is provided by a Laplace's equation for the piezometric head and the following boundary conditions (fig. 2):

- AB $\rightarrow \dfrac{\partial u}{\partial n} = 0$
- BC $\rightarrow u = h$
- CD $\rightarrow u = y$
- DE $\rightarrow \begin{cases} \dfrac{\partial u}{\partial n} = 0 \\ u = y \end{cases}$ *(free surface)*
- EA $\rightarrow u = H$

Figure 2 shows one of the discretizations that have been used in the numerical analysis. It can be noticed that there are double nodes at points A, B, C and E because of the discontinuity of the potential derivative. This flux discontinuity is caused by a discontinuity in geometry at points A, B and E, and by a singularity at point C [8].The problem has been treated, from the B-spline point of view, using triple knots at these points to impose C^0 continuity.

It is important to notice the tangency between the free surface and the seepage surface at the downstream face [9]. At the tangency point (point D),

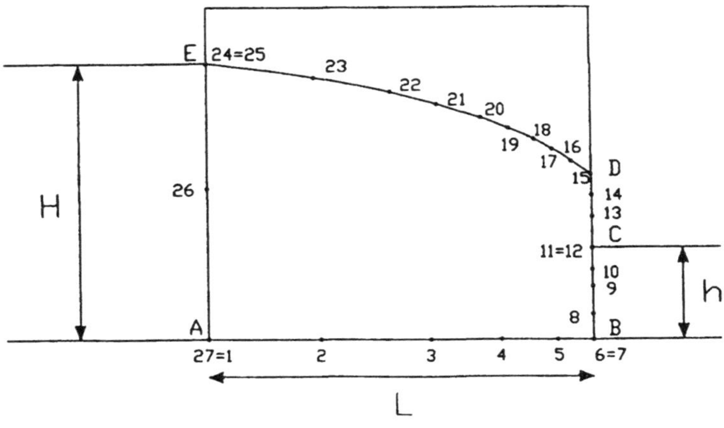

Figure 2: Flow through earth dam with vertical faces

sometimes called separation point, there is continuity of the slope and curvature, *i.e.* C^2 continuity [10]. Cubic B-spline elements can provide a good representation of these features using a single knot at the separation point.

From the BEM point of view, the consideration of a double node at point C is not straightforward. While at points A,B and E there is in fact only one unknown since the values of u and one of its normal derivatives are prescribed, at point C only u is known and the two flux values (before and after the node) are required as part of the solution.

The singularity of matrix **G** can be avoided using the so-called *Interpolated Element* [11] in which the source point is assumed somewhere inside the element instead of at the extreme points as usual.

The steady flow problem is solved iteratively since the free surface position is not initially known, and is computed as part of the solution process. We assume an initial position for the free surface, apply the boundary condition $\partial u/\partial n = 0$ and compute the potential u . Next, the computed values of u are compared with the assumed values of y. If there are some differences in their values, the free surface is adjusted to a new position according to the results for the potential and a new iteration is carried out.

For the rectangular dam, an analytical solution was derived by Polubarinova-Kochina [12]. Hornung and Krueger [13] have presented numerical results for Polubarinova-Kochina's formulation, for the height of seepage surface and the shape of the phreatic surface for some values of length (L), upstream height (H) and downstream height (h). Table 1 compares their results with those obtained by the B-spline BEM formulation.

L=0.5 H=1.0 h=0.5		
x	y	
	B-spline BEM	Analytical
0	1.000	1.000
0.1	0.966	0.967
0.2	0.920	0.920
0.3	0.859	0.860
0.4	0.782	0.782
0.45	0.732	0.733
0.5	0.664	0.662

L=1.0 H=1.0 h=0.0		
x	y	
	B-spline BEM	Analytical
0	1.000	1.000
0.2	0.936	0.938
0.4	0.850	0.850
0.6	0.736	0.738
0.7	0.671	0.671
0.8	0.593	0.595
0.9	0.502	0.503
1.0	0.365	0.368

Table 1: Results for homogeneous dam with vertical faces

The agreement between the two solutions is remarkable in spite of the coarse BEM discretization. The problem was also solved using a single node at point C; the results only changed for the flux in the neighbourhood of C, and did not affect the solution as a whole.

TRANSIENT FLOW

The boundary conditions for the transient flow problem generally specify the value of u or q at each node. On the free surface, however, both values are unknown but the following mathematical relations can be stated, using dimensionless variables [14]

$$u = \eta(x,t) \tag{12}$$

$$\left[\frac{\partial u}{\partial t}\right]_x = -\frac{1}{\cos\beta}\,q + W \qquad on \quad y = \eta \tag{13}$$

where η is the elevation above a reference datum, W is the recharge intensity and β is the angle between the free surface profile and the x-axis.

Applying a finite difference approximation to represent the free surface motion one can state, following Liggett and Liu [14]

$$u^{t+\Delta t} = u^t - \frac{\Delta t \cos\alpha}{\cos(\alpha+\beta^t)}[\theta q^{t+\Delta t} + (1-\theta)q^t] + W \tag{14}$$

where Δt is the nondimensionalized time interval ($= K\Delta t/\epsilon L$), ϵ is the effective porosity and α is the angle of displacement direction (fig. 3). This equation involves a linearization since the cosine of β is calculated at time t although the equation is written at time $t + \Delta t$. The use of small time steps is generally sufficient to produce good accuracy without iteration. Equation (14) provides an extra relation for the free surface nodes which completes the definition of the problem.

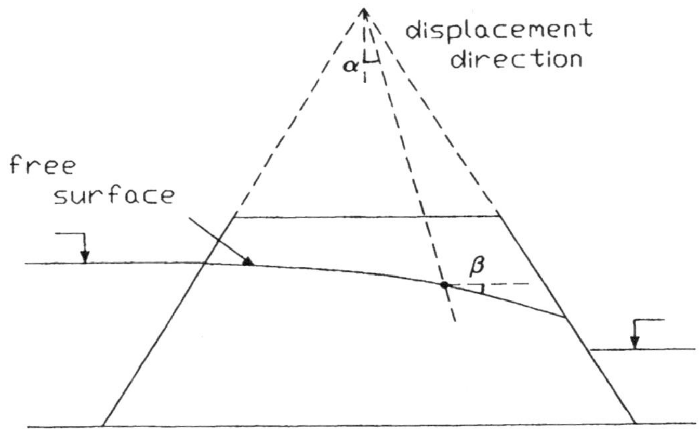

Figure 3: Flow through trapezoidal dam

Boundary conditions are then applied to equation (11). At fixed boundaries, either u or q is known at each point; at the free surface, expression (14) is substituted for $u^{t+\Delta t}$ and $q^{t+\Delta t}$ determined as part of the solution. Then, equation (14) can be applied to compute $u^{t+\Delta t}$ at each node explicitly.

After solution of the BEM equation (11) at each time step, the free surface nodes are displaced according to the following relation [14]

$$\eta^{t+\Delta t} = \eta^t - \frac{\Delta t \cos\alpha}{\cos(\alpha + \beta^t)}[\theta q^{t+\Delta t} + (1-\theta)q^t] + W \tag{15}$$

Flow through rectangular dam after drawdown on downstream face

This problem of practical importance has been analysed by several different numerical schemes [15],[16]. Here, the same values employed by Bruch [16] are used, *i.e.* dam width of 16 m, upstream water level at 24 m above the horizontal impervious base and a sudden drawdown from 24 m to 4 m at the downstream face. Other numerical data are hydraulic conductivity $K = 0.01$ cm/s, effective porosity $\epsilon = 0.4$, time relaxation factor $\theta = 0.69$ and time step $\Delta t = 0.445$ h.

At $t = 0$ the horizontal free surface violates the tangency condition. To avoid this problem, Liggett [15] used a time series solution to obtain the free surface profile at dimensionless time $t = 0.6$ which was then taken as the initial position. Cubic B-splines automatically follow the tangency condition by assuming a small curvature for the elements in the vicinity of the separation point at $t = 0$.

Table 2 shows the final position of the separation point above the dam base obtained by different methods. Polubarinova-Kochina [12] developed an analytical solution using the hodograph and complex potential planes which was later adjusted by Cryer [17]. Ozis [18] has used Cryer's formulation and improved

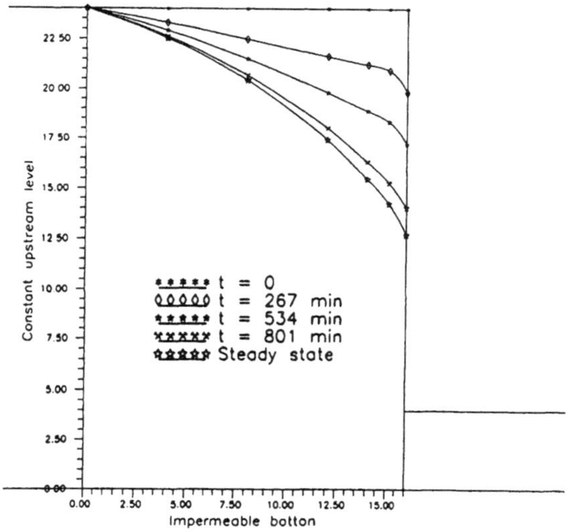

Figure 4: Vertical dam. Free surface positions at different times

Polubarinova-Kochina [12]	12.95
Cryer [17]	12.7132
Ozis [18]	12.7070
Bruch [16]	12.98
B-spline BEM	12.74

Table 2: Final position of the separation point above the dam base

the integrals evaluation using Gauss' quadrature with 64 integration points. Table 2 also shows results from Bruch [16] using linear boundary elements and an iterative procedure. Figure 4 shows free surface positions at different times.

NON-HOMOGENEOUS REGIONS

In practical problems, specially regarding earth dams, the water flows through zones of different materials. In this case the BEM formulation is applied to each zone independently and compatibility and equilibrium conditions used to relate the potential and flux between neighbouring zones [7].

For each node at the interface between zone i and zone $i+1$, the following compatibility and equilibrium conditions are generally used:

$$u^i = u^{i+1} \tag{16}$$

$$q^i = -\frac{K^{i+1}}{K^i} q^{i+1} \tag{17}$$

where K^i and K^{i+1} are the hydraulic conductivity of each zone.

x	B-spline BEM Two zones	Analytical [13]
0	1.000	1.000
0.2	0.936	0.938
0.5	0.796	0.796
0.6	0.736	0.738
0.9	0.499	0.503
1.0	0.370	0.368

Table 3: Validation test of multizone B-spline BEM computational code

The final system of equations for the whole region is obtained by adding the set of equations (11) for each subregion together with equations (16) and (17) for the interface nodes [7]. At the free surface of each zone equation (14) is applied.

After each time step, the free surface nodes were moved along lines radiating from a pole located at the intersection of lines passing along the lateral sides (fig. 3). In this way, one can avoid that the ratio between the size of neighbouring elements becomes very large.

In case of existence of an internal seepage surface, its rate of discharge has to be added to the next zone as a free surface recharge. In this work the recharge from the internal seepage surface was assumed as linearly applied over the first neighbour element.

Another point that has required careful modelization was the toe of the internal seepage surface. Figure 5 shows the four nodes assigned at the same geometrical position. Nodes C_2 and C_3 follow the interface compatibility and equilibrium conditions. Node C follows the free surface condition. The potential at node C_4 is equal to that of C, C_2 and C_3 but is still unknown, as well as the flux in C_4.

So the elements of matrix $\mathbf{H}$ related to node C_4 are added to the elements related to node C_3 (as the unknown potential has the same value for both) and the flux remains as a degree of freedom in C_4. In this way two unknown fluxes with similar equations are obtained for source points C_3 and C_4. To avoid the singularity of the system matrix, interpolated elements [11] are used for the elements that contain C_3 and C_4.

Two-zone rectangular dam with $K_1 = K_2$

For validation of the computational code, a two-zone rectangular dam with equal value of the hydraulic conductivity in both zones has been analysed. Table 3 compares the results with those published by Hornung and Krueger [13] for a homogeneous rectangular dam, with upstream height equal to 1.0, length equal to 1.0 and downstream height equal to 0.0. It can be seen that the accuracy of the B-spline BEM formulation is basically the same as for a single zone, table 1.

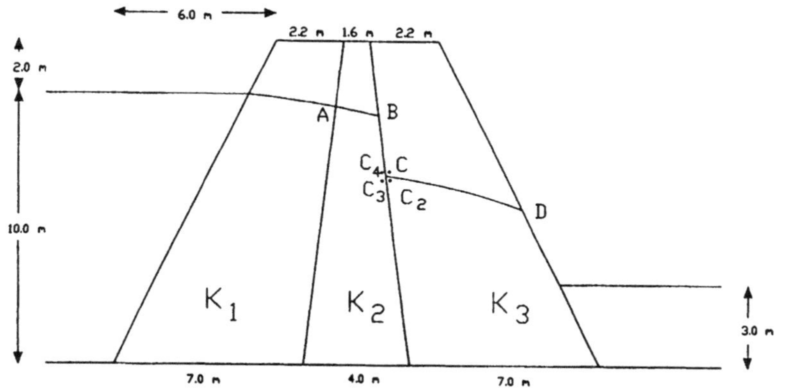

Figure 5: Vertical section of three zone dam: $K_1 = K_3 = 10^{-3} m/min$ and $K_2 = 10^{-2} m/min$

	B-spline BEM steady analysis	B-spline BEM transient analysis	BEM [19] quadratic elements
Number of Nodes	54	54	100
H_A	9.705	9.706	9.71
H_B	8.059	8.043	7.90
H_C	5.403	5.526	5.48
H_D	3.146	3.158	3.11

Table 4: Three-zone dam with internal seepage surface (see figure 5)

Three-zone dam with internal seepage surface

A steady-state free surface was obtained for a zoned dam with similar characteristics to that presented by Bruch *et al.* [19] and sketched in figure 5. Table 4 compares the present results with those of Bruch *et al.* [19], showing good agreement.

SALT WATER INTRUSION

In many practical cases the thickness of the transition zone between freshwater and saltwater is relatively small compared to the aquifer dimensions. It is thus a common approximation to neglect the transition zone and assume a sharp interface separating the two regions. Bear [9] has given some examples of real aquifers in which experimental measurements have shown the salt concentration to vary sharply at a determined location, clearly establishing a region of low salt concentration (freshwater) and a region of high salt concentration (saltwater).

In the BEM sharp interface model, equation (11) is applied for the fresh water and the salt water regions and interface conditions are imposed to combine both zones. For each node along the interface between the freshwater and saltwater zones the following conditions are used:

- Equilibrium of pressure on both sides of the interface, leading to

$$u_s = \frac{u_f}{\chi} + \frac{\chi - 1}{\chi} y \tag{18}$$

  where $\chi = \rho_s/\rho_f$, ρ is the density and the subscripts f and s are related to the fresh and salt waters.

- Compatibility of fluxes, which gives [14]

$$q_f = -\frac{\chi}{\alpha} q_s \tag{19}$$

  where $\alpha = \mu_s/\mu_f$ and μ is the dynamic viscosity.

After solving the BEM system of equations, the normal derivatives at the interface are used to compute the interface motion [14]. Assuming $x = \lambda(y,t)$, a finite difference approximation on time, similar to that of equation (14), is applied to compute λ after each time step.

Segregation by gravity action

This simple theoretical problem initially considers a vertical barrier separating fresh and salt waters in a horizontal confined aquifer of constant thickness. At time $t = 0$, the barrier is removed and the interface begins to tilt in a wedge-like shape due to the saltwater motion, forcing the freshwater upwards.

Several researchers have studied this problem applying theoretical, experimental and numerical techniques. Keulegan [20] presented an analytical solution based on the Dupuit approximation which is valid for large times, in the form

$$\ell(t) = \left(\frac{\Delta\rho\, K_f\, H}{\epsilon\, \rho_f}\right)^{\frac{1}{2}} t^{\frac{1}{2}}$$

in which ℓ is the location of the interface toe, H is the aquifer thickness and $\Delta\rho = \rho_s - \rho_f$. Gardner [21] modified Keulegan's expression to better represent the motion at small times, obtaining the equation

$$\ell(t) = \left(\frac{3t^{*2}}{4 + 3t^*}\right)^{\frac{1}{2}} H$$

where t^* is a dimensionless time defined as

$$t^* = \frac{\Delta\rho\, K_f}{\epsilon\, \rho_f\, H} t$$

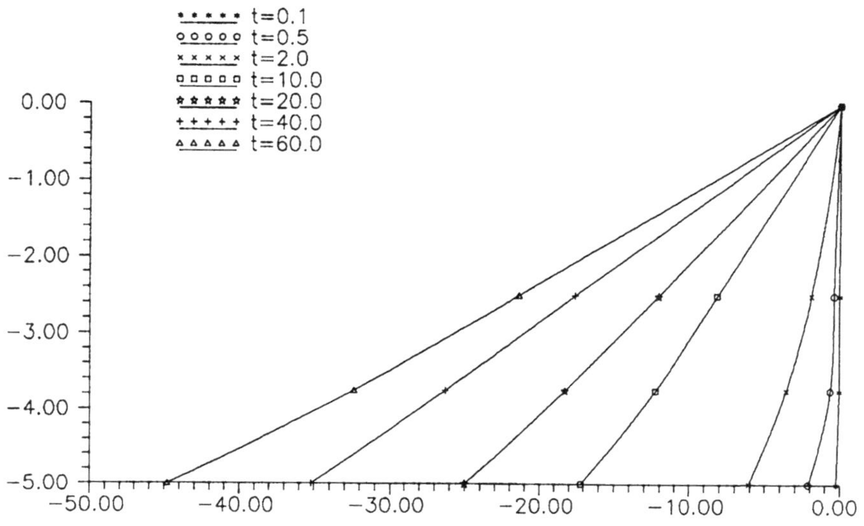

Figure 6: Segregation by gravity action: interface position at different times

Figure 6 shows the interface position at different times, as obtained by the present BEM formulation. The physical and geometrical characteristics of the aquifer are: $\rho_f = 1.0g/cm^3, \rho_s = 1.025g/cm^3, \mu_f = \mu_s = 1.2\times10^{-3}g/(cm\,s), K_f = 39.024m/day, K_s = 40m/day$ and $H = 10m$. The boundary element discretization employed 28 nodes along the external boundary and 7 nodes to model the interface. The initial time step was equal to $0.1day$; during the motion, this value was changed to $0.2day, 0.5day$ and finally $2.0day$.

Figure 7 compares the BEM results for the toe position with those obtained from the formulas of Keulegan [20] and Gardner [21]. Very good agreement between the solutions can be observed in the figure.

An interesting feature of this problem is that a Neumann boundary condition ($q = 0$) is applied all along the external boundaries, so the solution is not uniquely defined. To complete the problem specification and in order not to destroy the symmetry of the solution, a prescribed potential ($u = 0$) is applied at the middle point of the interface.

CONCLUSIONS

B-splines are very well suited for application in the BEM, both to represent the geometrical shape of the boundary and as interpolation functions for the potential and its normal derivative. Continuity up to the second derivative between adjacent elements can be provided by cubic B-splines and physical discontinuities

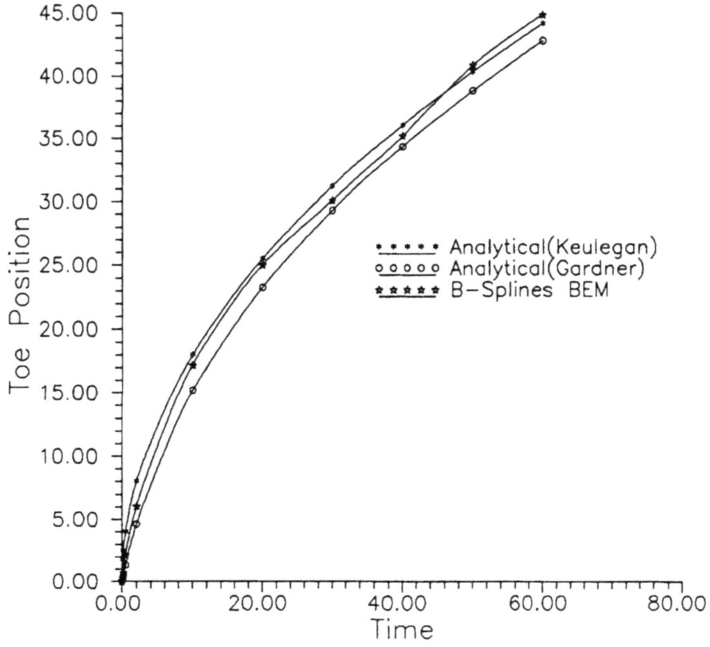

Figure 7: Segregation by gravity action: toe position at different times

can be handled by using multiple knots.

Two unknown fluxes at a corner or at a point presenting a discontinuity in the boundary conditions can be treated with B-splines using the so-called interpolated element.

For unconfined transient flow in porous media, the normal derivative at points along the moving boundary can be accurately computed using elements that provide derivative continuity. The results obtained suggest that, for the same level of accuracy, one can generally use far less degrees of freedom with B-spline elements than other kinds of elements.

Practical groundwater problems involving complexities such as zones of different materials, several moving boundaries and interfaces, external and internal seepage surfaces, etc., can be solved in an efficient, accurate and reliable way using B-spline boundary elements.

ACKNOWLEDGMENT

The first author would like to acknowledge the financial support of CAPES/Ministry of Education, Brazil.

References

[1] Chang,C.S., Boundary element analysis for unconfined seepage problems, *Journal of the Geotechnical Division, ASCE*, Vol. 114, pp. 556-572, 1988.

[2] Bruch,E. and Grilli,S., Computation of the transient seepage problem in anisotropic porous media by the BEM, *Boundary Elements IX*, Computational Mechanics Publications, Southampton and Springer-Verlag, Berlin, 1987.

[3] Cabral,J.J.S.P., Wrobel,L.C. and Brebbia,C.A., A BEM formulation using B-splines: I- uniform blending functions, *Engineering Analysis*, Vol. 7, pp. 136-144, 1990.

[4] Cabral,J.J.S.P., Wrobel,L.C. and Brebbia,C.A., A BEM formulation using B-splines: II- multiple knots and non-uniform blending functions, *Engineering Analysis*, Vol. 8, pp. 51-55, 1991.

[5] Yamaguchi,F., A new curve fitting method using a CRT computer display, *Computer Graphics and Image Processing*, Vol. 7, pp. 425-437, 1978.

[6] Bartels,R.H., Beatty,J.C. and Barsky,B.A., *An Introdution to Splines for Use in Computer Graphics and Geometric Modelling*, Morgan Kaufmann, Los Altos, California, 1987.

[7] Brebbia,C.A., Telles,J.C.F. and Wrobel,L.C., *Boundary Element Techniques*, Springer-Verlag, Berlin, 1984.

[8] Lafe,O.E., Montes,J.S., Cheng,A.H.P., Liggett,J.A. and Liu,P.L-F., Singularities in Darcy flow through porous media, *Journal of Hydraulic Engineering, ASCE*, Vol. 106, pp. 977-997, 1980.

[9] Bear,J. , *Hydraulics of Groundwater*, McGraw-Hill, New York, 1979.

[10] Crank,J., *Free and Moving Boundary Problems*, Clarendon Press, Oxford, 1984.

[11] Marques,E., Coupling of the Finite Element Method and the Boundary Element Method: An Application to Potential Problems, *M.Sc. Thesis*, COPPE/UFRJ, Rio de Janeiro, 1986 (in Portuguese).

[12] Polubarinova-Kochina,P.Y., *Theory of Groundwater Movement*, translated from russian by J.M.R. de Wiest, Princeton University Press, Princeton, New Jersey, 1962.

[13] Hornung,U. and Krueger,T., Evaluation of the Polubarinova-Kochina formula for the dam problem, *Water Resources Research*, Vol. 21, pp. 395-398, 1985.

[14] Liggett,J.A. and Liu,P.L-F., *The Boundary Integral Equation Method for Porous Media Flow*, Allen & Unwin Publishers, London, 1983.

[15] Liggett,J.A., Location of free surface in porous media, *Journal of Hydraulic Engineering, ASCE*, Vol. 103, pp. 353-365, 1977.

[16] Bruch,E., Application of the BEM to the steady and unsteady two dimensional phreatic groundwater flow, *Boundary Elements X*, Computational Mechanics Publications, Southampton and Springer-Verlag, Berlin, 1988.

[17] Cryer,C.W., *A survey of steady-state porous flow free boundary problems*, MRC Technical Summary Report 1657, Univ. of Wisconsin-Madison, USA, 1976.

[18] Ozis,T., *Ph.D. Thesis*, Brunel University, England, 1981.

[19] Bruch,E., Grilli,S. and Lejeune,A., Computation of the fluid flow in zoned anisotropic porous media and determination of the free surface seepage, *Boundary Elements VIII*, Computational Mechanics Publications, Southampton and Springer-Verlag, Berlin, 1986.

[20] Keulegan,G.H., Ninth progress report on model laws for density currents: An example of density current flow in permeable media, *Report No. 3411*, National Bureau of Standards, Washington, D.C., 1954.

[21] Gardner,G.H.F., Downie,J. and Kendall,H.A., Gravity segregation of miscible fluids in linear model, *Society of Petroleum Engineers Journal, Transactions AIME*, Vol. 225, pp. 95-104, 1962.

Domain Decomposition for Free Boundary Seepage

C.A. Papadopoulos (*), K.-P. Wang (**),
J.M. Sloss (*), J.C. Bruch, Jr. (**)
(*) Department of Mathematics,
(**) Department of Mechanical and Environmental Engineering, University of California, Santa Barbara, CA 93106, USA

ABSTRACT

Two domain decomposition techniques for solving a problem of flow through a porous dam in which there is a free surface are considered. An iterative procedure is proposed that reduces the problem to solving a sequence of two sets of problems. One set consists of solving independent Dirichlet problems on disjoint subdomains and the other set consists of solving independent Neumann problems on disjoint subdomains. Every Dirichlet problem subdomain has a common boundary with at least one Neumann problem subdomain. Along the common boundaries, Dirichlet data from the Neumann problems and Neumann data from the Dirichlet problems are used to generate a sequence of iterations. A relaxation is performed at the subdomain interfaces at each iteration as well as on each subdomain.

INTRODUCTION

For the past decade, parallel algorithms for solving problems using Multi-CPU computers have been studied intensively. Domain decomposition and multisplitting methods become an important subject for obtaining parallelism when solving a partial differential equation (PDE) on a known domain in which boundary conditions are assigned. The general idea to be carried out in domain decomposition is to subdivide the domain of definition of the PDE into a set of subdomains and then to obtain solutions of related PDE problems on each subdomain. The solutions on all the subdomains are then combined to obtain the approximate solution on the whole domain. Domain decomposition is advantageous in that it allows the efficient use of parallel processing in an obvious manner.

Many multisplitting techniques require overlaps between two adjoining subdomains. For example, the Parallel Schwarz alternating procedure (see references of [1]) equates the information on the overlapped region of two subdomains and solves each subdomain successively with data exchange between two adjacent subdomains. White [2] proposed several multisplitting techniques to divide the control domain into vertical, horizontal, or block subdomains with overlapped regions between adjacent subdomains. In addition, Rodrigue and Shah [1] imposed pseudo boundary conditions to accelerate the Parallel Schwarz method. Neumann and Dirichlet boundary conditions were imposed on each subdomain to ensure the continuities of the solutions. When trying to apply overlap methods to free boundary value problems a disadvantage arises since the free surface a priori is unknown, and one needs to assume that the free surface does not lie in the overlapped domain. It cannot be determined whether or not the free surface actually does lie in the overlap region.

However, there are techniques that do not require the overlapping of two neighboring subdomains when a free surface is not involved. For example, Farhat and Wilson [3] presented substructural concepts in finite element analysis using concurrent processing. Their basic approach was the automatic splitting of an arbitrary spatial domain, and processors are dynamically reassigned during the several phases of an analysis. Another example for domain decomposition without overlapping boundaries was proposed by Funaro et al. [4]. An interface relaxation factor θ was introduced in the iterative procedure. Iterative processes are then performed between two adjacent subdomains by imposing in one of them the continuity of the solution, and in the other one, the continuity of the normal derivative in a weak sense. At the limit of the convergence process, the transmission conditions at the interface are therefore satisfied. Marini and Quarteroni [5] proved the convergence of this iterative scheme for both the differential problem and its finite element approximation.

Papadopoulos et al. [6] have suggested a technique for domain decomposition on a problem involving a free surface which does not require an overlap region. In these seepage problems involving a free surface, the Baiocchi technique used to obtain a variational inequality can only be applied to regions with certain simple geometries. Domain decomposition of such problems can reduce the region containing the free surface to a simple one that allows a variational formulation, leaving the difficult geometry and boundary conditions to the regions which do not contain a free surface. A free surface seepage problem will be investigated here.

FORMULATION OF THE PROBLEM

Consider the following free boundary value problem: find the free surface in a steady, two-dimensional seepage through a rectangular dam. For simplicity, the soil in the flowfield is assumed to be homogeneous and isotropic, and the capillary and evaporation effects are neglected. In addition, the flow follows Darcy's Law:

$$\vec{q} = -k\ \nabla h = -k\ \nabla\left[\left(\frac{p}{\rho g}\right) + y\right] , \tag{1}$$

where $\vec{q}$ is the velocity vector, p is the pressure, k is the permeability of the soil, ρ is the density of the fluid, g is the gravitational acceleration, y is the vertical coordinate (positive upward), and h is the piezometric head. For homogeneous, isotropic soil, the permeability k is constant. The seepage velocity has a potential:

$$\phi = k\left[\left(\frac{p}{\rho g}\right) + y\right] . \tag{2}$$

In this study, the free surface, whose position is not known a priori, is to be found. On the free surface two boundary conditions have to be satisfied:

$$\phi = ky \tag{3}$$

and

$$\phi_\eta = 0 , \tag{4}$$

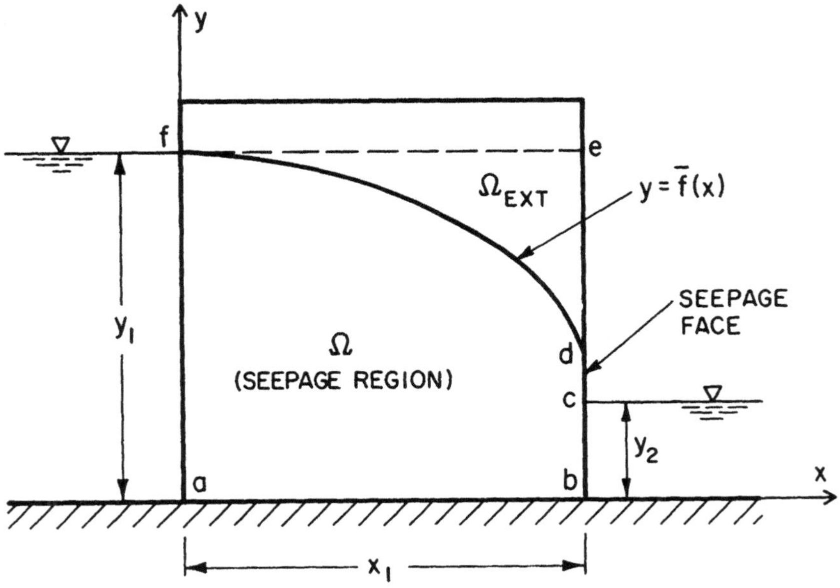

Figure 1. Free boundary seepage problem.

where η is the outward normal direction. Either Neumann or Dirichlet data are given on the remainder of the boundaries. The location of the free surface $y = \bar{f}(x)$ and the seepage domain Ω need to be found. $\bar{f}(x)$ satisfies

$$\bar{f}(0) = y_1, \quad \frac{d\bar{f}(0)}{dx} = 0, \quad \bar{f}(x_1) > y_2, \quad \text{and} \quad \frac{d\bar{f}(x_1)}{dx} = -\infty \ .$$

As shown in Fig. 1, the seepage region is defined as:

$$\Omega = \{ (x, y) \mid 0 < x < x_1, \ \ 0 < y < \bar{f}(x) \}$$

where x_1 is the horizontal distance from point a to b.

In the domain Ω, with the permeability k set equal to 1 for simplicity, the following conditions must hold:

$\Delta\phi = 0$	in Ω ,	(5)
$\phi = y_1$	on [af] ,	(5a)
$\phi = y_2$	on [bc] ,	(5b)
$\phi = y$	on [cd] ,	(5c)
$\phi = y$	on $\widehat{fd}$,	(5d)
$\phi_y = 0$	on [ab] ,	(5e)
$\phi_\eta = 0$	on $\widehat{fd}$,	(5f)
$\phi_\eta \leq 0$	on]cd[.	(5g)

The flow domain is not known a priori since the location of the free surface is unknown. A new known region D is defined as:

$$D = \{ (x,y) \mid 0 < x < x_1, \quad 0 < y < y_1 \} .$$

A new variable $\tilde{\phi}$ is defined:

$$\tilde{\phi}(x,y) = \begin{cases} \phi(x,y) & \text{in } \overline{\Omega} \\ y & \text{in } \overline{D} - \overline{\Omega} = \Omega_{ext} \end{cases} \tag{6}$$

which extends $\phi(x,y)$ into D. It follows that

$$-\Delta\tilde{\phi} = \left(\chi_\Omega\right)_y \tag{7}$$

in the sense of distributions on D, where $\chi_\Omega = 1$ in Ω, and $\chi_\Omega = 0$ elsewhere.

By using the Baiocchi transformation, a new dependent variable w is defined as:

$$w(x,y) = \int_y^{y_1} \left[\tilde{\phi}(x,\bar{\eta}) - \bar{\eta}\right] d\bar{\eta} \ . \tag{8}$$

Then w satisfies:

$$\Delta w = \chi_\Omega \qquad \text{in } D , \tag{9}$$

$$w(0,y) = \frac{1}{2}(y_1 - y)^2 \qquad \text{on } [af] , \tag{9a}$$

$$w(x,0) = \frac{y_1^2}{2} - \frac{y_1^2 - y_2^2}{2x_1}\,x \qquad \text{on } [ab] , \tag{9b}$$

$$w(x_1,y) = \frac{1}{2}\,(y_2 - y)^2 \qquad \text{on } [bc] , \tag{9c}$$

$$w = 0 \qquad \text{in } \overline{D} - \overline{\Omega} . \tag{9d}$$

If a w is found satisfying $\Delta w = \chi_\Omega$ in D subject to the above conditions, the following quantities can be obtained: $\Omega = \{ (x,y) \mid (x,y) \text{ in } D, w > 0 \}$; graph $\bar{f} = \partial\Omega - \partial D \equiv$ points of $\partial\Omega$ not in ∂D; and $\phi = y - w_y$ in Ω.

DOMAIN DECOMPOSITION

For simplicity, consider the differential equation:

$$\Delta w = f(x,y)\,\chi_\Omega \tag{10}$$

with homogeneous Dirichlet data. Decompose D into two non-overlapping regions D_1, D_2 with common boundary Γ such that D_2 is a region not containing the free surface.

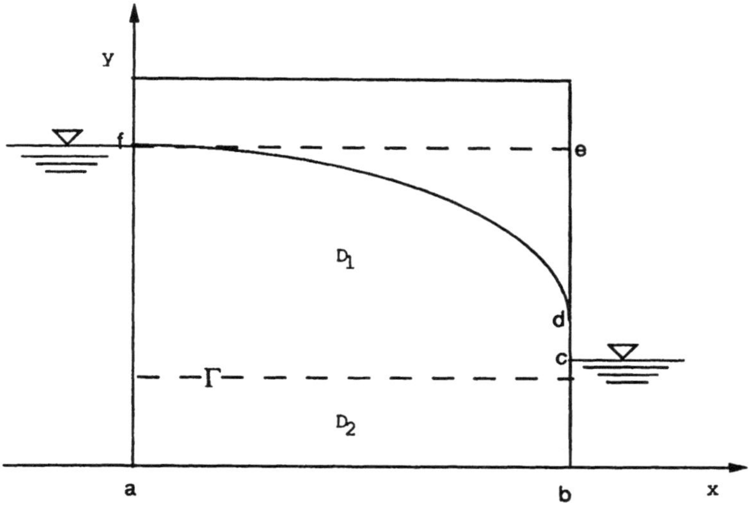

Figure 2. Theoretical domain decomposition.

Let $\Phi = \{v|_\Gamma \mid v \in H^1(D)\}$. For $\bar{\phi} \in H^{1/2}(\Gamma)$ let R_k: $H^{1/2}(\Gamma) \to H^1(D_k)$, k = 1,2 by

$$\int_{D_k} \nabla R_k \bar{\phi} \cdot \nabla v = 0 \quad \forall\, v \in H^1(D_k),\ \ v = 0 \text{ on } \partial D \cap \partial D_k, \ldots\ k = 1,2 \tag{11}$$

and consider the following pair of coupled problems:

Problem I: Given w_2 find $\{w_1, \Omega_1\}$ such that for $w_1 \in K = \{v \mid v \in H_1(D_1),\ v \geq 0$ on D_1, $v = w_2$ on $\Gamma\}$

$$\Delta w_1 = f(x,y)\, \chi_{\Omega_1},\quad w_1 \in K \cap H^2(D_1), \tag{12}$$

where

$$\Omega_1 = \{(x,y) \in D_1 \mid w_1(x,y) > 0\}\ . \tag{13}$$

Problem II: Given $\frac{\partial w_1}{\partial \eta}$ find w_2 such that

$$\Delta w_2 = f(x,y) \text{ in } D_2,\ \frac{\partial w_2}{\partial \eta} = \frac{\partial w_1}{\partial \eta},\ w_2 = 0 \text{ on } \partial D \cap \partial D_2\ , \tag{14}$$

where η is the exterior normal to D_2 on Γ.

Then it was shown by Papadopoulos et al. [6] that

Theorem 1: *If $\{w_1, w_2, \Omega_1\}$ solves problems I and II, then*

$$w = \begin{cases} w_1 & in\, D_1 \\ w_2 & in\, D_2 \end{cases}$$

solves $\Delta w = f(x,y)\, \chi_\Omega$ in D with $w \geq 0$ in D and $w = 0$ on ∂D.

Now consider the following iterative scheme.

Let $g \in \Phi$, $g \geq 0$ be given on D_1 and let

$$U_{ad}^{(1)}(g) = \{v \mid v \in H^1(D_1),\ v = 0 \text{ on } \partial D \cap \partial D_1,\ v = g \text{ on } \Gamma,\ v \geq \sigma g \text{ on } D_1\} \quad (15)$$

(in practice σg dies off quickly as we move away from Γ, e.g. less than a given tolerance.)

Given g^n, construct $w_1^n \in H^1(D_1)$, $w_2^n \in H^1(D_2)$ by:

Find $\{w_1^n, \Omega_1^n\}$, $w_1^n \in U_{ad}^{(1)}(g^n)$ and $\Omega_1^n = \{(x,y) \mid w_1^n(x,y) > 0\}$ such that:

$$\int_{D_1} \nabla w_1^n \cdot \nabla(v - w_1^n) \geq \int_{D_1} f(v - w_1^n) \quad \forall\ v \in U_{ad}^{(1)}(g^n) \quad (16)$$

and find $w_2^n \in U_{ad}^{(2)} = \{v \mid v \in H^1(D_2),\ \ v = 0 \text{ on } \partial D \cap \partial D_2\}$ such that:

$$\int_{D_2} \nabla w_2^n \cdot \nabla(v - w_2^n) = \int_{D_2} f(v - w_2^n) - \int_{D_1} \nabla w_1^n \cdot \nabla(R_1 \gamma_0(v - w_2^n)) + \int_{\Omega_1^n} f\, R_1 \gamma_0 (v - w_2^n)$$

(17)

from $g^{n+1} = \gamma_0(w_1^{n+1}) = \theta_n \gamma_0(w_2^n) + (1 - \theta_n)\, g^n$

with $0 < \theta_n < 1$ and γ_0 is the trace along Γ and proceed to construct $\{w_1^{n+1}, w_2^{n+1}, \Omega_1^{n+1}\}$ with the above scheme. From Papadopoulos et al. [6]

Theorem 2: Assume $\exists\ \theta_{min} > 0$ such that $\theta_n > \theta_{min}\ \forall\ n \geq 1$. If the sequence $\{\gamma_0\ w_1^n\}$ converges as $n \to \infty$, then the whole sequence $\{w_1^n, w_2^n\}$ converges to the solution $\{w_1, w_2\}$ of Problems I and II.

It can be shown that the following numerical scheme with homogeneous Dirichlet data on D is equivalent to the one above with certain regularity conditions. For simplicity, we let $\theta_n = \theta$, $n \geq 1$, be a constant.

$$\Delta w_1^n(x,y) = f(x,y)\chi_{\Omega_1} \qquad \text{in } D_1\,, \quad (18)$$

$$w_1^n(x,y) = g^n \qquad \text{on } \Gamma \quad (19)$$

and

$$\Delta w_2^n(x,y) = f(x,y) \text{ in } D_2 , \qquad (20)$$

$$\frac{\partial w_2^n}{\partial \eta}(x,y) = \frac{\partial w_1^n}{\partial \eta}(x,y) \qquad \text{on } \Gamma , \qquad (21)$$

where

$$g^{n+1} = \theta w_2^n(x,y) + (1-\theta)\, g^n . \qquad (22)$$

Equation (19) is the $\tilde{D}$ boundary condition, and Equation (21) is the N boundary condition. We denote this type of splitting as $\tilde{D}N$. The theory dictates that we need to have the Dirichlet boundary condition on the domain containing the free surface. In numerical implementation of this method for domain decomposition, it was found that we could create a variation of the scheme which formulates the problem as a $N\tilde{D}$ problem with the free surface domain containing Neumann data.

In cases of subdividing the whole domain into more than two subdomains, formulations are carried out in a similar manner. For each subdomain, Equation (10) (or (18) or (20)) holds inside and $\tilde{D}$ and N types of boundary conditions are assigned in pairs along the interface between two adjacent subdomains. For example, there will be two interfaces in the case of splitting the whole domain into three subdomains. Therefore, we can assign the boundary conditions with $N\tilde{D}$-$\tilde{D}N$ or $\tilde{D}N$-$N\tilde{D}$ combinations at the interfaces. In general if M subdomains are obtained, there will exist M-1 interfaces. Thus, several patterns of combinations of $\tilde{D}N$ pair interface boundary conditions will be studied.

NUMERICAL SCHEME

<u>Governing Equation</u>

The SOR (Successive over-relaxation) method combined with projection for the one domain case is one of the most successful iterative methods for solving systems of linear equations for this type of problem. During the computation using the SOR scheme, values at a new iteration as available are used along with values at the former iteration in order to speed up the convergence. Bruch [7] showed that this scheme can be used effectively in this study of the seepage problem.

When applying the SOR scheme, Equation (9) becomes:

$$w_{i,j}^{(m+1/2)} = \frac{\Delta x^2\, \Delta y^2}{2\left(\Delta x^2 + \Delta y^2\right)} \left[\frac{1}{\Delta x^2}\left(w_{i+1,j}^{(m)} + w_{i-1,j}^{(m+1)}\right) + \frac{1}{\Delta y^2}\left(w_{i,j+1}^{(m)} + w_{i,j-1}^{(m+1)}\right) - 1 \right] \qquad (23)$$

and

$$w_{i,j}^{(m+1)} = \max\left(0,\; w_{i,j}^{(m)} + \overline{\alpha}\left(w_{i,j}^{(m+1/2)} - w_{i,j}^{(m)}\right)\right) , \qquad (24)$$

where $\overline{\alpha}$ is the relaxation factor and the i and j are the column and row mesh point numbers, respectively. The superscript m in the parentheses refers to the iteration number within each of the two numerical approaches presented. The boundary data, Equations (9a)-(9c), must be taken into account.

This formulation, Equations (23) and (24), is for the one domain case. Two types of domain decomposition schemes will be employed in solving the seepage problem. The first involves splitting the region into two regions, D_1, D_2, with the same restrictions as the theoretical problem, i.e. the free surface must lie in D_1. An initial guess, g^1, for Dirichlet data on Γ is given. Then applying an SOR scheme with projection by implementing Equations (23) and (24) on D_1. This results in an approximate solution w_1^1 for D_1. Using this solution we calculate an approximation for the normal derivative on Γ using the three point second order scheme:

$$\frac{\partial (w_1^n)_{i,1}^{(m)}}{\partial \eta} = \frac{1}{2\Delta y}\left[3(w_1^n)_{i,1}^{(m)} - 4(w_1^n)_{i,2}^{(m)} + (w_1^n)_{i,3}^{(m)}\right] . \tag{25}$$

The above approximation will then be used as Neumann data at each grid point for solving the problem in region D_2, where a second order two point approximation is used for the normal derivative. An additional equation for w on Γ is then required. This involves assuming that the differential equation is satisfied along Γ, which is justified since Γ lies in D. This results in solving the following additional equation for the solution along the boundary where Neumann data is given in the SOR scheme.

$$\begin{aligned}(w_2^n)_{i,MAXROW2}^{(m+1)} = (1-\overline{\alpha})\,(w_2^n)_{i,MAXROW2}^{(m)} + \overline{\alpha}\,\frac{\Delta x^2\,\Delta y^2}{2(\Delta x^2+\Delta y^2)}\Bigg[&\frac{2}{\Delta y^2}\,(w_2^n)_{i,MAXROW2-1}^{(m+1)} \\ &+ \frac{1}{\Delta y^2}\left(4(w_1^n)_{i,2}^{(m+1)} - 3(w_1^n)_{i,1}^{(m+1)} - (w_1^n)_{i,3}^{(m+1)}\right) \\ &+ \frac{1}{\Delta x^2}\left((w_2^n)_{i+1,MAXROW2}^{(m)} + (w_2^n)_{i-1,MAXROW2}^{(m+1)}\right) - 1\Bigg]\end{aligned} \tag{26}$$

where MAXROW2 is the number of rows in D_2. This gives an approximation for w_2^1 for D_2. An interface parameter is used in Equation (22) to be used as Dirichlet data for the next iteration. As a result each iteration involves solving a system of equations in D_1 and D_2 using an SOR scheme with projection with a given tolerance for $w_1^2, w_2^2, w_1^3, w_2^3, \ldots$

Convergence on D_1 and D_2 is determined when

$$\max_{i,j}\left|\left(w_k^n\right)_{i,j}^{(m+1)} - \left(w_k^n\right)_{i,j}^{(m)}\right| < \varepsilon \tag{27}$$

for k = 1,2, respectively. Convergence of the entire scheme is determined on the interface as follows:

$$\max_i \left| (w_1^n)_i - (w_2^n)_i \right| < \varepsilon \qquad \text{on } \Gamma \ , \tag{28}$$

where $(w_1^n)_i = (w_1^n)_i^{(m+1)}$ and $(w_2^n)_i = (w_2^n)_i^{(m+1)}$. The scheme is stopped when a given tolerance, ε, is met along the interface and convergence of each SOR iteration has been successful.

The second method is a variation of the above method which may involve splitting the region into columns, rows, or blocks of subdomains. Inside each subdomain, Equations (23) and (24) still hold. In addition, the boundary conditions are selected in $\tilde{D}N$ pairs and are imposed on the interface between two adjacent subdomains. This method differs from the first in that the subdomains have no restriction to where the free surface lies. Also only a single iteration of the SOR method with projection is completed before data is exchanged along the interfaces, i.e. after each pass through the mesh in one domain using the algorithm, Equations (23) and (24), there is an exchange of data at an interface or interfaces and then the algorithm is applied to the next domain and data is again exchanged at an interface or interfaces, etc. Convergence is determined when equations similar to Equations (27) and (28) are met throughout D for a given iteration.

RESULTS

The numerical example problem investigated used the following data: $y_1 = 20.0$, $y_2 = 10.0$ or 5.0, $x_1 = 15.0$, $\overline{\alpha} = 1.85$, $\varepsilon = 0.001$, $\Delta x = \Delta y = 1/3$ and θ's as stated. The first case analyzed is that shown in Fig. 3 where $y_2 = 10.0$, and the extended solution domain $D = \{(x,y) \mid 0 < x < 15.0, 0 < y < 20.0\}$ is subdivided as shown where $D_1 = \{(x,y) \mid 0 < x < 15.0, 10 < y < 20.0\}$, $D_2 = \{(x,y) \mid 0 < x < 15.0, 0 < y < 10.0\}$ and $\Gamma = \{(x,y) \mid 0 < x < 15.0, y = 10.0\}$. Also shown is the free surface location that was obtained which was the same for the two iteration schemes presented. The free surface is taken as the first mesh point with a value of w that is less than 0.01 when you move in the vertical direction for a fixed x. The θ for the first iteration scheme (the one which iterated to convergence in a region before passing data on Γ) was 0.47 while θ for the second iteration scheme (a modified version of the first, i.e. passing of data after each iteration through the mesh using the SOR scheme with projection) was 0.45. Since convergence in these domain decomposition schemes is a function of θ, numerical experiments were performed to determine the best value to use. Tables 1 and 2 list the results of these experiments for the two approaches. The heading 'Iterations' in the Tables stands for the total number of iterations to obtain overall convergence. The second iteration scheme requires approximately half the number of iterations as the first scheme, who's implementation was suggested by the theoretical decomposition. Thus there is a considerable improvement in cost as well as more freedom in how to split the domains.

Table 1. Iteration information obtained by varying θ in the first iteration scheme.

θ	0.30	0.40	0.46	0.47	0.48	0.50	0.60
Iteration	502.5	396	360.5	348.5	349	371	525.5

Table 2. Iteration information obtained by varying θ in the second iteration scheme.

θ	0.10	0.20	0.40	0.45	0.50	0.53	0.56
Iterations	175	167	150	149	150	190	642

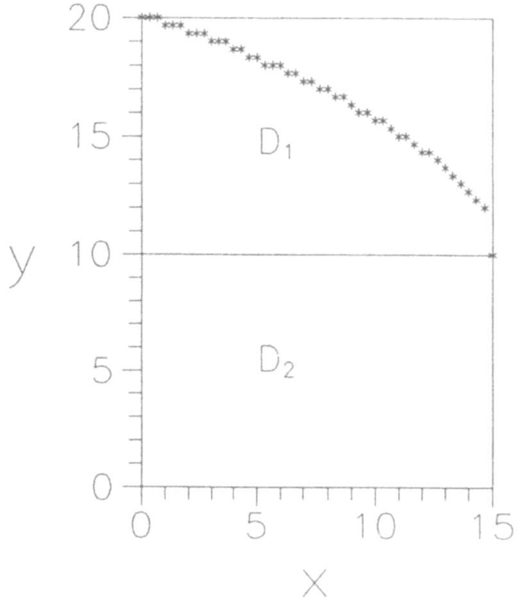

Figure 3. Horizontal decomposition.

Figures 4 - 6 demonstrate three different domain decompositions using the case for y_2 = 5.0, θ = 0.1 and the second iteration scheme. The final free surface location in all three examples was the same. The number of iterations for convergence in each case was 116, 122, and 146, respectively.

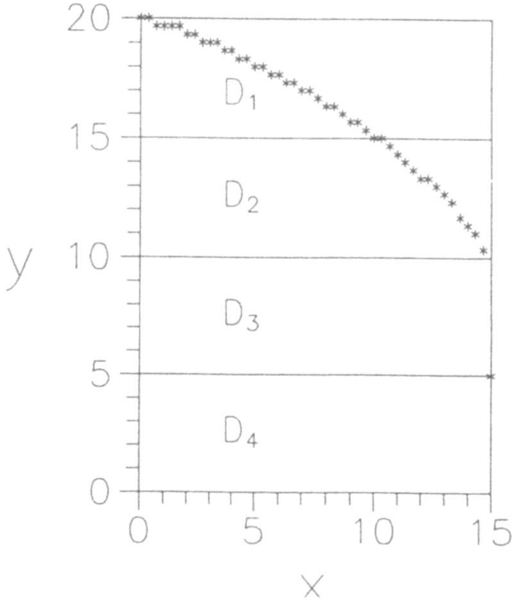

Figure 4. Horizontal decomposition.

ACKNOWLEDGEMENT

Some of this material is based upon work performed by the second and fourth authors who were supported by the National Science Foundation under Award No. ECS-9006516.

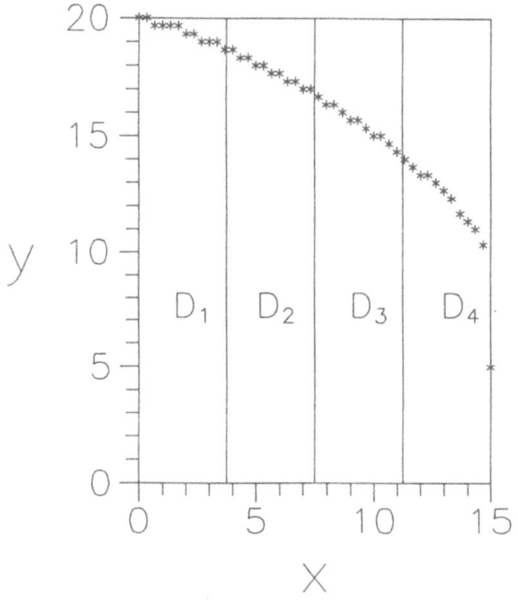

Figure 5. Vertical decomposition.

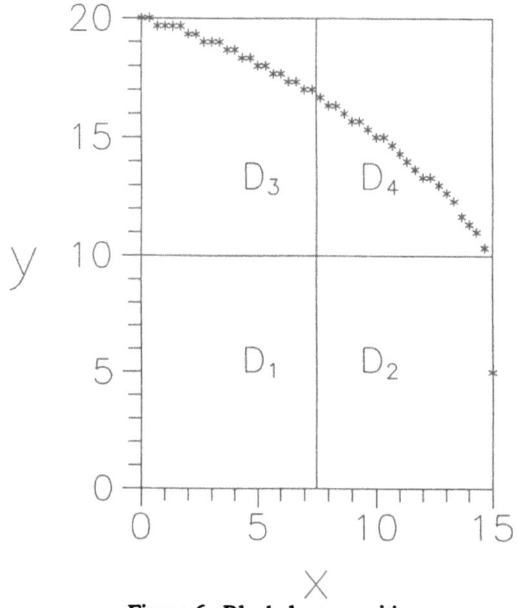

Figure 6. Block decomposition.

Note: The boundary data on the interfaces in Figs. 4-6 was: For Fig. 4: between regions D_1 and D_2 it was a $N\tilde{D}$ combination; between regions D_2 and D_3 it was a $\tilde{D}N$ combination; and between regions D_3 and D_4 it was a $N\tilde{D}$ combination. For Fig. 5: there were $\tilde{D}N$, $N\tilde{D}$, and $\tilde{D}N$ combinations from left to right interfaces, respectively. For Fig. 6: regions D_1 and D_4 had $\tilde{D}$ boundary data while regions D_3 and D_2 had N boundary data on the interfaces.

REFERENCES

1. Rodrigue, G. and Shah, S. Parallel Supercomputing: Method, Algorithms and Applications, John Wiley and Sons, New York, pp. 77-88, 1989.

2. White, R.E. Multisplittings and Parallel Iterative Methods, Comp. Meth. Appl. Mech. Engng., Vol. 64, pp. 567-577, 1987.

3. Farhat, C. and Wilson, E. A New Finite Element Concurrent Computer Program Architecture, Int. J. Num. Meth. Engng., Vol. 24, pp. 1771-1792, 1987.

4. Funaro, D., Quarteroni, A., and Zanolli, P. An Iterative Procedure with Interface Relaxation for Domain Decomposition Methods, SIAM, J. Numer. Anal. Vol. 25, pp. 1213-1236, 1988.

5. Marini, L.D. and Quarteroni, A. A Relaxation Procedure for Domain Decomposition Methods Using Finite Element, Numer. Math., Vol. 55, pp. 575-598, 1989.

6. Papadopoulos, C.A., Sloss, J.M. and Bruch, J.C. Jr. A Domain Decomposition Technique Applied to a Free Surface Problem, to appear in Proceedings of the Conference on Numerical Methods for Free Boundary Problems, (Ed. Neittaanmaki, P.), Jyvaskyla, Finland, 1990.

7. Bruch, Jr., J.C. A Survey of Free Boundary Value Problems in the Theory of Flow Through Porous Media: Variational Inequality Approach, Advances in Water Resources, Part I, Vol. 3, pp. 65-80, 1980.

Spreadsheet Solution for Transient Fluid Flow in a Porous Media with a Moving Drainage Boundary

D.J. Kerkes

Department of Civil Engineering, University of Connecticut, Storrs, Connecticut 06269-3037, USA

ABSTRACT

This is an application oriented paper that describes how the problem of two-dimensional drainage with a moving boundary can be solved simply and efficiently with a desk top personal computer and commercially available spreadsheet software, using the numerical method of finite differences and an explicit form of the solution. The built-in features of spreadsheet software can be combined to perform operations which once required that an engineer formulate a solution using a specific computer language, such as BASIC or FORTRAN, and the spreadsheet's graphics capabilities greatly facilitate the presentation of results. To illustrate how the solution algorithm is developed, a detailed example is provided from the field of geotechnical engineering. The geotechnical problem of soil consolidation is one of transient fluid flow in a porous media, and the two-dimensional form of the diffusion equation can be used effectively to model the phenomenon. The consolidation problem is one of many encountered in the construction of an earth dam, with the rising fill constituting a moving drainage boundary. From this example, the reader will see that it is a simple matter to extend the solution to more than one moving boundary.

The material in the paper has been presented so that the logical sequencing of operations, as well as the general manner in which spreadsheets function, is clear enough to the reader who has no experience with spreadsheets; however, a working knowledge of spreadsheets on the part of the reader is necessary for a thorough understanding of the discussion.

DESCRIPTION OF THE PROBLEM

The diffusion equation is used to model a variety of engineering problems; therefore, to assist the reader in following the specific application discussed herein, and applying the method to their particular problem, a brief description of the geotechnical problem is presented in the following paragraphs to acquaint the reader with some of the terminology and variables that will be used throughout the paper.

In its most abstract form, an earth or rockfill dam is simply a stable structure, constructed of naturally occurring geologic materials, that restricts the flow of water at its location to the point of creating an impoundment. An earth dam typically consists of two basic zones of material. One zone's primary function is to retard the flow of water through the embankment sufficiently to create an impoundment; therefore, the essential criteria of the material in this zone is that of low permeability or hydraulic conductivity. This zone is typically referred to as the core or core zone. The other basic zone provides support to the core zone on both the upstream and downstream sides to insure the structure's stability against catastrophic release of the impoundment; thus, the essential criteria of the material in this portion of the dam is strength. This zone is referred to as the shell or upstream and downstream shoulders.

The moist soil, commonly clay, used as the construction material for the core zone experiences a volumetric reduction under the weight of the rising fill, and the water in the soil pore space experiences a pressure increase in response to the load. The pore water pressures generated in this way can have an adverse effect on the stability of the earth structure during and immediately subsequent to construction. Therefore, it is essential that the designer be able to estimate the magnitude of these pore pressures throughout the structure, as well as the rate at which these pressures dissipate.

The problem of predicting pore pressures in the core of a zoned embankment can be thought of in terms of essentially three interdependent components. The first involves establishing the distribution of stresses within the embankment, which is a function of the embankment geometry and construction material properties. Second is the matter of defining the pore pressure response of the soil to the stress changes that occur as the embankment is constructed. Third is the consolidation process, which addresses the dissipation of excess pore pressures and attendant volumetric strain of the core material. The first two components of the problem are beyond the scope of this paper and have been discussed by the author elsewhere [2]. This paper will focus on the third component, which involves defining the pore pressure as a function of two spatial variables and time, and the nature of the problem is one of fluid flow in a porous media, with the rising fill constituting a moving drainage boundary.

SPREADSHEETS

Before proceeding with a discussion of the solution algorithms, a brief explanation will be given of what a spreadsheet is and how it operates. A spreadsheet, or electronic worksheet, is simply an array of rows and columns, the intersections of which make up the spreadsheet elements that are referred to as cells. The columns are typically designated by capital letters and the rows are designated by numbers; thus, the intersection of the particular column and row defines the location of a specific cell and constitutes that cell's address: for example, A1, J200, etc. A cell may contain text, numbers, formulas or logical statements. When formulas or logical statements are contained in a cell, the formula or statement is hidden and only the result is displayed on both the monitor and a print-out. Consequently, the manner in which a particular spreadsheet actually functions is not apparent from simply viewing the display or a printout. Placing the cursor over any cell, however, reveals the true nature of its contents on the computer monitor.

The real value of spreadsheet software as an analytical tool lies in the features that enable the cells within the spreadsheet to interact, and the speed at which such interactions or computations are carried out. Formulas within the spreadsheet are automatically recalculated whenever the contents of any cell in the spreadsheet is changed. Thus, by linking formulas to the contents of other cells, through the use of relative and/or absolute cell addresses, the spreadsheet can be modified by simply changing the value of any controlling parameters contained in appropriately identified cells. In addition to the interactive cells, the spreadsheet possesses various software features that are executed through the use of numerous preprogrammed commands, which the user accesses and directs by specifying the cells upon which the desired operations are to be performed. These features include a complete set of trigonometric, exponential and statistical functions, as well as matrix operations, along with an assortment of operations that permit the contents of cells to be moved, copied or otherwise manipulated. By properly sequencing the spreadsheet's various features, the user can perform mathematical operations that once required the formulation of a program in a specific computer language. While a detailed discussion of the myriad features of spreadsheet software is beyond the scope of this paper, a brief discussion of the algorithms quintessential commands will be given. The spreadsheet discussed in this paper was developed using Lotus 1-2-3 (copyright Lotus Development Corporation) [4], and while the various functions are described herein using the Lotus format, comparable functions exist in the various other quality spreadsheet packages available on the market.

Through user specified "macros," a spreadsheet can be made to function in the same way as a conventional computer program.

A macro is a sequence of keystrokes that are entered into a cell, as a label, in the same order that they would be executed from the keyboard for any sequence of operations that can be performed by the spreadsheet. Directing the cursor to that cell, using the appropriate keystroke, causes the operations to be performed. Coupled with the interactive cells, macros enable spreadsheets to execute complicated mathematical operations, subroutines, loops, etc., in the same way as programs written in BASIC or FORTRAN.

THE SOLUTION ALGORITHM

As previously noted, the consolidation problem can be modeled effectively using the two-dimensional form of the diffusion equation, which is:

$$c_v\left\{\frac{\partial^2 u}{\partial z^2} + \frac{\partial^2 u}{\partial y^2}\right\} = \frac{\partial u}{\partial t} - \frac{\partial \sigma}{\partial t} \tag{1}$$

in which c_v is the coefficient of consolidation, u is the excess pore water pressure, t is time, and σ is total stress. The explicit form of the finite difference approximation to this equation is:

$$\begin{aligned} u(z_0,y_0,t+\Delta t) &= u(z_0,y_0,t) + \sigma(t+\Delta t) - \sigma(t) + \\ &\frac{c_v\Delta t}{(\Delta z)^2}\{u(z_2,y_0,t) + u(z_4,y_0,t) - 2u(z_0,y_0,t)\} + \\ &\frac{c_v\Delta t}{(\Delta y)^2}\{u(z_0,y_1,t) + u(z_0,y_3,t) - 2u(z_0,y_0,t)\} \end{aligned} \tag{2}$$

where (z_0,y_0,t) corresponds to point 0 in Figure 1 at time t, etc. Equation (2) simply states that the value of u at any point in the consolidating zone subsequent to an incremental change Δt in time, can be determined from the values of u prior to the change Δt at that point and four points immediately adjacent to that point, located at increments Δz above and below, and Δy to the left and right, as illustrated in Figure 1. Thus, the determination of the values of u throughout the core zone during the consolidation process is a straightforward matter, provided the initial values of u are known throughout the zone and the values of Δz, Δy and Δt are properly chosen.

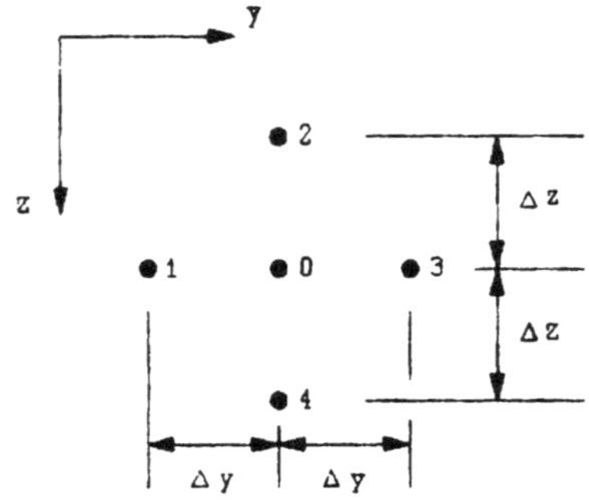

FIGURE 1: Finite difference grid reference system

The advantage of the explicit form of the approximation to the governing partial differential equation is that it involves a very simple algorithm in the spreadsheet, despite the fact that it imposes certain restrictions on both the spatial and time increments in order to ensure convergence and stability of the approximate solution. It has been shown that both convergence and stability of equation (2) will be assured if the following criteria is satisfied:

$$\frac{c_v \Delta t}{(\Delta z)^2 + (\Delta y)^2} \leq \frac{1}{8} \tag{3}$$

While this restriction does increase the execution time, the author has not found this to be a serious limitation to the method.

The second and third terms on the right side of equation (2) $[\sigma(t+\Delta t) - \sigma(t)]$ simply define the change in total stress $(\Delta\sigma)$ over the time increment (Δt), and the change in pore pressure is a function of this change in total stress; therefore, the term can be interpreted as a pore pressure increment. The total stress change and associated pore pressure increase is due to the addition of successive layers, or lifts, of material onto the embankment. Application of the equation in the model implies that each lift is applied to its full height instantaneously at the end of the time increment, which is a reasonable assumption provided the lift thickness is small, since time does elapse between lifts as the contractor works his equipment over the fill to raise it uniformly. Thus, equation (2) describes the pore pressure dissipation that occurs between application of successive lifts, over the time step Δt, and applies a pore pressure increment in response to the application of the next lift placed on the fill. As the embankment rises, however, the length of the drainage paths change, particularly in the vertical direction, which must be accounted for in the analysis.

In its most basic form, the analysis is performed with a spreadsheet that consists of (a) an algorithm to define the pore pressure increment, (b) an algorithm, using equation (2), to model the pore pressure dissipation, and (c) a conditional statement to model the moving upper drainage boundary of the advancing fill.

To develop the spreadsheet, the core geometry is first divided into nodes, defined in terms of their elevation and offset from the centerline of the dam. Figure 2 illustrates the finite difference grid adopted for the dam selected for this example. Within the spreadsheet, a block of columns and rows, referred to as a range, are respectively assigned centerline offsets and elevations, and specific cells within that block correspond to the nodes of the grid. It is often convenient to "name" the range so that the range can be specified subsequently within a macro by its name. Names also make it easier to follow the macro operations, in much the same way as comment statements do in a conventional computer program. Since core zones are not typically rectangular, certain cells within the designated range will not

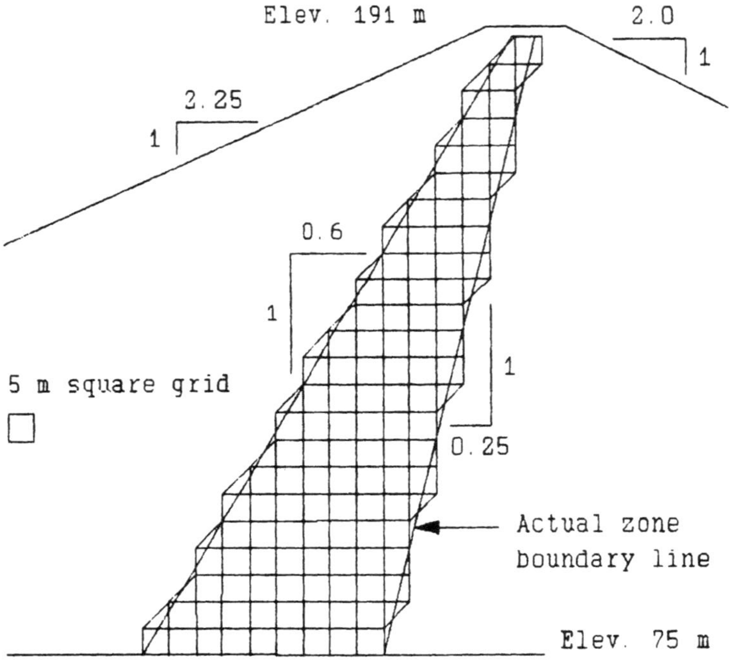

FIGURE 2: Core zone finite difference grid

correspond to nodes and are simply ignored by the spreadsheet in the solution. Figure 3 illustrates the way in which the core zone finite difference grid, shown in Figure 2, is defined in the actual spreadsheet. The row of letters at the top of Figure 3 and the column of numbers at the left, illustrate how the columns and rows appear in the actual spreadsheet, with the columns defined by the letters and the rows by the numbers. The zeros are resident in the cells that represent the nodal points of the finite difference grid, and the interior lines illustrate how a range of cells would be defined. Only rectangular ranges or blocks can be specified, which are defined by the two cells at the corners of the rectangle's diagonal; for example, the range \U1 in Figure 3 would be defined simply as D25..S48, which specifies all the cells from column D row 25 through column S row 48 inclusive. Each spreadsheet will have a specific series of steps for the user to follow to "name" the range, in order to store that piece of information in the spreadsheet's memory so that the those respective cells can subsequently be identified or accessed simply by the name, in this case \U1. The \ in the range name prevents the spreadsheet from interpreting the name as a single cell address, rather than a block of cells.

While the problem is three-dimensional in space and time, the spreadsheet is only two-dimensional. This limitation is overcome by using macros to create a loop, comparable to a DO LOOP in a conventional computer program, where each execution of the loop corresponds to an increment in time. Figure 4 presents a flow diagram for the loop and will assist in following the discussion. The actual macros are quite simple and the details will be given subsequently.

	A	B	C	D	E	F	G	H	I	J	K	L	M	N	O	P	Q	R	S
20	CENTERLINE			71.2		61.2		51.2		41.2		31.2		21.2		11.2		1.2	
21	OFFSET (m)				66.2		56.2		46.2		36.2		26.2		16.2		6.2		-3.8
22																			
23	ELEV. (m)																		
24					Range Name: \U1 (Pore pressures at time t)														
25	190																	0	0
26	185																0	0	0
27	180															0	0	0	
28	175															0	0	0	
29	170														0	0	0	0	
30	165														0	0	0	0	
31	160													0	0	0	0		
32	155												0	0	0	0	0		
33	150												0	0	0	0	0		
34	145											0	0	0	0	0	0		
35	140											0	0	0	0	0			
36	135										0	0	0	0	0	0			
37	130									0	0	0	0	0	0	0			
38	125									0	0	0	0	0	0	0			
39	120								0	0	0	0	0	0	0				
40	115								0	0	0	0	0	0	0				
41	110							0	0	0	0	0	0	0	0				
42	105						0	0	0	0	0	0	0	0	0				
43	100						0	0	0	0	0	0	0	0					
44	95					0	0	0	0	0	0	0	0	0					
45	90					0	0	0	0	0	0	0	0	0					
46	85				0	0	0	0	0	0	0	0	0	0					
47	80			0	0	0	0	0	0	0	0	0	0						
48	75			0	0	0	0	0	0	0	0	0	0						

FIGURE 3: Typical range of nodal values in the spreadsheet

To create the loop, one block or range of cells in the spreadsheet is assigned the pore pressure values at time t for the nodes corresponding to the respective cells. The cells within this range contain numerical values, and the range was assigned the name \U1 for this analysis. Another range of cells within the same spreadsheet, given the name \U2, contains the finite difference formulas for the pore pressure at time $t+\Delta t$, for each corresponding node in \U1. These formulas are simply the repeated application of equation (2), the details of which will be discussed subsequently. The variables for the nodal formulas in \U2 (the second range) that change with time are contained in \U1 (the first range). The loop consists of copying the values obtained from the nodal formulas in \U2 over the values in \U1, after which the spreadsheet automatically re-executes the nodal formulas in \U2 using the new values in \U1. Thus, at the end of the first loop \U1 displays the pore pressures at time t+Δt and \U2 displays the pore pressures at time t+2Δt. In this manner, each cycle of the loop corresponds to an increment in time.

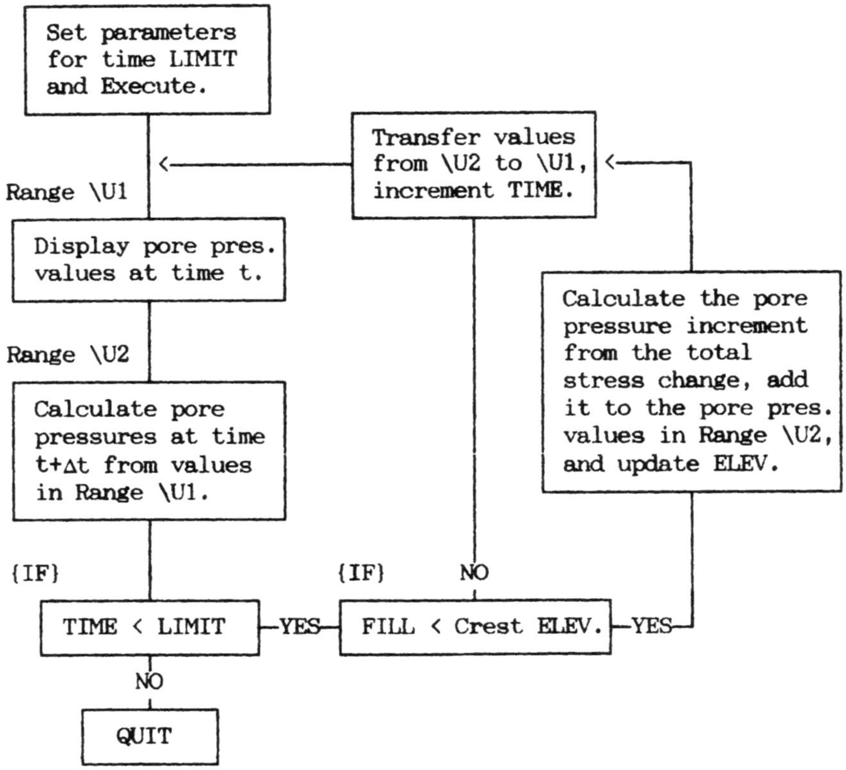

FIGURE 4: Spreadsheet flow diagram

The heart of the solution and the key to modeling the moving boundary is found in the formulas that reside in the cells of range \U2, and the relative ease in which these formulas are entered into the spreadsheet. These formulas (a) model the moving drainage boundary, (b) apply an appropriate pore pressure increment, and (c) model the dissipation of pore pressures over the time period represented by the analysis. The formulas use logical statements of the form IF(condition,x,y) which cause the spreadsheet to process the argument x if the condition is true or the argument y if it is false. In the field, the top of the fill constitutes a free draining boundary, as do the upstream and downstream sides of the core zone. In the model, a free draining boundary is defined by specifying a constant value of zero for the excess pore pressure at the nodes along the boundary. Therefore, in the model all nodes are assigned an excess pore pressure value of zero until the elevation of the fill exceeds that of the node, after which a pore pressure increment is applied to the node. This is accomplished with a conditional statement of the form IF(Node Elev.>=Fill Elev., 0, Eq. (2)), which assigns a zero pore pressure value to the node if it is above the present elevation of the fill thereby making it a drainage boundary for the node below it, provided the elevation of that node is below the elevation of the fill. The finite difference equation (2) is used once the elevation of the fill exceeds that of the node. Referring to Figure 1, the actual expression corresponding to equation (2) in the IF statement above will be of the form:

$$u(z_0,y_0,t+\Delta t) = u(z_0,y_0,t) + \Delta u + \mathrm{CRIT}\{u(z_2,y_0,t) + u(z_4,y_0,t) + u(z_0,y_1,t) + u(z_0,y_3,t) - 4u(z_0,y_0,t)\} \qquad (4)$$

In equation (4), CRIT is a range name, with an absolute cell address, which contains the following expression, from equation (2):

$$\mathrm{CRIT} = \frac{c_v \Delta t}{(\Delta z)^2} \qquad (5)$$

where Δz, Δy and Δt are chosen so as to satisfy the stability and convergence criteria for the finite difference approximation, given by equation (3). In equation (5), the vertical and horizontal increments (Δz and Δy) were set equal so that terms could be conveniently combined to yield equation (4).

Consider, for example, the node at elevation 100 m and centerline offset 31.2 m in Figure 3. This cell, which corresponds to $u(z_0,y_0,t)$ in equation (4), is contained in range \U1 and has L43 as its cell address. Similarly, $u(z_0,y_1,t)$ would have K43 as its cell address, and $u(z_4,y_0,t)$ would have L44 for a cell address, both in range \U1. Now, if cell L105 in another range contains the value of Δu appropriate to the node at elevation 100 m and centerline offset 31.2 m, then the spreadsheet formula for that same node in range \U2 would be:

```
@IF($A43>=FILL,0,L43+L105+CRIT*(L42+L44+K43+M43-4*L43))    (6)
```

The @ sign before the IF is part of the spreadsheet command, and the asterisk (*) in the expression is the multiplication operator. Thus, if the elevation of the fill (contained in range FILL) is less than the elevation of the node (contained in cell A43), the node is assigned a zero value, because the dam does not yet exist at that point; however, the finite difference formula is applied once the elevation of the fill exceeds that of the node.

The use of relative and absolute cell addresses in formula (6) permits the formulas to be physically written only once, then copied into as many cells as necessary to define the nodal points for the core zone finite difference grid by using the spreadsheet's copy command. Note in Figure 3 that column A contains the elevations of the nodal points. The $ in the A43 cell address ties all formulas to the A column when they are copied into their respective locations in range \U2. This is known as a mixed cell address. The cells corresponding to the upstream and downstream drainage boundaries in range \U2 simply contain the value zero; however, a conditional statement could be written into the cells corresponding to the upstream drainage boundary that would increase the water pressures at these locations as a function of time to simulate reservoir impoundment.

As previously noted, the details of how the pore pressure change is computed from the total stress change associated with application of each successive lift have been presented by the author elsewhere. Suffice it to say that a third range, which contains cell L105 in formula (6), is defined within the spreadsheet, the cells of which correspond to the nodes of the finite difference grid in the same way as the cells in ranges \U1 and \U2. The cells in this third range contain formulas of the form $\Delta u = f(\Delta\sigma)$, where the total stress change ($\Delta\sigma$) is computed for the position of each respective nodal point within the core zone from the dimensions of the applied lift and unit weight of the construction material. In this manner the distribution of stresses within the embankment as the fill elevation increases is accounted for in the value of the pore pressure increment used in the analysis.

This analysis used two macros, that were named \A and \B, to model the time element. The macros were linked to several parameters stored in cells that were identified by specific range names, where the ranges were simply defined B20..B20, etc., for a single cell. These ranges are given in upper case letters throughout this paper. The macros were basically defined as follows.

```
\A {WINDOWSOFF}{PANELOFF}
   {IF FILL<134}{LET LIFT,0.176*Dt}{LET FILL,FILL+LIFT}/XG\B~
   {IF FILL<136}{LET LIFT,0.016*Dt}{LET FILL,FILL+LIFT}/XG\B~
   {IF FILL<191}{LET LIFT,0.154*Dt}{LET FILL,FILL+LIFT}/XG\B~
   {LET FILL,191}{LET LIFT,0}/XG\B~
\B /RV\U2~\U1~{LET TIME,TIME+Dt}{IF TIME>=LIMIT}{QUIT}
   /XG\A~
```

The expressions in brackets { } are specific macro command options that are part of the features of the spreadsheet, while the expressions preceded by a / are instructions that are also executable by the user from the keyboard. The ~ in the macro is equivalent to striking the enter key on the keyboard. The {WINDOWSOFF} and {PANELOFF} commands greatly reduce execution time, but are not required. The macro is accessed from the keyboard by holding down the ALT key, then simultaneously depressing the key defining the macro name, which in this case would simply be the A key. The \ key is not used to access the macro from the keyboard even though the \ appears in the name.

Macro \A calculates the thickness of each lift placed on the fill, based on the rate of fill placement anticipated during construction, and computes the current elevation of the fill, stored in range FILL. When the rate of fill placement is not constant, a series of IF statements can be written and stacked to model any desired construction sequence. For the purpose of this example, three fill placement rates were used: a rate of 0.176 meters per day (m/day) was used up to elevation 134 m; 0.016 m/day between elevations 134 m and 136 m; and 0.154 m/day between elevations 136 m and the top of the dam at elevation 191 m. The spreadsheet reads the IF statement and continues program execution on the same line if the statement is true, but drops to the line below if the statement is false, where it encounters another IF statement, and so on. The IF statements in macro \A compare the elevation of the fill (FILL) to the elevation at which the placement rate changes. If the current elevation of the fill is below that elevation, program control passes to the LET statement, which computes the current lift thickness, stored in range LIFT. The lift thickness is simply the product of the fill placement rate and the time increment Δt, which is a user specified value stored in range Dt. The next LET statement increases the elevation of the fill (FILL) by adding the lift thickness (LIFT). The final statement in the row (/XG\B~) passes control to macro \B. When the fill elevation (FILL) is at or above the elevation specified in the IF statement, control is passed to the statement in the cell below. The final statement in the macro fixes the fill elevation at the design crest elevation, since the final lift will probably bring the embankment a fraction of a meter over this elevation, and all subsequent lifts are set at zero thickness. The last statement in this row (/XG\B~) again passes control to macro \B.

Macro \B transfers the contents of range \U2 as values, rather than the resident formulas, into range \U1 by means of the Range~Value commands (/RV\U2~\U1~). The spreadsheet keeps track of time by incrementing a specified cell, assigned the range name TIME, by Δt each time the loop is executed. If the spatial increments (Δz and Δy) for the finite difference grid are selected by the designer, then the time step (Δt) must be established so as to satisfy the stability and convergence criteria given by equation (3). The macro then compares the elapsed time, in range TIME, to the user specified time limit, which is stored in range LIMIT. The {QUIT} command terminates macro execution when the elapsed time is greater than or equal to the specified limit, but until then control passes to the next line, which instructs the computer to re-execute macro \A, thereby re-initiating the loop.

The nodal values that appear in range \U1 are the pore pressures corresponding to the time and embankment elevation displayed in the ranges named TIME and FILL respectively.

Therefore, the designer can examine the excess pore pressure distribution at any time after the start of construction by simply specifying the value in TIME. To continue the model, the designer need only redefine the value in range TIME and re-initiate the loop by invoking macro \A. In this way the spreadsheet can model the process of pore pressure development during construction, and pore pressure dissipation both during and subsequent to construction. By expanding the macros or adding another, the spreadsheet can be programmed to store the contents of any cells that the designer desires during macro execution. This data, coupled with the spreadsheet's graphics capability enables the designer to prepare figures such as Figure 5, which compares the pore pressures predicted by the model against actual field data obtained from one of many piezometers that were placed in the embankment during construction.

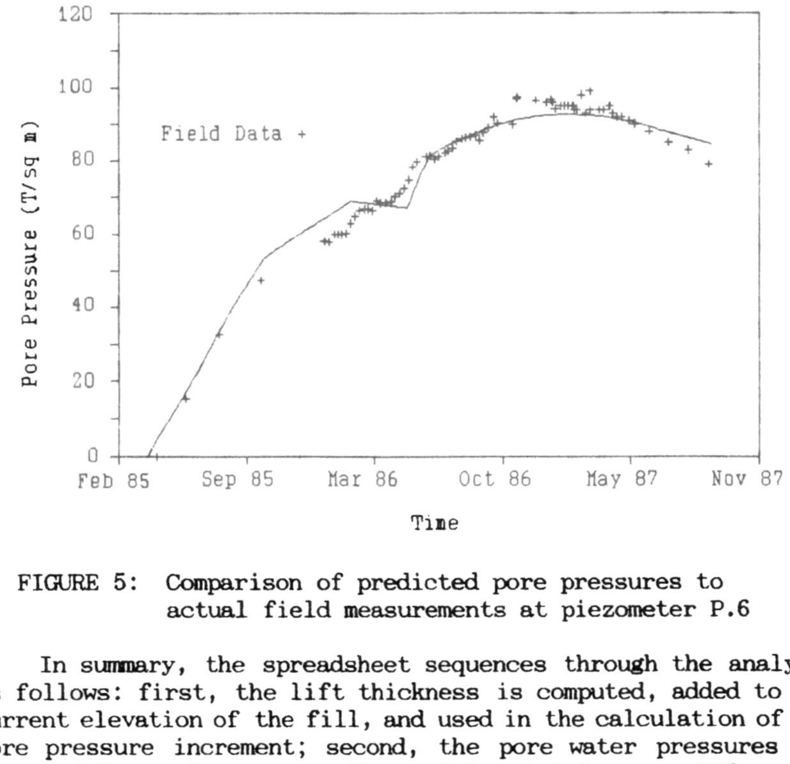

FIGURE 5: Comparison of predicted pore pressures to actual field measurements at piezometer P.6

In summary, the spreadsheet sequences through the analysis as follows: first, the lift thickness is computed, added to the current elevation of the fill, and used in the calculation of the pore pressure increment; second, the pore water pressures are computed for each corresponding nodal point in range \U2, using the appropriate pore pressure increment and the previous pore water pressures from range \U1; and third, the pore water pressures in range \U2 are transferred as values into range \U1, the current time is incremented by Δt, and the sequence is re-initiated until the specified time limit is reached. When execution begins, the pore pressure values are initially zero in \U1 and consequently zero in \U2, until the first lift of material is placed, after which positive values are applied to the formulas in \U2 as the elevation of the fill rises to the maximum crest elevation, with no increases thereafter. When execution is terminated, the pore water pressures contained in range \U1 are those appropriate to the time and embankment elevation indicated in ranges TIME and FILL respectively.

The main components of the spreadsheet formula (6) used in the preceding example and typical of the formulas contained in range \U2, can be summarized as follows:

(a) The expression given as formula (6) is the conditional statement, resident in the cells of range \U2, that considers the moving upper drainage boundary of the fill by comparing the elevation of the nodal point to the current elevation of the fill.
(b) The second argument in the conditional statement of formula (6), of the form L43+L105+CRIT*(L42+L44+K43+M43-4*L43), is the typical form of the finite difference equation for two-dimensional consolidation, which contains a pore pressure incrementing term.
(c) The pore pressure is incremented after each lift is applied to the fill, and this increment is represented by cell L105 in the conditional statement of formula (6).

CONCLUSIONS

With the ever increasing use of desk top personal computers (PC's) in engineering firms comes the need for computational software suited to the PC; however, this does not necessarily mean that special application software must be either developed or purchased. Problems that can be modeled using the method of finite differences can be analyzed simply and efficiently using a PC and commercially available spreadsheet software, thus offering the practicing engineer an excellent alternative to the more conventional computerized solutions. In the preceding pages, the author has attempted to demonstrate the way in which a spreadsheet can be used to model the problem of transient fluid flow in a porous media with a moving drainage boundary. An example from the field of geotechnical engineering was used to illustrate the primary components of the solution algorithms in what the author hopes is sufficient detail to enable another engineer to develop a similar spreadsheet for his or her particular application. Extending a similar spreadsheet model to consider more than one moving boundary would be equally as straightforward as the example presented herein. While the author did not discuss the treatment of impervious boundaries, the appropriate equation is no more complicated than the equation for a drainage boundary used herein.

The built-in features of a spreadsheet lend themselves well to the analysis of a transient fluid flow problem with a moving boundary that is three-dimensional in space and time. The spreadsheet's rows and columns can be used to keep track of the two spacial variables, and the development of a simple macro enables the spreadsheet to model the time variable. The computational power of the spreadsheet's interactive cells, coupled with the use of a simple conditional statement, enables the spreadsheet to model the moving boundary. By the use of additional conditional statements, the spreadsheet can be made to vary the physical conditions along any particular boundary. For example, a simple conditional statement linked to the time variable could be used to model the effect of reservoir impoundment on post construction pore pressure dissipation for the problem presented herein.

The use of a spreadsheet presents several advantages, not the least of which are the relative ease with which problems can be input and the speed in which the analysis is executed. The spreadsheet is particularly well suited to sensitivity analyses or "what if" scenarios, since the entire spreadsheet is recomputed whenever a change is made to a cell whose address is linked to the formulas in the spreadsheet. As if these advantages were not enough, the spreadsheet's graphics capabilities greatly facilitate the presentation of results. An indirect advantage associated with the use of a spreadsheet is that it requires a better understanding of the theoretical aspects of the basic problem than is sometimes necessary with special application software; consequently, the responsibility for engineering is placed on the designer, where it belongs, rather than the computer.

REFERENCES

1. Gerald, C. F. & Wheatly, P. O. Applied Numerical Analysis, Menlo Park, California, Addison-Wesley Publishing Co., 1985

2. Kerkes, David J., "Analysis and Prediction of Stresses and Pore Pressures Associated with Wet Core Construction for Embankment Dams," Ph.D. Thesis, University of Colorado, May 1990

3. Kerkes, David J., "Application of Spreadsheet Software to the Solution of Problems in Consolidation," University of Connecticut Department of Civil Engineering, Report CE 90-187, October 1990

4. Lotus 1-2-3 Reference Manual, Cambridge, Mass., Lotus Development Corporation, 1985

5. Scott, R. F. Principles Of Soil Mechanics, Menlo Park, California, Addison-Wesley Publishing Co., 1963

Water Table Aquifers and Finite Element Method: Analysis and Presentation of a Case Study

A. Dassargues

Laboratoires de Géologie de l'Ingénieur, d'Hydrogéologie et de Prospection géophysique (L.G.I.H.), University of Liège, Bat. B19, 4000 Liège, Belgium

ABSTRACT

In groundwater modelling of water table aquifers the finite element codes searching for the free (or water table) surface and adapting the mesh at each time step can be considered now as obsolete. Different ways of modelling the free surfaces in transient conditions and with a fixed mesh have been found and used. The main concepts of the methods, based on the introduction in the constitutive laws of a storage or permeability variation law, are recalled and discussed. Non-linear codes are required but CPU time is spared because the geometry of the mesh has not to be adapted in function of the new position of the water table. One of the methods developed at the University of Liege and based on a storage variation law has been used to simulate the water table alluvial aquifer of the River Meuse located downstream the city of Liege (Belgium). In this simulation a very detailed 3D mesh of 8-nodes brick finite elements represents the discretized aquifer and a very accurate calibration has been completed fitting the model results on more than 200 measurement piezometers.

INTRODUCTION

Classically, the simulation of groundwater problems by the finite element method is realized using linear codes.However, in the case of water table aquifers, the free surface corresponds to an unknown seepage boundary and the mesh must be adjusted after each time step. As the discretization is changed, the solution of a new equation system has to be found, based on the new boundary conditions. If the mesh adjustments are made by an automatic meshing procedure, some distorted finite element could appear in the neighbourhood of the new free surface boundary.

New methods, based mainly on the works of Bathe and Khoshgoftaar [1], have been developed using non-linear codes. Despite the fact that normally, non-linear codes consumes more CPU time than linear ones, for this particular application, it is certainly not the case considering the mesh adjustment procedure which takes a very long CPU time and increases the risk of instability or non-convergence by introduction of uncontrolled distorted elements.

FLOW EQUATION IN CONFINED AND UNCONFINED AQUIFERS

The usual form of the equation describing the flow in a porous medium is written :

$$\frac{\partial}{\partial x_i} (K_{ij} \frac{\partial h}{\partial x_j}) - Q = S_s . \frac{\partial f}{\partial t} \qquad (1)$$

where general index notations are adopted, x_j are the coordinates, K_{ij} is the permeability tensor, h is the piezometric head, Q is the external sink/source term and S_s is the specific storage coefficient.

For confined aquifers, S_s is deduced (De Marsily [2]) from :

$$S_s = \rho . g . m_v \qquad (2)$$

where the water compressibility and the solid grain compressibility are neglected in regard to the compressibility of the porous medium (m_v), ρ is the water volumic mass.

The storage coefficient of the confined aquifer is written :

$$S = \int_o^d S_s . dz \qquad (3)$$

where d is the thickness of the confined aquifer and z the vertical coordinate.

For unconfined or water table aquifers, the storage coefficient, noted S, is usually approximated by the effective porosity (n_e) of the geological formation, corresponding to the quantity of water that could be obtained by the total drainage of the porous medium.
In some cases of intensive water withdrawal, a confined aquifer may become a free or water table aquifer by lowering the piezometric head under the top of the upperlying aquitard. In such particular conditions, the storage coefficient passes from its value in equation (3) to the value of the effective porosity (n_e), i.e. there is a discontinuity.

METHOD USING THE NON-LINEARITY OF THE PERMEABILITY

The permeability coefficient is introduced in the non-linear code, as a function of the pore pressure (figure 1). Different relationships between K and the pore pressure (p) can be chosen depending on the nature of the studied porous medium.

In the non-saturated part ($p \leq 0$), this approach can take into account the fact that the medium situated above the water table is not completely impervious. As K is not equal to zero, a (reduced) unsaturated flow is simulated and, of course, it can influence the flow in the saturated zone.

As mentioned by Dysli and Rybisar [3], from an hydrogeological point of view, the unsaturated zone can be characterized by a relationship linking the permeability coefficient to the suction pressure (figure 2). This one corresponds to the negative pressure energetically equivalent to the pressure which is applied to the medium in order to mobilize the capillarity water. The evolution of water content of the porous formation in function of the suction pressure varies also from a geological medium to another (figure 2).

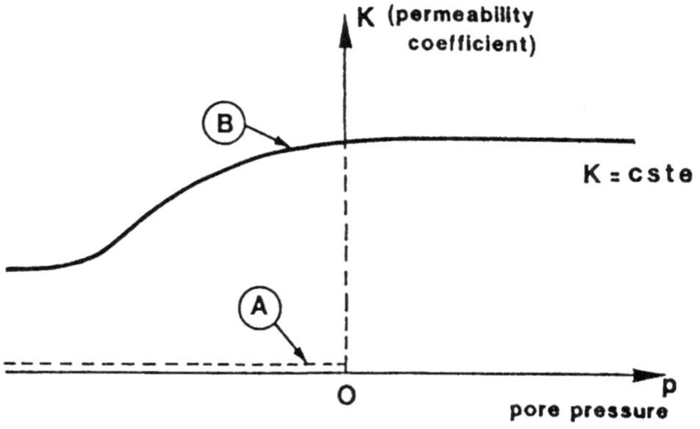

Fig.1. Permeability coefficient in function of the pore pressure :
A) theoretical relationship used in common FEM codes with adjustment of the mesh to the water table
B) relationship used in the non-linear code

(after [3])

In flow-compaction problems linked to the lowering of a water table aquifer, a coupling relation linking the permeability coefficient to the porosity or the void-ratio has to be introduced in a second finite element code. The coupling could be achieved by an interprocessor (Dysli and Rybisar [3]), based on a relationship between the variation of pore pressure and the variation of effective stress i.e. the Terzaghi [4] principle.

In flow-compaction problems induced by lowering of the pore pressure in confined aquifers (most cases of land subsidence are of this type), a second coupling affecting the storage coefficient is necessary as expressed by equation (2).

METHOD USING THE NON-LINEARITY OF THE STORAGE COEFFICIENT

The storage coefficient is introduced in the non-linear code, as a function of the pore pressure (Dassargues, Radu and Charlier [5]). The method has been imagined from the modelling by "enthalpic technique" of the phase changes in heat conduction problems : the phase changes occur at constant temperature, with heat storage. In our case, the geological porous medium passes from unsaturated to saturated state at a constant zero pressure with storage of water. This storage corresponds mainly to the effective porosity of the medium. The variation of the storage in function of the pore pressure is shown for a sandy aquifer on figure 3.

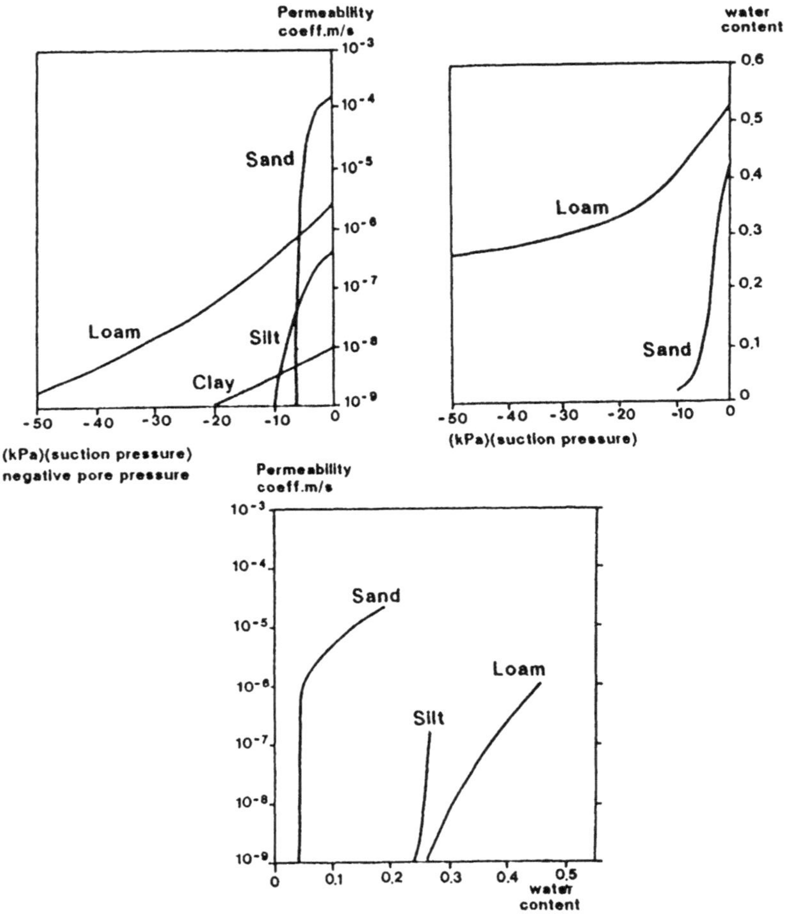

Fig.2. Evolution of the permeability and the water content in function of the suction pressure, for different porous media.

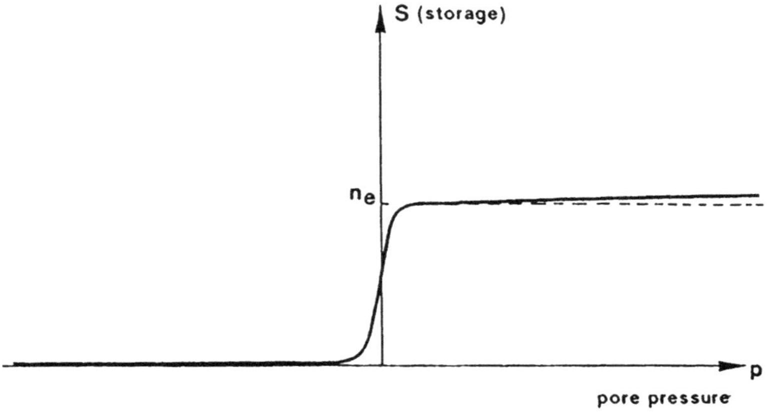

Fig.3. The real storage law.

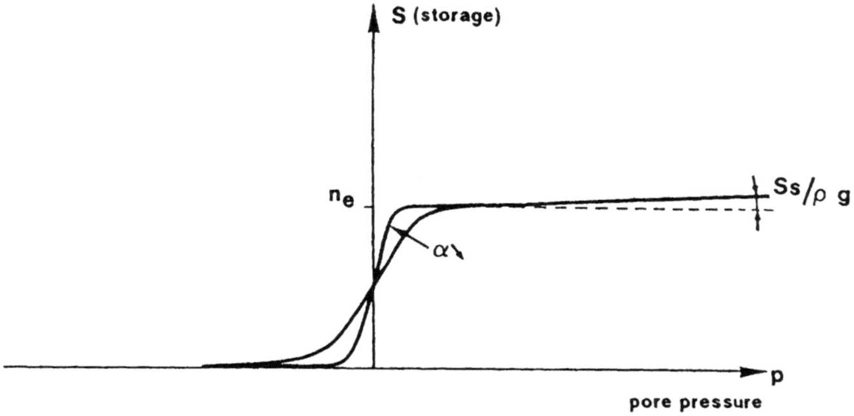

Fig.4. The storage law introduced in the non-linear code LAGAMINE.

In the non-linear code, the storage law which approximates the real variation is an arc-tangent function (figure 4) (Charlier, Radu and Dassargues [6]) :

$$S = n_e \left(\frac{1}{\pi} \operatorname{arctg} \left(\frac{p}{\alpha} \right) + \frac{1}{2} \right) + \frac{S_s}{\rho . g} \cdot \langle p \rangle \tag{4}$$

where S_s = specific storage coefficient

$\langle p \rangle = p$ if $p > 0$

$\langle p \rangle = 0$ if $p \leq 0$

α = coefficient influencing the shape of the function (figure 4).

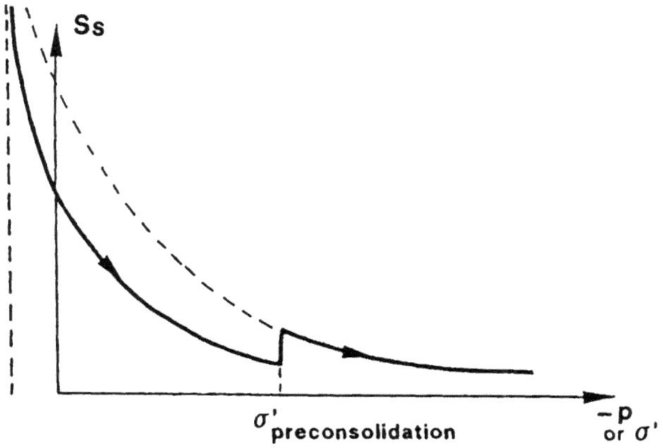

Fig.5. Evolution of the S_s value in function of the pore pressure (or minus the effective stress, by the Terzaghi principle and with constant total stress).

In flow-compaction problems linked to the lowering of a water table, any coupling relation linking the permeability coefficient to the porosity or void-ratio can be chosen without problem. This new non-linearity can be introduced in the same FEM code.

In flow-compaction problems induced by the lowering of the pore pressure in confined aquifers, the variation of S_s (figure 5) is introduced by the following coupling relation (Dassargues [7] :

$$f^V = S_s \cdot \dot{p} / \rho . g \tag{5}$$

where f^V is the total quantity of water expelled during the compaction and $\dot{p}$ is the time derivation of the pore pressure.

Assuming the Terzaghi principle with a constant total stress, this equation can be written :

$$- f^V = S_s . \dot{\sigma}' / \rho . g = -\dot{\varepsilon} \tag{6}$$

where $\dot{\varepsilon}$ is the time derivation of the strain.

APPLICATION

Our non-linear code,called LAGAMINE (Charlier [8]), has been already used to model different kinds of problems :

- model of the Ekofisk oil-field subsidence (Schroeder et al. [9])
- groundwater model of a regional water table aquifer,called "Hesbaye aquifer", in Belgium (Bolly et al. [10])
- model of the subsidence of the central area of Shanghai (Dassargues et al. [11] and Dassargues & Li [12]).

The case study that will be illustrated hereafter consists of the modelling of a regional water table aquifer: the alluvial aquifer in the valley of the River Meuse, downstream to the city of Liege (Belgium). The LAGAMINE code has been applied with the non-linear storage law in order to simulate the water table conditions.

A very detailed model of the alluvial plain has been built, including a complex spatial distribution of the permeability (more than 25 different values are used). Additionally, all the geometrical features and constraints are physically represented in the model : uniform recharge, leakage from a canal, infiltration from the hills feeding the alluvial plain, uplift infiltration from a confined aquifer in karstified limestones underlying the alluvial plain, impervious banks of the River Meuse in some places, pumping and recharge. In such a complex case study, the non-linear code with implementation of the storage law for modelling the water table, is very useful avoiding the need for an automatic meshing procedure on a regional domain with highly heterogeneous conditions and more than 2350 elements disposed in four layers.

The calibration with such a detailed spatial distribution of the permeability was particularly uneasy, but the final results are fitted with measured piezometric maps very accurately (less than 15 cm of mean error on the whole domain). One of the computed piezometric map is shown on figure 6. On basis of these calibrated results, fluxes maps have been computed and drawn (figure 7) showing the main trends in the alluvial aquifer.

ACKNOWLEDGEMENTS

The case study described here has been carried out under a contract with the Public Works Ministry of Belgium.

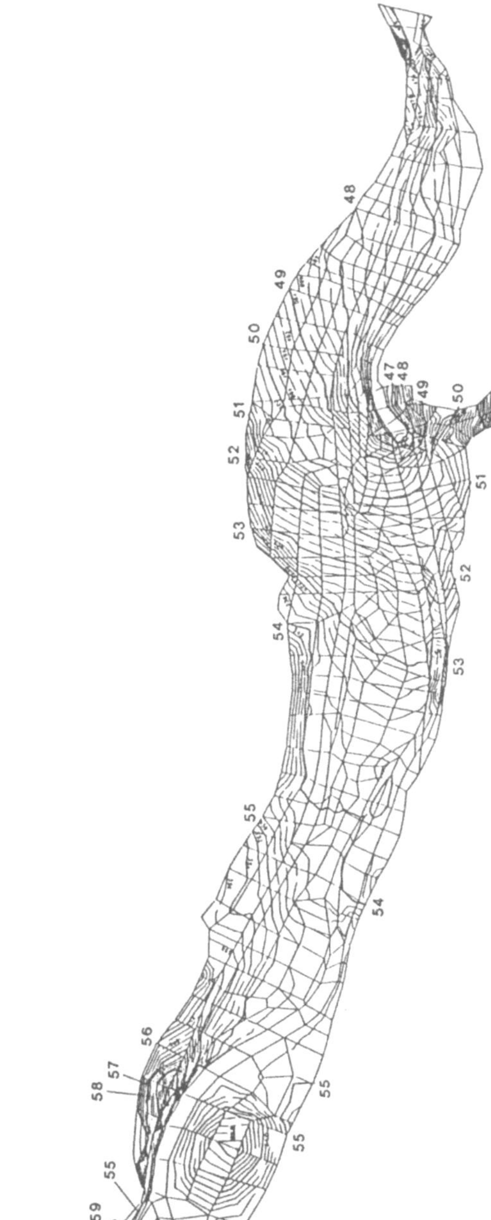

Fig.6. Computed piezometric map corresponding to the situation of July 1989.

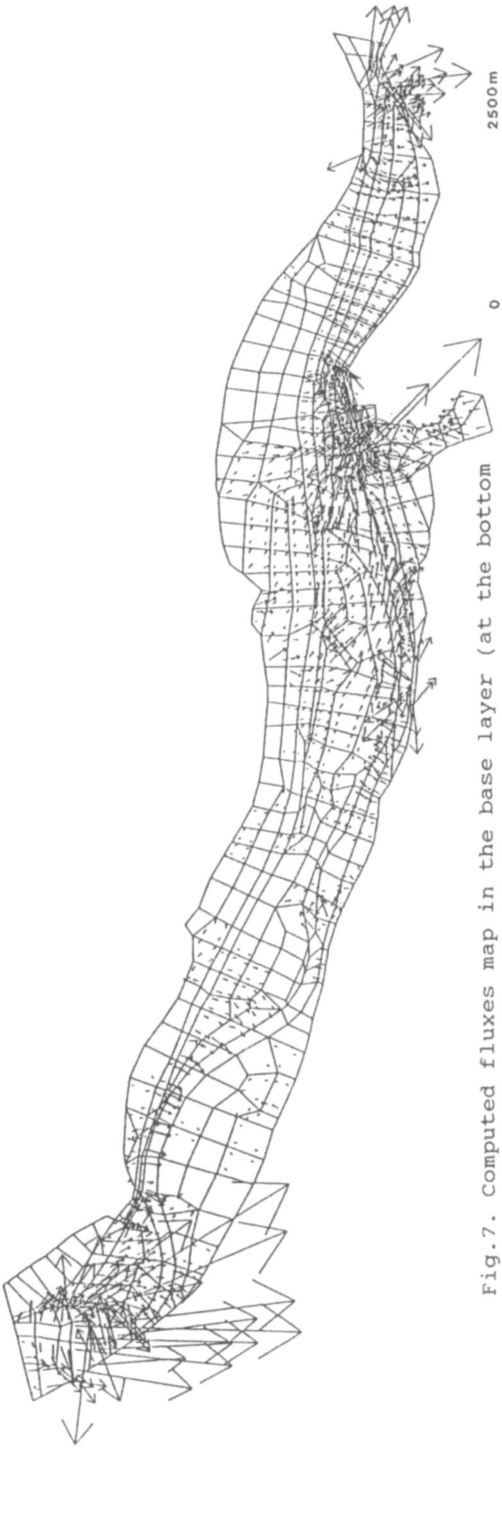

Fig.7. Computed fluxes map in the base layer (at the bottom of the model).

REFERENCES

1. Bathe, K.J. and Khoshgoftaar, M.R. Finite element free surface seepage analysis without iteration, Int. J. Num. and Anal. Meth. in Geomechanics, Vol. 3, pp. 13-22, 1979.
2. de Marsily, G.,Quantitative Hydrogeology, Academic Press, 1986.
3. Dysli, M. and Rybisar, J. Coupled models and free-surface seepage analysis without mesh iteration, in ICONMIG/88 (Ed. Swoboda, G.), pp. 791-795, Proceedings of the sixth Int.Conf.on Numerical Methods in Geomechanics, Innsbruck, Austria, 1988.
4. Terzaghi, K. and Peck, R., Soil mechanics in engineering practice, Wiley and Sons, New-York, 1948.
5. Dassargues, A., Radu, J.P.and Charlier, R. Finite element modelling of a large water table aquifer in transient conditions, Adv. in Water Resources, Vol. 11, June, pp. 58-66, 1988.
6. Charlier, R., Radu, J.P. and Dassargues, A. Numerical simulation of transient unconfined seepage problems, in Computer and Water Resources : Groundwater and aquifer modelling (Ed. Ouazar D. and Brebbia C.A.), pp. 143-155, Proceedings of the 1st Int. Conf. in Africa on Computer Methods and Water Resources, Vol. 1, Rabat, Morocco, 1988.
7. Dassargues, A. Non-linear analysis of the parameter in a coupled flow compaction model, submitted to Adv. in Water Resources in october 1990.
8. Charlier, R. Approche unifiée de quelques problèmes non-linéaires de mécanique des milieux continus par la méthode des éléments finis, Doctorat or Ph.Thesis,Faculté des Sciences appliquées, Université de Liège, Collection des publications n° 109, Belgium, 1987.
9. Schroeder, Ch., Monjoie, A., Radu, J.P., Charlier, R. and Fonder, G. Ekofisk subsidence compaction, mathematical modelisation synthesis report, internal report L.G.I.H.-M.S.M. for Fina Exploration Norway, Fina 881, 1988 (unpublished).
10. Bolly, P.Y., Dassargues A. and Monjoie, A. Finite difference and finite element modelling of an aquifer in Cretaceous chalk, in ICONMIG/88 (Ed. Swoboda G.), pp.747-756, Proceedings of the 6th Int. Conf.on Numerical Methods in Geomechanics, Innsbruck, Austria, 1988.
11. Dassargues, A.,Schroeder, Ch. and Monjoie, A. Engineering geology in the central area of Shanghai : Preparation of the data for subsidence modelling, IAEG/90, Proceedings of the 6th Int. IAEG Congress, Amsterdam, 1990.
12. Dassargues, A. and Li, X.L. Computing the land subsidence of Shanghai by Finite Element Method, submitted for the 4th Int. Symp. on Land Subsidence, Houston, May 1991.

SECTION 2: WAVE PROPAGATION

Wave Motion and Overturning Induced by Moving Bodies: Application to Slender Ship Wave Resistance

S.T. Grilli

Department of Ocean Engineering, The University of Rhode Island, Kingston-RI 02881-0814, USA

ABSTRACT

A Numerical method developed for two-dimensional fully nonlinear water waves using Boundary Integral Equations (BIE), is used to model wave motion and wave overturning induced by moving bodies. The method is applied to the computation of wave resistance of high speed slender ships. Due to slenderness, the steady three-dimensional wave resistance problem can be approximated by an unsteady two-dimensional problem.
A higher-order Boundary Element Method (BEM) is used for solving the BIE's in the physical space. All boundaries are discretized and, in particular, arbitrary bottom geometry can be modeled. Time integration is based on an explicit Taylor expansion in an Eulerian-Lagrangian formulation. Continuity and compatibility of both geometry and field variables are imposed at the intersection between free surface and moving bodies. High accuracy of the integrations is achieved with a distance-adaptive numerical integration. This is particularly needed at boundary corners and in narrow regions of the free surface like in the tip of overturning waves.

1 INTRODUCTION

Over the last decade, numerical solution of the exact nonlinear equations for the two-dimensional inviscid water wave problem, using a Boundary Integral Equation Method (BIEM), has become extremely successful. In these solutions, full nonlinearity is kept in the kinematic and dynamic free surface boundary conditions, and the problem is generally solved in a mixed Eulerian-Lagrangian formulation, i.e., following water particles in time along the free surface. Hence, no approximation other than the restrictions inherent to potential flow theory is made on the wave motion and, in particular, both deep and shallow water cases can be solved with the same model and computations carried on until a wave overturns and hits the free surface.

Most of the first contributions were restricted to space periodic waves and solved the Integral Equations in a conformally mapped space, in which the equivalent of the free surface was the only part left of the boundary. There are many advantages of this approach which has been extensively explored by many authors. Important steps in its development are the first computation of deep water overturning waves by Longuet-Higgins & Cokelet 1976 [13], the computation of wave interaction with submerged structures in finite depth by Vinje & Brevig 1981 [18], and the development of a more accurate and stable time integration by Dold & Peregrine 1986 [5], which allowed to follow wave trains and to compute nonlinear interactions between wave components over hundreds of periods. A more complete literature survey can be found in Grilli, *et al.* 1989 [7].

Space periodicity, however, makes it difficult or impossible to analyze wave motion induced by a moving body or to model wave propagation over a bottom of complex geometry. To be able to address these general water wave problems, Grilli, *et al.* 1989 [7] developed a method without this restriction. In their model, arbitrary wave motion is analyzed and waves are generated reflected or radiated at parts of the boundary. Laplace's equation is solved in the physical space by a Boundary Element Method (BEM). All boundaries are discretized using higher-order elements, which allows for arbitrary bottom geometry to be modeled. Time integration of the two nonlinear free surface boundary conditions is based on Dold & Peregrine's [5] explicit Taylor expansion, in an Eulerian-Lagrangian formulation. Continuity and compatibility of both geometry and field variables are imposed at the intersection between free surface and moving bodies. High accuracy of the integrations is achieved with a distance-adaptive numerical integration (Grilli & Svendsen 1990b [10]).

Nonlinear wave generation, propagation and runup on plane slopes and on various coastal structures have successfully been modeled with this method. Various ways of generating waves included numerical piston or flap wavemakers (Grilli, *et al.* [7, 10]), internal sources (Grilli & Svendsen [8]), and the imposition of an initial wave potential on the free surface (Grilli & Svendsen [11]). Nonlinear wave absorption was also addressed (Grilli, *et al.* [7]). Many comparisons of computational results with experimental and other numerical or analytical results have been presented (Grilli & Svendsen [9, 11]; Svendsen & Grilli [16]). In all cases, the method was found to be accurate and efficient as compared to other methods. The agreement of computational results with available experimental results was found surprisingly good, down to very small scale oscillations and over extensive periods of time. This confirmed the applicability of potential flow theory for these cases.

Korsmeyer & Tuck 1989 [12] presented ideas of a theory for calculating slender ship wave resistance. Due to slenderness, computation of the steady three-dimensional wave resistance problem approximately reduces to solving an unsteady fully nonlinear two-dimensional wave generation problem, in which the ship hull acts as a wavemaker and the coordinate y along the ship becomes a time-like coordinate t. Diverging waves only (lying behind the ship) can be calculated this way. Davidson 1989 [3] and Davidson & Tuck 1991 [4] presented a deep water solution to this problem. They further showed that, to maintain consistency in the orders of magnitude of the geometric and field variables following from the slender ship hypothesis, the method was only valid for large enough Froude numbers, i.e., for large enough ship speeds. In their numerical model, they used a BIE formulation for the complex velocity, identical to Roberts 1983 [14]. They avoided, however, representing the intersection between the free surface and the ship's hull and simply extrapolated results in the corners.

In this paper, nonlinear two-dimensional wavemaking problems are solved in arbitrary depth, using the model by Grilli, *et al.* [7, 10]. In this model, the corner solution is accurately represented and, due to solving in the physical space, it is in principle straightforward to implement wave generation by a wavemaker of arbitrary shape in a numerical wave tank of complex bottom geometry. Though a wave absorption condition has earlier

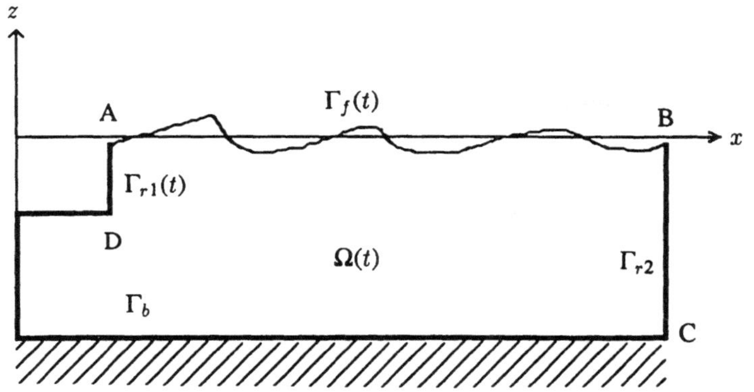

Figure 1: Computational domain $\Omega(t)$ with four corners A to D. $\Gamma_f(t)$ is the free surface, Γ_b the bottom, $\Gamma_{r1}(t)$ the moving body boundary and Γ_{r2} a reflective boundary.

been introduced in this model, a reflective condition at the end of the tank will simply be used in the present applications, and reflection will be made negligible by using a long enough tank.

The mathematical formulation is presented in Section 2 and the numerical implementation, briefly, in Section 3. Generation of overturning waves by a piston wavemaker is analyzed in Section 4, and the computation of slender ship wave resistance in Section 5.

2 MATHEMATICAL FORMULATION

2.1 Governing equations and boundary conditions

We study two-dimensional irrotational flows of an inviscid fluid whose motion is described by a velocity potential $\phi(\mathbf{x}, t)$. The velocity field is given by $\mathbf{u} = \nabla\phi = (u, w)$ in the plane (x, z). The continuity equation in the fluid domain $\Omega(t)$ with boundary $\Gamma(t)$ is a Laplace equation for ϕ (Figure 1),

$$\nabla^2\phi = 0 \qquad \text{in } \Omega(t) \qquad (1)$$

On the free surface $\Gamma_f(t)$, ϕ satisfies nonlinear kinematic and dynamic boundary conditions,

$$\frac{D\mathbf{r}}{Dt} = (\frac{\partial}{\partial t} + \mathbf{u}\cdot\nabla)\mathbf{r} = \mathbf{u} = \nabla\phi \qquad \text{on } \Gamma_f(t) \qquad (2)$$

$$\frac{D\phi}{Dt} = -gz + \frac{1}{2}|\nabla\phi|^2 - \frac{p_a}{\rho} \qquad \text{on } \Gamma_f(t) \qquad (3)$$

with $\mathbf{r}$ the position vector of a free surface fluid particle, g the acceleration due to gravity, z the vertical coordinate (positive upwards and $z = 0$ at the undisturbed free surface), p_a the atmospheric pressure and ρ the fluid density.

In the present applications, waves are generated by a moving body on boundary $\Gamma_{r1}(t)$. The fluid normal velocity is specified over the body surface by,

$$\nabla\phi\cdot\mathbf{n} \equiv \overline{\frac{\partial\phi}{\partial n}} = \mathbf{u}_b\cdot\mathbf{n} \qquad \text{on } \Gamma_{r1}(t) \qquad (4)$$

where the overline denotes a specified value, $(\mathbf{x}_b(t), \mathbf{u}_b(\mathbf{x}_b(t), t))$ are prescribed body motion and velocity respectively, and $\mathbf{n}$ is the unit outward normal vector.

Along the stationary bottom Γ_b and the domain reflective extremity Γ_{r2}, we have,

$$\nabla\phi \cdot \mathbf{n} \equiv \overline{\frac{\partial \phi}{\partial n}} = 0 \qquad \text{on } \Gamma_b \text{ and } \Gamma_{r2} \qquad (5)$$

2.2 The time integration

The time integration follows the mixed Eulerian-Lagrangian approach introduced by Dold & Peregrine [5]. The procedure, described in detail in Grilli, *et al.* [7] in the scope of the present model, uses Taylor expansions in terms of the Lagrangian or material time derivative (as defined in (2)) and a small time increment Δt, to establish at time $t + \Delta t$, both the new position $\overline{\mathbf{r}}(t+\Delta t)$ and the potential $\overline{\phi}(t+\Delta t)$ on the free surface. This actually corresponds to following in time the pathline of a fluid particle.

Both free surface conditions (2),(3) provide first-order terms in the Taylor series. Higher-order terms are obtained by material time derivation of these. Hence, all coefficients in the series are functions of normal and tangential derivatives of the potential and of its time derivatives. All time derivatives $(\frac{\partial \phi}{\partial t}, \frac{\partial^2 \phi}{\partial t^2}, ...)$ satisfy Laplace's equations which can be successively solved, each problem solution providing boundary conditions for the next one. Series have been limited to second-order in Δt, which corresponds to solving two Laplace problems at each time step to get the coefficients in the Taylor series. Notice, however, full nonlinearity has been kept in (2),(3) and the truncation error is only function of $\mathcal{O}(\Delta t^3)$ terms.

After solving the first Laplace problem for ϕ at time t, using boundary conditions (4),(5), boundary conditions are determined for the $\frac{\partial \phi}{\partial t}$problem. Along the free surface, Bernoulli's equation yields,

$$\overline{\frac{\partial \phi}{\partial t}} = -gz - \frac{1}{2} | \nabla \phi |^2 - \frac{p_a}{\rho} \qquad \text{on } \Gamma_f(t) \qquad (6)$$

For a body of arbitrary shape moving along $\Gamma_{r1}(t)$,

$$\overline{\frac{\partial^2 \phi}{\partial t \partial n}} = [\frac{d\,(\mathbf{u}_b \cdot \mathbf{n})}{d\,t} - \mathbf{u}_b \cdot \nabla(\mathbf{u}_b \cdot \mathbf{n})] \qquad \text{on } \Gamma_f(t) \qquad (7)$$

where $\frac{d}{dt}$ denotes a time derivative following the body motion.

Along the fixed boundaries, we get,

$$\overline{\frac{\partial^2 \phi}{\partial t \partial n}} = 0 \qquad \text{on } \Gamma_b \text{ and } \Gamma_{r2} \qquad (8)$$

2.3 Detailed boundary conditions for a moving body

Cointe [2] developed (7) for a two-dimensional rigid body of arbitrary shape, moving with velocity $\mathbf{u}_g = (u_g, w_g)$ at point $\mathbf{x}_g$, and angular velocity $\dot{\theta}$ around it (where dots denote material time derivatives). For a point $\mathbf{x}_b = (x_b, z_b)$ of the body surface,Cointe's expression reads,

$$\overline{\frac{\partial^2 \phi}{\partial t \partial n}} = (\dot{\mathbf{u}}_b \cdot \mathbf{n})\, [(\mathbf{u}_b \cdot \mathbf{s}) - \frac{\partial \phi}{\partial s}]\, \dot{\theta} \; - \; [\frac{\partial^2 \phi}{\partial n \partial s} + \frac{1}{R}\frac{\partial \phi}{\partial s}]\, (\mathbf{u}_b \cdot \mathbf{s}) \; + \; [\frac{\partial^2 \phi}{\partial s^2} - \frac{1}{R}\frac{\partial \phi}{\partial n}]\, (\mathbf{u}_b \cdot \mathbf{n}) \qquad (9)$$

where $\mathbf{u}_b = (u_b, w_b) = \dot{\mathbf{x}}_b$, $s(\mathbf{x}_b)$ is the curvilinear abscissa and $\frac{1}{R}(\mathbf{x}_b)$ the curvature along the body boundary curve with,

$$x_b = x_g + r_g \cos\theta \; ; \; z_b = z_g + r_g \sin\theta$$

$$u_b = u_g - (z_b - z_g)\dot{\theta} \; ; \; w_b = w_g + (x_b - x_g)\dot{\theta}$$

$$\dot{u}_b = \dot{u}_g - (x_b - x_g)\dot{\theta}^2 - (z_b - z_g)\ddot{\theta} \; ; \; \dot{w}_b = \dot{w}_g - (z_b - z_g)\dot{\theta}^2 + (x_b - x_g)\ddot{\theta} \quad (10)$$

where r_g is the distance between $\mathbf{x}_b$ and $\mathbf{x}_g$.

For a plane rigid body like a piston or flap wavemaker, (9) reduces to,

$$\overline{\frac{\partial^2\phi}{\partial t\partial n}} = (\dot{\mathbf{u}}_b \cdot \mathbf{n}) + [(\mathbf{u}_b \cdot \mathbf{s}) - \frac{\partial\phi}{\partial s}]\dot{\theta} - \frac{\partial^2\phi}{\partial n\partial s}(\mathbf{u}_b \cdot \mathbf{s}) + \frac{\partial^2\phi}{\partial s^2}(\mathbf{u}_b \cdot \mathbf{n}) \quad (11)$$

For a piston wavemaker at the leftward end of the domain, consisting in a vertical plate moving horizontally over the full depth h of the fluid domain, with motion $\mathbf{x}_b = (x_b(t), 0)$ and velocity $\mathbf{u}_b = (u_b(x_b(t), t), 0)$, we get from (4),(11) and after some calculations, complete boundary conditions over the paddle,

$$\overline{\frac{\partial\phi}{\partial n}} = -u_b(t), \qquad \overline{\frac{\partial^2\phi}{\partial t\partial n}} = -\dot{u}_b(t) - u_b(t)\frac{\partial^2\phi}{\partial z^2} \qquad \text{on } \Gamma_{r1}(t) \quad (12)$$

For moving bodies of complex shape or motion, full expressions (4) and (10), (9) or (11) should be used.

3 NUMERICAL IMPLEMENTATION

3.1 Solution by a Boundary Element Method

Using free space Green's function $G(\mathbf{x}, \mathbf{x}_l) = -\frac{1}{2\pi}\log|\,\mathbf{x} - \mathbf{x}_l\,|$, Laplace's equation (1) for ϕ becomes the Boundary Integral Equation (BIE),

$$\alpha(\mathbf{x}_l)\phi(\mathbf{x}_l) = \int_{\Gamma(\mathbf{x})} [\frac{\partial\phi}{\partial n}(\mathbf{x})G(\mathbf{x}, \mathbf{x}_l) - \phi(\mathbf{x})\frac{\partial G(\mathbf{x}, \mathbf{x}_l)}{\partial n}]\, d\Gamma(\mathbf{x}) \quad (13)$$

where $\mathbf{x} = (x, z)$ and $\mathbf{x}_l = (x_l, z_l)$ are position vectors for points on boundary $\Gamma(\mathbf{x})$ and $\alpha(\mathbf{x}_l)$ is a geometric coefficient.

The BIE (13) for ϕ and its equivalent for $\frac{\partial\phi}{\partial t}$ are solved by a Boundary Element Method (BEM, Brebbia [1] 1978) using a set of collocation nodes on the boundary and higher-order elements to interpolate between the collocation nodes. Quasi-spline elements (Grilli & Svendsen [10]) are used on the free surface, and isoparametric elements elsewhere (linear to quartic) (Grilli, *et al.* [7]).

Each integral in (13) is transformed into a sum of integrals over each boundary element. Non-singular integrals are calculated by standard Gauss quadrature rules. A kernel transformation is applied to weakly singular integrals, which are then integrated by a numerical quadrature exact for the logarithmic singularity. An adaptive numerical integration is used for improving the accuracy of the regular integrations near corners (A-D, Fig. 1) and other narrow regions of the boundary, like the overturning jet in breaking waves (as, e.g., in Fig. 2c), where elements on different parts of the boundary get close to each other.

Boundary conditions (4)-(12), as well as the potential on the free surface specified by the Taylor series are used to solve both BIE's, for each time t. Tangential derivatives of the boundary solution are calculated by sliding 4th-order polynomial interpolation. Discretization and types of boundary conditions are identical for both Laplace problems, which leads to a unique algebraic system matrix to be solved for each time for two load vectors. This makes the method computationally quite efficient.

The numerical implementation is detailed in Grilli, *et al.* [7, 8, 10].

3.2 Corner treatments

Mathematical aspects When waves are generated by a body motion, there is a corner on the boundary curve where body and free surface meet. The flow near the intersection between a free surface and a moving solid body has given rise to substantial concern in the literature (e.g., see Roberts [15], Cointe [2]). This has been reviewed by Grilli & Svendsen [10]. The conclusion is that, provided the initial acceleration of the moving body is small with respect to gravity, there will not be any strong singularity at the free surface corner.

Numerical aspects At the intersection between a free surface of imposed potential and a moving body, both the potential and its normal gradient are known on the body part of the free surface corner, whereas there is a different and unknown normal gradient on its free surface part. Double nodes are used at corners A to D (Figure 1), for which coordinates of the nodes are the same but normal vectors differ, reflecting, they belong to each side of the corner, with different boundary conditions. Furthermore $\frac{\partial \phi}{\partial n}$ and $\frac{\partial^2 \phi}{\partial t \partial n}$ are different on each side of a corner, and two algebraic BIE's are expressed for each corner. Continuity of ϕ (and $\frac{\partial \phi}{\partial t}$) is imposed explicitly in these equations.

The velocity vector must be unique at a double node. Although it should be satisfied in the numerical solution, such a kinematic condition in general is not exactly fulfilled, due to numerical errors. Since there is no damping in the model, errors may add up through time updating and lead to instability of the solution. Hence, a compatibility relation is explicitly imposed to ϕ's and $\frac{\partial \phi}{\partial n}$'s (or $\frac{\partial \phi}{\partial t}$\$ and $\frac{\partial^2 \phi}{\partial t \partial n}$\$) after each time step.

The numerical treatment of corners is detailed in Grilli & Svendsen [10].

3.3 Selection of time steps

Grilli & Svendsen [10] showed, for a given initial interval between nodes on the free surface $\Delta x' = \frac{\Delta x}{h}$, numerical errors are proportional to $\Delta t'^3$ when the Courant number $\mathcal{C} = \frac{\Delta t'}{\Delta x'} \geq 0.5$ (with $\Delta t' = \Delta t \sqrt{\frac{g}{h}}$). This is consistent with the second order accuracy of the time integration. When $\mathcal{C} < 0.5$, errors cannot be further reduced by decreasing $\Delta t'$, but become proportional to $\Delta x'$ only. This corresponds to the spatial discretization effect on the accuracy.

To account for Lagrangian time updating of the free surface nodes, a varying time step procedure is used to satisfy the Courant condition : At $t = t_o$, the initial Courant number $\mathcal{C}_o = \frac{\Delta t'_o}{\Delta x'_o}$ is selected at 0.5 or less and, at subsequent time t, time step is adjusted as,

$$\Delta t' = \Delta \mid \mathbf{r}' \mid^{min} \mathcal{C}_o \tag{14}$$

where $\Delta \mid \mathbf{r}' \mid^{min}$ represents the minimum distance between two successive nodes on the free surface.

3.4 Accuracy

Accuracy of the computations is checked by computing the maximum change in volume of the fluid domain during the time stepping procedure. Constancy of total energy can also be checked, but former results showed, errors in total energy closely fallow errors in volume. In the present applications, relative volume errors of less than to a few 0.01% are usually obtained.

4 GENERATION OF WAVES BY A WAVEMAKER

Generation of waves by a piston wavemaker in constant depth h has been thoroughly studied with this model (Grilli & Svendsen [9, 10, 11], Svendsen & Grilli [16]). First-order solitary waves were generated as in Goring [6], by using boundary conditions (12) with the piston motion,

$$x_b(t) = \frac{H}{\kappa}[\tanh \frac{\kappa}{h}(ct - x_b(t) - \lambda) + \tanh \frac{\kappa}{h}\lambda] \tag{15}$$

where H is the wave height, $c = \sqrt{gh(1 + \frac{H}{h})}$ its celerity, and $\kappa = \sqrt{\frac{3H}{4h}}$ and λ are constant.

In accordance with the discussion in Section 3.2 results showed, the free surface elevation stayed finite, provided the acceleration was small during the first few time steps at the start of the wavemaker motion. For large but finite initial accelerations, a thin jet of water of large amplitude could be produced close to the wavemaker (as also observed in, e.g., Davidson [3]) . Details of u_b and $\dot{u}_b$ corresponding to (15) can be found in [10] where it is also showed, the initial acceleration is controlled by λ. For $\frac{H}{h} = 0.5$ and $\frac{\kappa}{h}\lambda = 3.8$, for instance, the initial acceleration is less than $0.002g$. In the following, primes denote dimensionless variables in accordance with classical long wave theory : lengths are divided by depth h, times by $\sqrt{\frac{h}{g}}$, velocities by $\sqrt{gh}$ and accelerations by g.

Before considering the ship problem, generation of overturning waves is studied using piston motion (15). With right selection of λ this indeed leads to a smooth start, with very small initial acceleration of the piston, even for very high waves, which eliminates both the singularity problem at the corner and the jet generation. Solitary waves, theoretically, become unstable for heights above $H' \simeq 0.78$. Therefore, introducing higher H' in piston motion (15) creates waves which rapidly overturn quite close to the wavemaker.

Figure 2 shows piston motions $x_b'(t')$ for different H' between 1.0 and 2.2, with $\frac{\kappa}{h}\lambda = 3.8$, and Figure 3 shows computational results for H'=1.0, 1.6 and 2.0, in a tank of length $\ell' = 10$. The initial discretization of the free surface at time $t_o' = 0$ has horizontal node intervals $\Delta x' = 0.25$ and cubic spline elements are used for the interpolation between nodes. Linear elements are used elsewhere. Figures 3abc show dimensionless free surface elevations η' as function of x' for different times t' (of non-constant intervals) of the wavemaker motion. The wavemaker, always, is at the leftward extremity of each curve. Though for $H' = 2$ the wavemaker accelerates up to almost one g, initial acceleration is only $0.03g$, which accounts for the observed regularity of the free surface geometry around its intersection with the wavemaker. Notice, ϕ, $\frac{\partial \phi}{\partial n}$,$\frac{\partial \phi}{\partial t}$ and $\frac{\partial^2 \phi}{\partial t \partial n}$ also vary smoothly along the free surface.

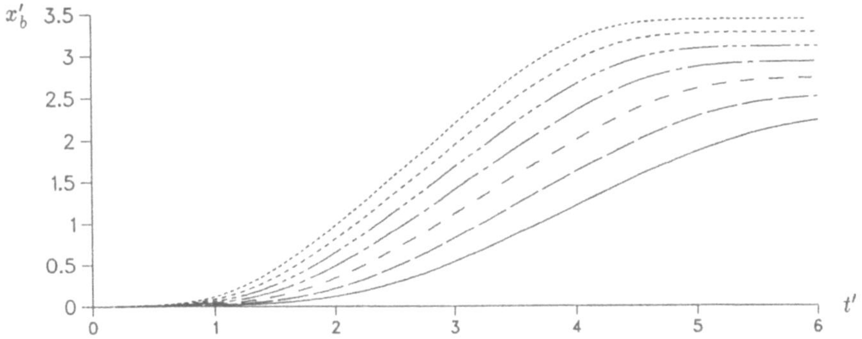

Figure 2: Wave overturning induced by a piston wavemaker : horizontal piston motion $x'_b(t') = \frac{x_b}{h}$, for $H' = \frac{H}{h}$=1.0 (——) to 2.2 (- - - -) by steps 0.2, with parameter $\frac{\kappa}{h}\lambda$ = 3.8.

5 SLENDER SHIP WAVE RESISTANCE

Equations for the steady three-dimensional ship wave resistance problem in finite depth are also given by (1)-(5) for the potential $\phi(x, y, z)$. The ship hull boundary $\Gamma_{r1}(t)$ is moving forward at constant speed U in the y direction, and Γ_{r2} is a lateral cylindrical boundary where reflective or radiation conditions are imposed. Davidson 1989 [3] and Davidson & Tuck 1991 [4] presented a deep water solution to this problem showing that for slender ships, it can be reduced to solving an unsteady two-dimensional problem similar to the problem of wave generation by a wavemaker described above, in which the ship hull acts as the wavemaker and the coordinate y along the hull becomes a time-like coordinate t. Their analysis is briefly outlined in the following Section.

5.1 Approximate two-dimensional problem

In a coordinate system (x, y, z) fixed to the ship, the velocity potential reads,

$$\phi(x, y, z) = \varphi(x, y, z) + Uy \tag{16}$$

where φ is an unknown disturbance velocity potential, vanishing at infinity, superimposed to a uniform flow. With (16), continuity equation (1) reads,

$$\nabla^2 \varphi = 0 \qquad \text{in } \Omega \tag{17}$$

and free surface boundary conditions (2),(3) become, after some transformations,

$$\frac{\partial \varphi}{\partial z} = \frac{\partial \varphi}{\partial x}\frac{\partial \eta}{\partial x} + (\frac{\partial \varphi}{\partial y} + U)\frac{\partial \eta}{\partial y} \qquad \text{on } \Gamma_f \tag{18}$$

$$g\eta + U\frac{\partial \varphi}{\partial y} + \frac{1}{2} \mid \nabla \varphi \mid^2 = 0 \qquad \text{on } \Gamma_f \tag{19}$$

where η represents the free surface elevation above $z = 0$ and $p_a = 0$. Along the steady ship hull ($\mathbf{u}_b = 0$), the sea bottom and reflective lateral boundary, (4) and (5) simply become,

$$\nabla \varphi \cdot \mathbf{n} = -U \nabla y \cdot \mathbf{n} = -U n_y \qquad \text{on } \Gamma_{r1}, \Gamma_{r2} \text{ and } \Gamma_b \tag{20}$$

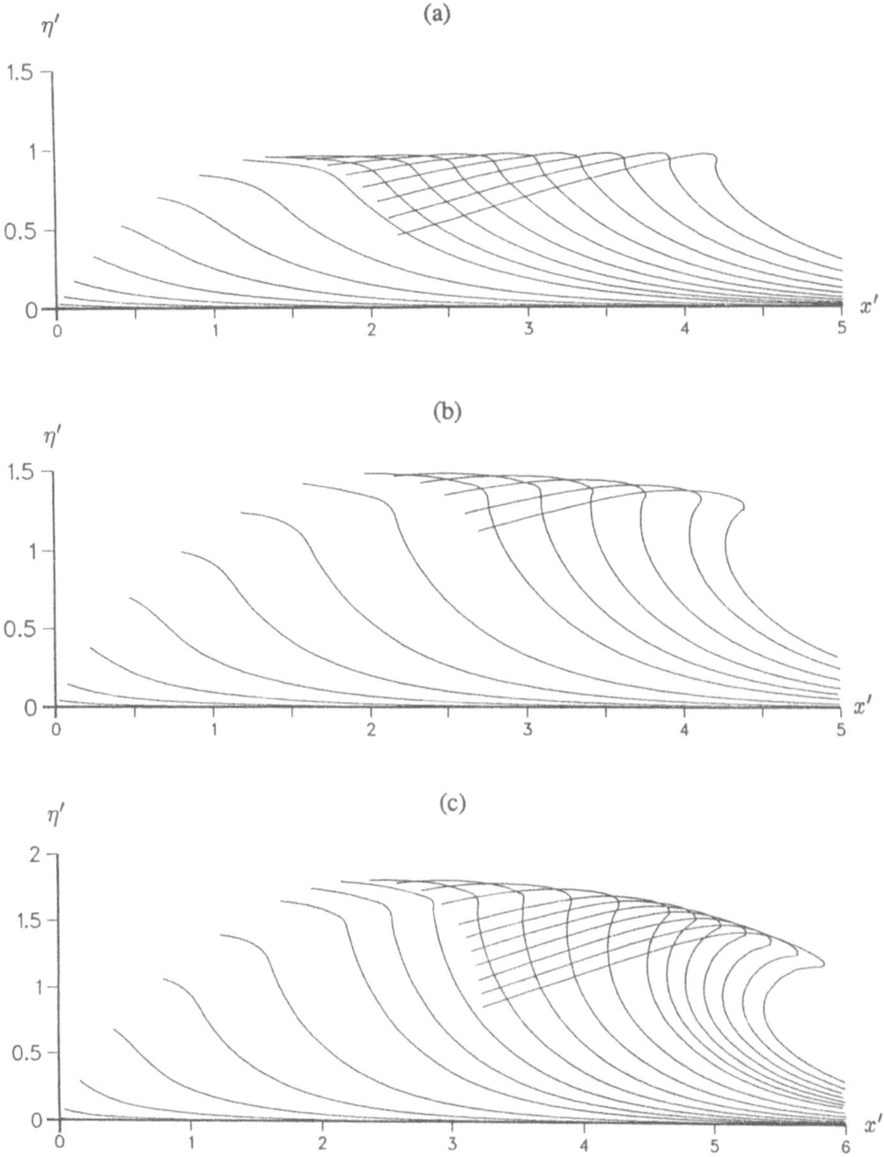

Figure 3: Wave overturning induced by a piston wavemaker : free surface elevations $\eta'(x') = \frac{\eta}{h}$ in a tank of length $10h$, for : (a) H'=1.0, (b) H' = 1.60, (c) H' = 2.0.

where n_y denotes the y component of **n**. Equations (16)-(20) define a three-dimensional steady wave resistance problem for a ship of arbitrary geometry. Introducing a time like variable $t = \frac{y}{U}$ along the ship length instead of y, and the slender ship hypothesis, i.e., that ship dimensions in the transverse directions (x, z) are much smaller than in the longitudinal one y, we get,

$$\frac{\partial}{\partial y} \ll \frac{\partial}{\partial x}, \frac{\partial}{\partial z}$$

$$\frac{\partial}{\partial y} = \frac{1}{U}\frac{\partial}{\partial t} \tag{21}$$

Using (21) to neglect some of the $\frac{\partial}{\partial y}$ terms in (17)-(19) and then, introducing the variable $t = \frac{y}{U}$, we get,

$$\frac{\partial^2 \varphi}{\partial x^2} + \frac{\partial^2 \varphi}{\partial z^2} = 0 \qquad \text{in } \Omega \tag{22}$$

$$\frac{\partial \varphi}{\partial z} = \frac{\partial \varphi}{\partial x}\frac{\partial \eta}{\partial x} + \frac{\partial \eta}{\partial t} \qquad \text{on } \Gamma_f \tag{23}$$

$$g\eta + \frac{\partial \varphi}{\partial t} + \frac{1}{2}(\frac{\partial \varphi}{\partial x}^2 + \frac{\partial \varphi}{\partial z}^2) = 0 \qquad \text{on } \Gamma_f \tag{24}$$

and (20) remains unchanged. Equations (23),(24) are equivalent to (2),(3) and, hence, (20), (22)-(24) define a complete set of equations for an unsteady two-dimensional problem in the plane (x, z).

5.2 Geometric modeling

To further simplify the problem, the slender ship is assumed to be wall-sided with constant draft D, half beam B and half length L, and the water depth h is constant. Along the y-axis, the ship half-width varies as $b(y)$. In this situation, lateral sides of the ship act like piston wavemakers, moving with horizontal motion $x_b(t) = b(y(t))$ over the ship draft, in a tank of depth h.

A simple variation for the ship half-width is to take it parabolic along y (or t). Parabolic-waterlined ships have been studied by Davidson [3] who observed, the first bow wave could break for large Froude numbers, hence terminating computations. To avoid this early breaking, Tuck [17] suggested using a geometry providing smoother variation for the half-width at bow and stern, of the form,

$$x_b(t) = B[1 - (1 - \frac{t}{t_f})^2]^2 \ , \ t_f = \frac{L}{U} \ , \ t = \frac{y}{U} \tag{25}$$

which corresponds to a cusp-ended ship.

Boundary conditions along the ship walls follow from (12) expressed in the half ship cross section $\gamma_h(t)$ with, by (20),(21),

$$u_b(t) = U n_y \ , \ \dot{u}_b(t) = U^2 \frac{\partial n_y}{\partial y} \tag{26}$$

It can easily be shown, by simple geometric considerations along the slender ship hull, $n_y = \frac{d\,b(y)}{dy}$. Hence from (25),(26),

$$u_b(t) = 4\varepsilon\,\sqrt{2gL}\,F\,(1-\frac{t}{t_f})\,[1-(1-\frac{t}{t_f})^2]$$

$$\dot{u}_b(t) = 8\varepsilon\,g\,F^2\,[3(1-\frac{t}{t_f})^2 - 1]\ ,\ \varepsilon = \frac{B}{L}\ ,\ F = \frac{U}{\sqrt{2gL}} \tag{27}$$

where ε and F are the ship slenderness and (length) Froude number respectively. Depth also can be expressed as a (depth) Froude number $\mathcal{F} = \frac{U}{\sqrt{gh}}$ (or $\mathcal{F} = F\sqrt{\frac{2B}{\varepsilon h}}$)

Only half the fluid domain is discretized in region $x > 0$. By symmetry, a no flow condition is imposed over a vertical boundary, under the ship axis, and the same is done over the ship bottom. On both these boundaries, regriding of the BEM discretization nodes is used after each time step, to account for the change in geometry in the ship cross section from bow to stern, i.e., when the wavemaker-ship wall is moving from $x_b = 0$ (for $y = t = 0$) to $x_b = B$ (for $y = L$; $t = t_f$) and to $x_b = 0$ (for $y = 2L$; $t = 2t_f$).

5.3 Wave resistance computation

The pressure on the ship is obtained as a function of time, i.e., in $\gamma_h(t)$, by Bernoulli's equation (6) as,

$$p = \rho\,[\frac{\partial\phi}{\partial t} + gz + \frac{1}{2}(\frac{\partial\phi}{\partial s}^2 + \frac{\partial\phi}{\partial n}^2)] \qquad \text{on } \gamma_h(t) \tag{28}$$

where $\frac{\partial\phi}{\partial n}$ and $\frac{\partial\phi}{\partial t}$ are directly provided by the BEM solution of the flow, and $\frac{\partial\phi}{\partial s}$ is calculated from ϕ, in the interpolation polynomials of each boundary element. Wave resistance is obtained by integrating the y component of the actual pressure over the ship hull as,

$$R_w = 2\int_0^{2L}\int_{\gamma_h(y)} p(x,y,z)\,n_y\,d\gamma\,dy$$

For a simple wall-sided slender ship of geometry described by (25), this is further simplified by (26) into,

$$R_w = 2\int_0^{2t_f} f_x(t)\,\frac{d\,b(y(t))}{dt}\,dt \simeq 2\sum_{t=0}^{2t_f} f_x(t)u_b(t)\Delta t \tag{29}$$

where $f_x(t)$ is the resultant horizontal force in cross section $\gamma_h(t)$, acting at $x = b(t)$, given by,

$$f_x(t) = \int_{-D}^{\eta(b(t))} p\,(b(t),t,z)dz \tag{30}$$

and computed by summation of contributions of each boundary element in $\gamma_h(t)$, in which the integral is evaluated numerically, for the pressure variation (28).

D'	B'	ℓ'	$\frac{\ell}{B}$	$\Delta x'_o$	$\Delta t'_o$	N	M
0.5	0.5	10	20	0.25	0.125	103	99
0.5	0.5	15	30	0.25	0.125	143	103
0.25	0.25	15	60	0.25	0.125	143	103
0.20	0.20	10	50	0.20	0.100	97	76
0.10	0.10	3	30	0.05	0.025	114	90

Table 1: Geometric and discretization data for slender ship wave resistance computation for $\varepsilon = 0.1$ with $D' = \frac{D}{h}$ the ship draft, $B' = \frac{B}{h}$ the ship half beam, $\ell' = \frac{\ell}{h}$ the length of computational domain, $\Delta x'_o = \frac{\Delta x_o}{h}$ the initial spatial discretization step on the free surface, $\Delta t'_o = \Delta t_o \sqrt{\frac{g}{h}}$ the initial time step and (N, M), the number of nodes and boundary elements of the discretization .

5.4 Applications

In the applications, for a given ship geometry (ε, B, D), and depth h, Froude number F is selected and it follows $t_f = \frac{1}{F}\sqrt{\frac{B}{2\varepsilon g}}$ and $\mathcal{F}$ (with $h = \frac{F^2}{\mathcal{F}^2}\frac{2B}{\varepsilon}$). Total time of computation is $2t_f$. Primes denote, otherwise mentioned, dimensionless variables defined as in Section 4.

A first example is computed for $\varepsilon = 0.1$, $B' = 0.5$ and $D' = 0.5$. Discretization data are identical to the example in Section 4. Figure 4 upper part (a) shows variations for the dimensionless horizontal force $f'_x(t')$ (i.e., divided by the hydrostatic pressure at still water level : $\frac{1}{2}\rho g D^2$). Lower part (b) of the figure shows corresponding half-width variations $x'_b(t')$. Three cases have been computed with Froude numbers F=(0.25, 0.50, 0.75) and corresponding $\mathcal{F}$=(0.79, 1.58, 2.37) respectively. These are very shallow water cases with transition over the critical depth Froude number. Dimensionless wave resistances for these cases are R'_w=(0.312, 0.155, 0.130) (i.e., divided by $\rho BDU^2 = \frac{1}{2}\rho g V F^2$, where V denotes volume $4BDL$). Notice, variations of f'_x with t' in Figure 4a, also approximately correspond to variations of water level $\eta'(b(y(t)))$, along ship length y. Discontinuities in the f'_x curves at the beginning and at the end of the wavemaker motion (i.e., at bow and stern) are believed to result from non-zero accelerations at times $t' = 0$ and $2t'_f$. By (27), these both are equal to $\dot{u}_b(0) = 16\varepsilon\, g\, F^2$ and vary, here, between 0.10 and $0.90g$.

In a second computational example, R'_w is computed as a function of F, for different values of $D' = B'$ while maintaining $\varepsilon = 0.1$. Discretization data vary for each case. The initial discretization of the free surface at time $t'_o = 0$ has horizontal node intervals $\Delta x'_o$ between 0.05 and 0.25 and cubic spline elements are used for the interpolation between nodes. Linear and quadratic elements are used elsewhere. Length of the computational domain varies between ℓ= 3 to $15h$ and $\frac{\ell}{B}$= 20 to 60. Detailed data are given in Table 1. Computations of R'_w are carried out for the five cases in Table 1, in the Froude number range F =0.1-0.5. Results are presented in Table 2 and in Figure 5. The first two cases have identical data, except for the length of computational domain, 50% larger in the second case. Wave resistances for these cases closely agree in Table 2 and Figure 5 above $F = 0.15$ and results for both cases show, R'_w decreases when F increases. For smaller Froude numbers, however, R'_w drops much faster in the shorter

F	$D' = 0.5, \frac{\ell}{B} = 20$	$D' = 0.5, \frac{\ell}{B} = 30$	$D' = 0.25$	$D' = 0.20$	$D' = 0.10$
0.10	0.0624	0.7297	0.5148	0.4526	0.0080
0.15	0.4803	0.5095	0.3542	0.3223	0.1221
0.20	0.3850	0.3862	0.2763	0.2546	0.1940
0.25	0.3124	0.3129	——	——	——
0.30	0.2643	0.2645	0.2014	0.1919	0.1872
0.35	0.2275	0.2279	——	——	——
0.40	0.1988	0.1987	0.1601	0.1565	0.1674
0.45	0.1739	0.1752	——	——	——
0.50	0.1549	0.1550	0.1348	0.1336	0.1433

Table 2: Dimensionless wave resistance $R'_w = R_w \frac{\varepsilon}{2\rho g D B^2 F^2}$ as a function of length Froude number F and relative draft $D' = \frac{D}{h} = 0.25$.(——) means not calculated

domain. One sees, here, consequences of interaction effects between ship side walls and the reflective boundary at the end of the computational domain. When the ship speed is lower, waves, relatively, have more time to travel and reflect between the ship and the domain far end, hence to modify wave resistance. Wave resistance has been computed for four different ratios $D' = \frac{D}{h}$. As could be expected, results in Table 2 and in Figure 5 show, R'_w, globally, decreases with D'. For $F > 0.3$, all results somehow converge towards identical values. Small local variations could be due to discretization effects. For Smaller F's, dimensionless wave resistance drops in the deepest water case, $D' = 0.1$, whereas it still increases for the others, down to $F = 0.1$. Shallow water effects have somewhat more influence on R'_w when the ship speed is low. Notice, as mentioned in the introduction, Davidson & Tuck [4] showed, consistency in the orders of magnitude in the method requires F to be large enough. Results, therefore, are closer to actual slender ship wave resistance in this latter case.

In a third and last example, one will simply re-analyze results of Section 4, in the light of the present slender ship wave resistance theory. Computations in Figure 3, indeed, qualitatively correspond to a situation identical to wave breaking induced by the bow of a slender ship whose cross section half-width would vary as $b(y(t)) = x_b(t)$, given by (15) and shown in Figure 2, with $\frac{D}{h} = 1$. Figure 3c, in particular, gives quite a realistic view of the roller created by such a ship bow, if one simply looks at it in a perspective way. This illustrates the ability of the present method to analyze and predict possible breaking induced by different geometry of a ship bow.

6 CONCLUSIONS

A model for fully nonlinear potential wave flows [7] has been extended [10] to accurately compute wave generation by moving bodies of arbitrary shape. An approximate theory [3, 4, 12], valid for high speed slender ships, has been used with the model for calculating wave resistance of wall-sided cusp-ended ships in water of constant depth, as a function of ship geometry and speed (Froude number). Wave overturning induced by the ship bow can also be analyzed for arbitrary bow shape. A reflective boundary has been used at the

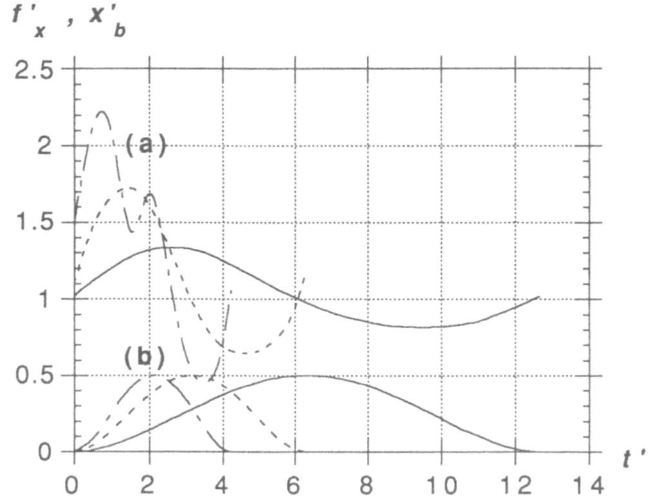

Figure 4: Horizontal wave force $f_x'(t') = \frac{f_x}{\frac{1}{2}\rho g D^2}$ (a) and ship half width $x_b'(t') = \frac{x_b}{h}$ (b), for F= 0.25 (——), 0.50 (- - - -) and 0.75 (— - —), with $\varepsilon = 0.1$, $B' = D' = 0.5$, and corresponding $2t_f'$= (12.65, 6.32, 4.22).

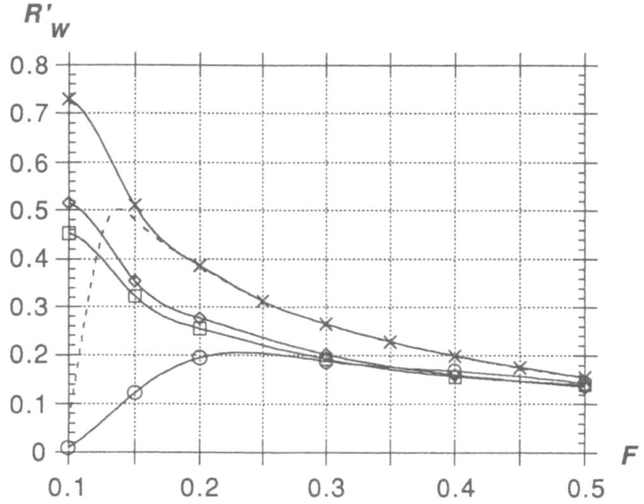

Figure 5: Wave resistance $R_w' = R_w \frac{\varepsilon}{2\rho g D B^2 F^2}$ as a function of Froude number F, for ε=0.1 and $B' = D' =$: 0.1 (—○—), 0.2 (—□—), 0.25 (—◇—), 0.5 ($\frac{\ell}{B}$ = 30,—×—) and ($\frac{\ell}{B}$ = 20, - - - -) (see Tables 1,2). Symbols denote computational results and lines are cubic spline interpolations between them.

end of the computational domain which, however, has been shown not to significantly influence results for large enough Froude numbers. Reduction of water depth has also been shown, as expected, to increase wave resistance. Possible extensions of the present computations could be the analysis of other half-width variations for wall-sided ships, and also to introduce slender ships of arbitrary cross section.

Acknowledgments

The author wishes to thank Dr. E.O. Tuck for giving him the initial theoretical ideas for solving slender ship wave resistance problems and also for providing him later with some more material, in the form of an unpublished manuscript. The author acknowledges financial support from the research grant "FNRS-Prix de Waele ex. 1990-91".

References

[1] Brebbia, C.A. *The Boundary Element Method for Engineers.* John Wiley & Sons, U.K., 1978.

[2] Cointe, R. Quelques aspects de la simulation numérique d'un canal à houle. *Thèse de Docteur de l'Ecole Nationale des Ponts et Chaussées,* 284 pps., 1989.

[3] Davidson, T.E. Numerical Simulation of the Water Waves Generated by a Floating Body. *Master Thesis,* University of Adelaide, South Australia, 1989.

[4] Davidson, T.E. & Tuck, E.O. Slender Ships at High and Moderate Speeds. To appear, 1991.

[5] Dold, J.W. & Peregrine, D.H. An Efficient Boundary Integral Method for Steep Unsteady water Waves. *Numerical methods for Fluid Dynamics II* (ed. K.W. Morton & M.J. Baines), pp. 671-679. Clarendon Press, Oxford, 1986.

[6] Goring D.G. Tsunamis - The Propagation of Long Waves onto a Shelf. *W.M. Keck Laboratory of Hydraulics and Water Resources, California Institute of Technology, Report No.* **KH-R-38,** 1978.

[7] Grilli, S., Skourup, J. & Svendsen, I.A. An Efficient Boundary Element Method for Nonlinear Water Waves. *Engineering Analysis with Boundary Elements* **6** (2), 97-107, 1989.

[8] Grilli, S. & Svendsen, I.A. The Modeling of Highly Nonlinear Waves. : Some Improvements to the Numerical Wave Tank. *Advances in Boundary Elements* (Proc. 11th Intl. Conf. on Boundary Elements, Cambridge, Massachusetts, USA), Vol. 2 (ed. C.A. Brebbia, J.J. Connor), pp. 269-281. Computational Mechanics Publication. Springer Verlag, Berlin, 1989.

[9] Grilli, S. & Svendsen, I.A. Computation of Nonlinear Wave Kinematics during Propagation and Runup on a Slope. *Water Wave Kinematics,* (Proc. NATO ARW, Molde, Norway, May 89) (ed. A. Torum & O.T. Gudmestad), NATO ASI Series E: Applied Sciences Vol. **178,** 387-412. Klüwer Academic Publishers, 1990a.

[10] Grilli, S. & Svendsen, I.A. Corner Problems and Global Accuracy in the Boundary Element Solution of Nonlinear Wave Flows. *Engineering Analysis with Boundary Elements,* **7** (4), 178-195, 1990b .

[11] Grilli, S. & Svendsen, I.A. Wave Interaction with Steeply Sloping Structures. In *Proc. 22nd Intl. Conf. on Coastal Engineering* (ICCE22, Delft, The Netherland, July 90) 14 pps. ASCE edition, 1990c (in press).

[12] Korsmeyer, F.T. & Tuck, E.O. Marching toward a Slender Ship Wave Resistance Theory. In *Proc. 4th Intl. Workshop on Water Waves and Floating Bodies* (Oystese, Norway, May 89) (ed. J. Grue), pp. 117-121, 1989.

[13] Longuet-Higgins, M.S. & Cokelet, E.D. The Deformation of Steep Surface Waves on Water - I. A Numerical Method of Computation. *Proc. R. Soc. Lond.* **A350,** 1-26, 1976.

[14] Roberts, A.J. Stable and Accurate Numerical Method to Calculate the Motion of a Sharp Interface between Fluids *IMA J. Appl. Math.* **31,** 13-35, 1983.

[15] Roberts, A.J. Transient Free-Surface Flows Generated by a Moving Vertical Plate. *Q.J. Mech. Appl. Math.* **40** (1), 129-158, 1987.

[16] Svendsen, I.A. & Grilli, S. Nonlinear Waves on Steep Slopes. *J. Coastal Research* **SI 7,** 185-202, 1990.

[17] Tuck, E.O. Personal communication, 1989.

[18] Vinje, T. & Brevig, P. Numerical Simulation of Breaking Waves. *Adv. Water Ressources* **4,** 77-82, 1981.

The Modelling of Extreme Gravity Waves: An Approach towards a Numerical Wave Channel

C. Haack, P. Gravert, V. Schlegel
Arbeitsbereich Meerestechnik II, Technische Universität Hamburg-Harburg, D-2100 Hamburg 90, Germany

ABSTRACT

The paper presents a combined boundary element and time stepping method for the numerical simulation of nonlinear gravity waves. The method uses a real velocity potential with time-dependent nonlinear free surface boundary conditions. The mathematical formulation of the moving free surface is utilized to simulate a deep water wave channel. The numerical implementation uses the Cauchy integral equation method for the discretization in space and a multistep predictor-corrector scheme for the time integration. The numerical stability of this technique is improved by spline elements for the free surface discretization, a weighted least squares method as solution scheme for the boundary integral equation, double-node representation of corners and semi-analytical treatment of singular integral kernels. Results for regular periodic and for overturning waves in deep and shallow water are presented. In addition numerical simulations of an existing wave channel are compared with the results of experimental measurements within that channel.

INTRODUCTION

An accurate and efficient calculation scheme to model and analyze the considerably complex behavior of ocean waves in a more general way is a valueable contribution in the subject of fluid dynamics of the upper ocean as well as for ocean engineering applications. Since the approach of Longuet-Higgins and Cokelet [6], the combined boundary element and time stepping technique became a well-known and reliable numerical method for this kind of problems. This primary approach was optimized and extended by several other authors, but however, even nowadays the numerical stability of the algorithm and computation time are the most restrictive factors for practical purpose. In this proposal the method of Longuet-Higgins and Cokelet is extended with regard to the following aspects — see Gravert [3]:

- analysis of deep and shallow water waves,
- use of a real velocity potential and solution in the physical space,
- improved numerical stability by the implementation of spline elements, double-nodes, etc.,
- numerical simulation of a deep water wave channel.

PROBLEM FORMULATION

Consider a two-dimensional domain Ω with the boundary Γ (Fig. 1) that includes a part of the free surface, the sea bottom, and two vertical boundaries.

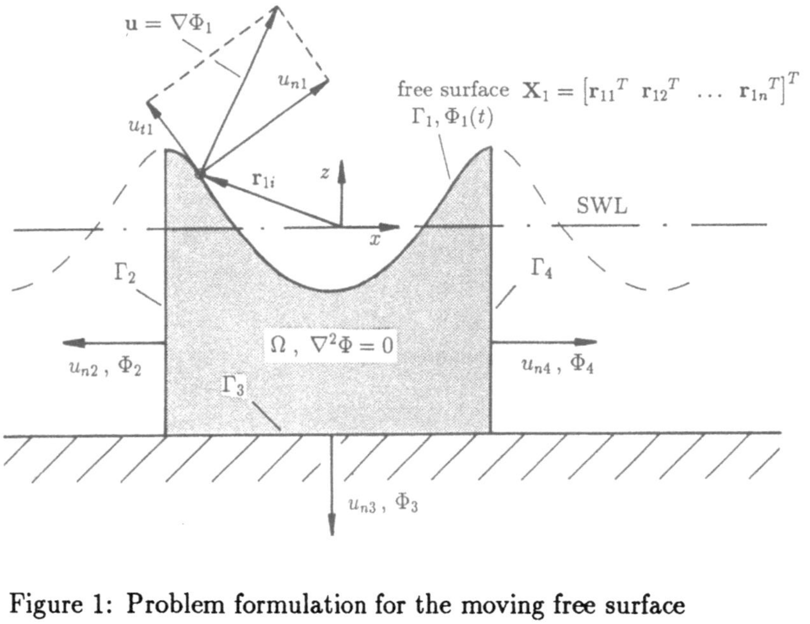

Figure 1: Problem formulation for the moving free surface

In the described domain Ω the continuity equation and the Navier-Stokes equations are valid for the fluid. The velocity field $\mathbf{u}$ can be expressed as the gradient of a scalar potential Φ, if an incompressible fluid and an irrotational flow is assumed:

$$\mathbf{u} \ = \ [u_n \ u_t]^T \ = \ \nabla\Phi \ . \tag{1}$$

With this assumption the continuity equation becomes a Laplace equation for Φ in the fluid domain:

$$\nabla^2\Phi = 0 \ . \tag{2}$$

By means of a direct boundary element method and by using Cauchy's principal value the Laplace equation (2) is transformed into a boundary integral equation (BIE) of the form

$$C(\boldsymbol{\xi})\,\Phi(\boldsymbol{\xi}) = \fint_{\Gamma} \left(\frac{\partial\Phi(\mathbf{x})}{\partial n(\mathbf{x})} G(\mathbf{x},\boldsymbol{\xi}) - \Phi(\mathbf{x}) \frac{\partial G(\mathbf{x},\boldsymbol{\xi})}{\partial n(\mathbf{x})} \Phi \right) d\Gamma(\mathbf{x}) \ , \tag{3}$$

with $\Gamma = \Gamma_1 + \Gamma_2 + \Gamma_3 + \Gamma_4$ (see Fig. 1). The position vectors for field and source points are denoted by $\mathbf{x}$ and $\boldsymbol{\xi}$ and the unit outward normal vector by $\mathbf{n}$. $C(\boldsymbol{\xi})$ ia a geometric constant and $G(\mathbf{x},\boldsymbol{\xi})$ the fundamental solution of the plane problem .

FREE SURFACE FLOW

In order to describe the motion of the fluid particles on the free surface Γ_1, a Lagrangian formulation is chosen. With the assumptions mentioned before the dynamic boundary condition becomes the Bernoulli equation. Thus the kinematic and dynamic boundary conditions on the free surface can be written as

$$\frac{d\mathbf{X}_1}{dt} = \nabla\Phi_1 \ , \tag{4}$$

$$\frac{d\Phi_1}{dt} = -\frac{1}{\rho}p - g\,(\mathbf{z}_1 - \bar{\mathbf{z}}_1) + \frac{1}{2}(\nabla\Phi_1)^2 \ , \tag{5}$$

with the position vector of the fluid particles on the free surface $\mathbf{X}_1$ (see Fig. 1). Here t denotes the time, p the pressure, ρ the density of the fluid, g the acceleration due to gravity and $\mathbf{z}_1$ the vertical reference coordinate of the fluid particles on the free surface ($\bar{\mathbf{z}}_1$ for the undisturbed surface — the still water line (SWL)). The motion of the fluid particles and therefore the unknown behavior of the free surface Γ_1 is derived by integrating the time-dependent nonlinear kinematic and dynamic boundary conditions with respect to time t.

The following boundary conditions are given for the fluid. On the free surface Γ_1 the potential Φ_1 is given as an initial value problem

$$\Phi_1 = \Phi_1(t) \ , \tag{6}$$

and on the sea bottom Γ_3 the normal velocity vanishes:

$$u_{n3} = 0 \ . \tag{7}$$

There are no general boundary conditions known on the vertical boundaries Γ_2 and Γ_4 , except for some special cases:

- in the linear theory of regular waves Sommerfeld's radiation condition is valid,
- in the shallow water theory a relation between the mean flux through the vertical boundaries and the elevation of the free surface can be applied (see Haugel [5] or Schaper [8]),
- in the case of two-dimensional flow the assumption of periodicity in space results in:

$$\Phi(x) = \Phi(x + \lambda) \ , \tag{8}$$

$$u_n(x) = u_n(x + \lambda) \ , \tag{9}$$

where λ is the wavelength.

Especially for nonlinear transient waves, there is no other condition known than the periodicity, even in the case of one-directional propagating waves. Therefore, along the vertical boundaries Γ_2 and Γ_4 the following conditions are chosen:

$$u_{n2} = -u_{n4} \ , \tag{10}$$

$$\Phi_2 = \Phi_4 \quad . \tag{11}$$

WAVE CHANNEL

To describe the propagation of waves in a wave channel the boundary conditions have to be defined in a modified way. A typical deep water channel (Fig. 2) consists of a flap as wave generator (e.g. for sinusoidal oscillations), a plane bottom and a slope construction in order to avoid wave reflections at the end of the wave channel. Such flume is widely used in laboratories for the experimental analysis of two-dimensional deep water waves and fluid structure interaction. The boundary conditions in the wave channel are given by:

$$\Phi_1 = \Phi_1(t) \ , \tag{12}$$

$$u_{n2} = v_{n,\text{flap}} , \tag{13}$$

$$u_{n3} = 0 , \tag{14}$$

$$u_{n4} = 0 , \tag{15}$$

where $v_{n,\text{flap}}$ is the normal velocity at the flap with respect to translation $\dot{x}$ and rotation $\dot{\alpha}$ (see Fig. 2). The physical damping of the slope is modelled by a modified dynamic boundary condition at the free surface Γ_1. The wave absorption is introduced by assuming Rayleigh-damping on the free surface in the region of the slope (see Betts and Mohamad [1]). Thus, the Bernoulli equation (5) is extended to

$$\frac{d\Phi_1}{dt} = -\frac{1}{\rho}p - g(z_1 - \bar{z}_1) + \frac{1}{2}(\nabla\Phi_1)^2 - \mu_R\Phi_1 , \tag{16}$$

where $\mu_R = f(x)$ represents the Rayleigh-damping coefficient as a function of the fluid particle's position. The additional term $\mu_R\Phi_1$ in the Bernoulli equation represents a force in opposite direction of the velocity and results therefore in the dissipation of energy.

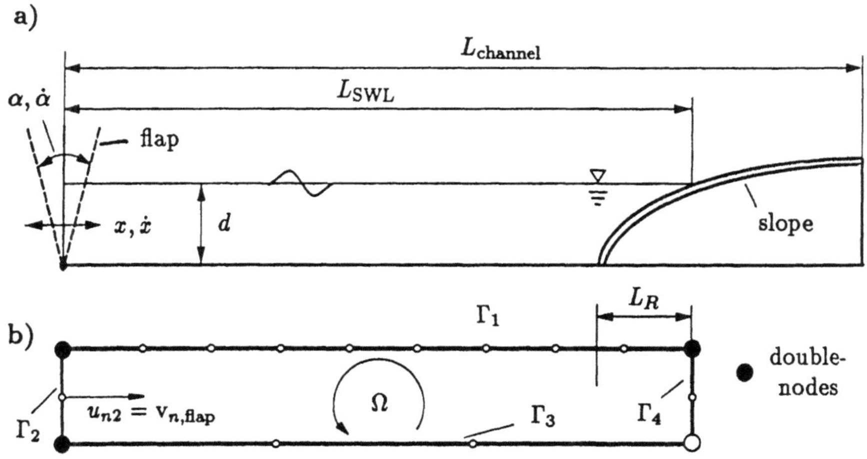

Figure 2: Deep water wave channel (a) and its discretization in the numerical model (b)

NUMERICAL IMPLEMENTATION

In order to carry out the numerical simulation of the moving free surface the boundaries of the domain Ω are discretized into finite boundary elements. In the Lagrangian description the position of the fluid particles is identical to the position of the element nodes on the free surface. By approximating the unknown boundary functions Φ and $\partial\Phi/\partial n$ in each node of the discretization, and by numerical integration, the solution of the boundary integral equation (3) is obtained. These approximations represent the solution with a residuum

$$\mathbf{D}(\boldsymbol{\xi}) = \fint_{\Gamma} \left(\frac{\partial\Phi(\mathbf{x})}{\partial n(\mathbf{x})} \mathrm{G}(\mathbf{x},\boldsymbol{\xi}) - \Phi(\mathbf{x}) \frac{\partial \mathrm{G}(\mathbf{x},\boldsymbol{\xi})}{\partial n(\mathbf{x})} \right) d\Gamma(\mathbf{x}) - \mathrm{C}(\boldsymbol{\xi})\,\Phi(\boldsymbol{\xi}) \,. \quad (17)$$

To minimize the defect a least squares method

$$\int_{\Gamma} \mathbf{D}^2(\boldsymbol{\xi})\, d\Gamma(\boldsymbol{\xi}) \;\Rightarrow\; \text{Min.} \quad (18)$$

is used. Thus, an additional numerical integration is required

$$\sum_{j=1}^{n_{\mathrm{GP}}} \mathbf{D}^2(\boldsymbol{\xi}_j)\, \mathbf{Q} \;=\; \mathbf{D}^T \mathbf{Q}\, \mathbf{D} \;\Rightarrow\; \text{Min.}\,, \quad (19)$$

with n_{GP} grid points $\boldsymbol{\xi}$. This results in the weighted Gaussian normal equations

$$\underbrace{\mathbf{A}^T \mathbf{Q}\, \mathbf{A}}_{\text{symmetric}} \; \mathbf{a} \;=\; \mathbf{A}^T \mathbf{Q}\, \mathbf{b}\,, \quad (20)$$

where $\mathbf{Q}$ is a diagonal matrix of the integration weights multiplied by the determinant of the Jacobian matrix. $\mathbf{A}$ is the matrix of coefficients, $\mathbf{a}$ is the vector of unknown and $\mathbf{b}$ the vector of known boundary conditions. This procedure leads to a symmetric matrix $\mathbf{A}^T \mathbf{Q}\, \mathbf{A}$ and achieves a high order of convergence for the Dirichlet condition at the free surface (see Wendland [10]). This is important for the numerical stability of the subsequent time integration of the nonlinear boundary conditions.

The solution of the linear equations (20) gives the normal velocity u_n for the fluid particles on the free surface. To determine the time-dependent nonlinear boundary conditions in equation (4) and (5) the tangential component of the velocity vector $\mathbf{u}$ in equation (1) has to be calculated as well. This is done by spline interpolation of the given potential Φ_1.

The accuracy of the proposed boundary element method is improved by the use of an isoparametric cubic spline representation of the geometry of the free surface, and higher order polygonial elements (up to fourth order) for the other boundaries. The computational costs of this approach are accepted with respect to the advantage of improved numerical stability. Another important computational aspect is the numerical integration for the assembly of the coefficient-matrices in equation (20). Conventional quadrature formulas give acceptable results in the case of significant distances between an element and a source point $\boldsymbol{\xi}_j$. Near the source point the order of integration schemes has to be increased. In the extreme case of singularities, modified or special integration schemes have to be used. The treatment of singular integrals depends mainly on the element type and on the position of the source point in the element. Therefore, different forms of semi-analytical integration formulas for certain singularities have been developed (see Gravert [3]). In order to reduce computation time the analytical representation of a singular integral is converted into a numerical integration scheme.

The corners of two boundaries are modelled by double-nodes at the end of the boundaries with an identical position vector (see Gravert [3]). Thus, the discontinuities of the normal velocities at the corners of the boundaries are taken into account. The importance of this aspect was recently pointed out and discussed in detail by Grilli and Svendsen [4].

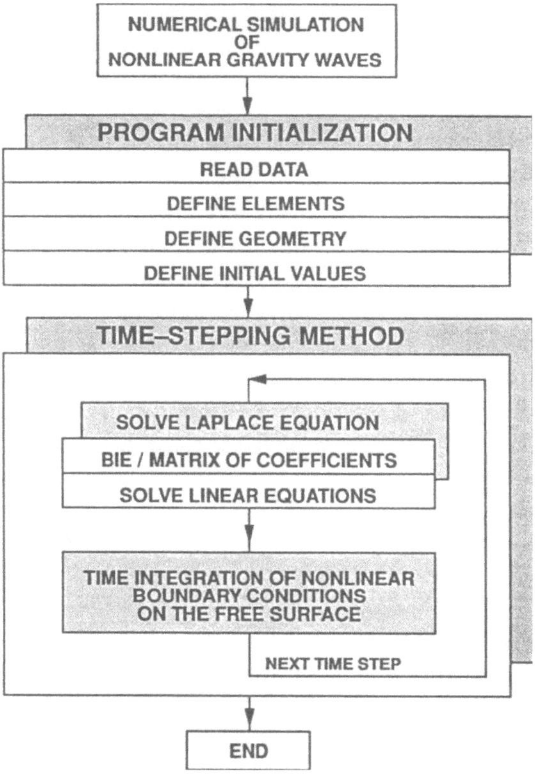

Figure 3: Flow-chart of the algorithm

To determine the position of the fluid particles as a function of time an initial value problem has to be solved by integrating the time-dependent nonlinear free surface boundary conditions. The integration of these ordinary differential equations (4) and (5) is carried out as usual by the multistep-method of Adams-Bashforth-Moulton. A classical Runge-Kutta method is applied as starting procedure for this predictor-corrector scheme (both fourth order schemes).

Figure 3 shows the simplified algorithm of the combined boundary element and time stepping technique that is used in this proposal.

NUMERICAL WAVE CHANNEL

Due to the restrictive assumption of spatial periodicity the proposed numerical simulation of extreme gravity waves allows only a qualitative understanding of the complex behavior of nonlinear wave motions, whereas a numerical wave channel model makes even quantitative comparisons with experimental measurement data feasible. The numerical implementation of the wave channel uses all the methods and aspects discussed before. Figure 2 presents the numerical modelling — the discretization of the channel. Even in this case the modelling of the contact points (free surface - flap, flap - bottom and slope - free surface) does not cause any difficulties because double-nodes are introduced. As explained before, the boundary condition at the flap is $\partial \mathbf{u}/\partial \mathbf{n} = u_n = \mathrm{v}_{n,\mathbf{flap}}$. On the bottom and on the slope the normal velocity u_n is equal to zero. The time-dependent translational and rotational motions of the flap, and therefore the normal velocity $\mathrm{v}_{n,\mathbf{flap}}$ is given either as an analytical function (e.g. time-periodic excitation) or as measured data. By assuming Rayleigh-damping on the free surface along the distance L_R (see Fig. 2) the dynamic boundary condition is used in the form of equation (16). The damping coefficient μ_R is assumed to be a linear or an exponential function with respect to the distance of the fluid particles from the contact point of the slope and the free surface. This mathematical modelling of the properties of a physical slope allows to treat the slope as a vertical boundary.

RESULTS

WAVE PROPAGATION

Several numerical experiments on propagating and overturning waves were carried out. In order to check the accuracy of the discussed methods the long time behavior of a regular periodic fifth order Stokes wave was analyzed. For this type of waves an analytical approximation is known (e.g. Rienecker and Fenton [2]) and because of the steadiness of a regular periodic wave the results of the numerical simulation of the propagating free surface can be compared in an accurate way.

Figure 4 shows a deep water fifth order Stokes wave with a wave steepness $H/\lambda = 0.1$ — the wave height is denoted by H, the wave length by λ — and a fraction $d/\lambda = 1$, where d is the water depth. In the case of steady deep water waves Williams [11] indicated an upper limit of $H/\lambda = 0.141063$ for the wave steepness. One wave length λ of the free surface is discretized into 50 cubic spline elements. Two polygonial elements of fourth order are used for each of the vertical boundaries and three for the sea bottom. For a qualitative check the time-dependent motion of a single fluid particle P_1 (Fig. 4) was computed. The time integration was carried out with a constant step size of approximately $T/12$, where T is the wave period. Due to the steadiness of the considered wave the fluid particle has to end up exactly at the position of the initial wave after each period (12 time steps)
The observed particle is moving along a helix (see Fig. 4), which is significant for the nonlinear behavior of fluid particles and the mass transport of Stokes waves.

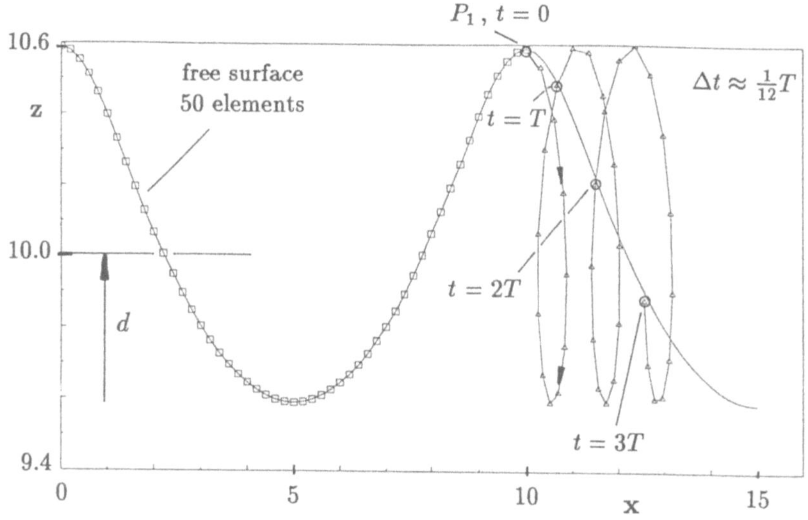

Figure 4: Nonlinear motion of a fluid particle P_1 in a fifth order Stokes wave

To control the combined boundary element and time stepping technique in a more accurate way some characteristic quantities were calculated. For each time step the flux through the boundary has to vanish:

$$\oint \frac{\partial \Phi}{\partial n} \, d\Gamma \overset{!}{=} 0 \,. \tag{21}$$

The water level of the undisturbed free surface (SWL) has to be constant

$$\bar{\zeta} = \frac{1}{\lambda} \int_0^\lambda \zeta \, dx \overset{!}{=} \text{const.}\,, \tag{22}$$

where ζ is the elevation of the fluid particles with respect to the undisturbed surface and x is the horizontal coordinate. In addition, the energy of the system under consideration has to be constant in time

$$\mathbf{E} = \mathbf{E}_{\text{kin}} + \mathbf{E}_{\text{pot}} = \frac{\rho}{2} \int_\Gamma \Phi \frac{\partial \Phi}{\partial n} \, d\Gamma + \frac{\rho}{2} \int_0^\lambda \zeta^2 \, dx \overset{!}{=} \text{const.}\,, \tag{23}$$

where $\mathbf{E}_{\text{kin}}$ is the kinetic and $\mathbf{E}_{\text{pot}}$ the potential energy. Figure 5 gives the relative error of these quantities with regard to the time t (for 10 periods T). Even the long time behavior shows acceptable results and no numerical instabilities occur in this example.

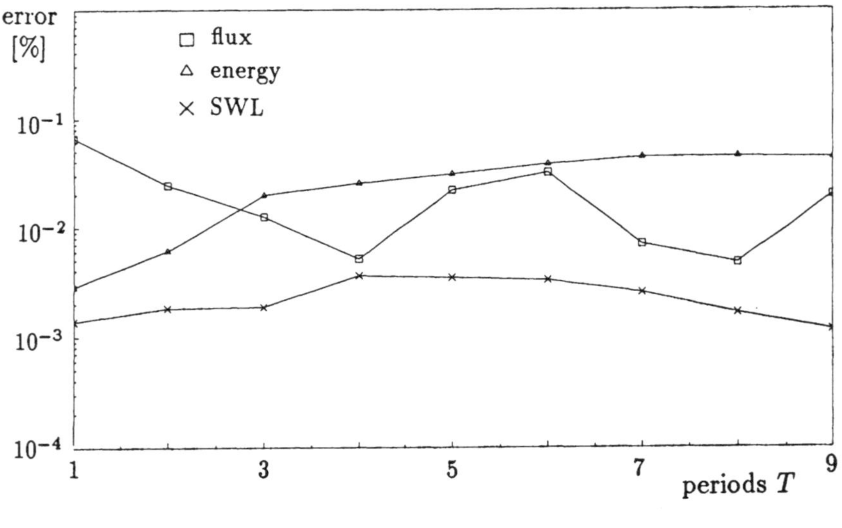

Figure 5: Relative error of the flux through the boundaries, the still water line (SWL), and the energy with respect to the simulation time

Another typical application of the proposed method is the numerical simulation of overturning and breaking waves.

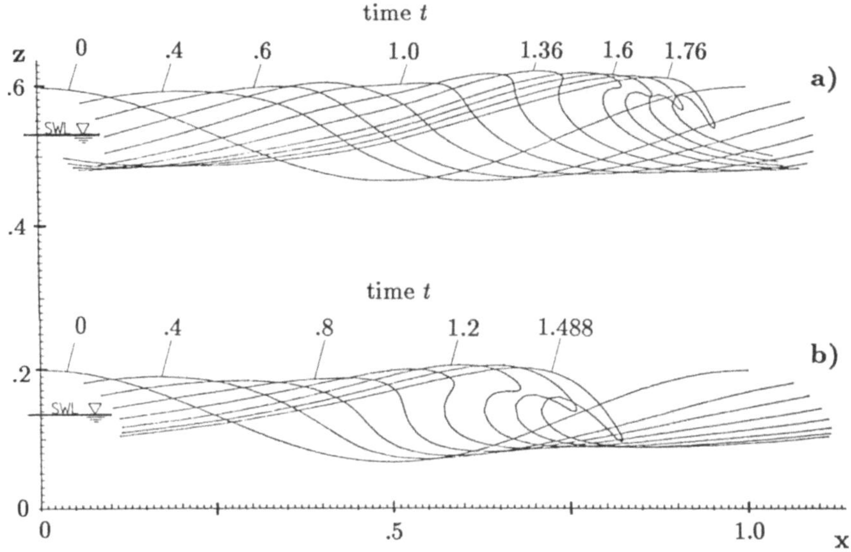

Figure 6: Overturning waves in deep (a) and shallow (b) water

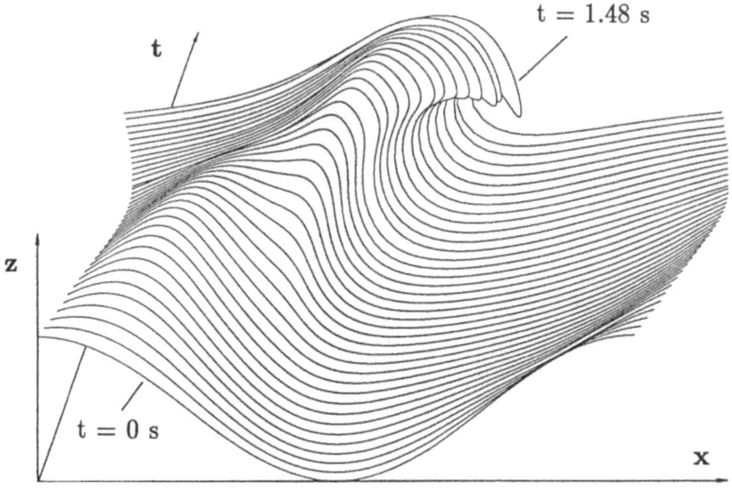

Figure 7: Time evolution of a shallow water wave

Figure 6 shows the results of the propagation of waves in shallow and deep water, respectively. For the computations the free surface was divided into 59 spline elements. Two polygonial elements of order four were used for the vertical boundaries, and three for the bottom. A wave steepness $H/\lambda = 0.133$, a fraction $d/\lambda = 0.133$ for the shallow water and $d/\lambda = 0.53$ for the deep water were supposed. As initial condition for the free surface a regular periodic sinusoidal wave was chosen (compare with New, Mc Iver and Peregrine [7] and Vinje and Brevig [9]). The simulation was carried out with a constant step size of $\Delta t = 0.01$ seconds for the time integration. The computation for the shallow water wave breaks down after 1.488 seconds – for the deep water wave after 1.76 seconds.

In Figure 7 the results of the simulation of a first order Stokes shallow water wave is given, using a different scaling than before, but the same values for the wave steepness H/λ and for the fraction d/λ. In this presentation the simulation time t is introduced as the third dimension. The simulation uses 100 spline elements for the free surface and a constant step size of $\Delta t = 0.04$ seconds. Apparently, the wave gets out of shape three times, until the overturning starts.

After 37 time steps the computation breaks down. On a CONVEX C220 the CPU time per time step was 54 seconds in this example.

WAVE CHANNEL APPROACH

Experiments were carried out in a test flume ($L_{\text{channel}} = 12.0$ meter, 1.5 meter width). The undisturbed surface has a length of $L_{\text{SWL}} = 9.82$ meter, where the water depth is $d = 1.0$ meter — see Figure 2. The flap is excited without any translational motion ($\dot{x} = 0$) by a sinusoidal oscillation with a frequency of 0.44 Hz and has a maximal angle-amplitude of $\alpha_{max} = 8.56°$ (compare Fig. 2). Starting from a resting position the elevation of the free surface was measured at a distance of 6.0 meters away from the flap.

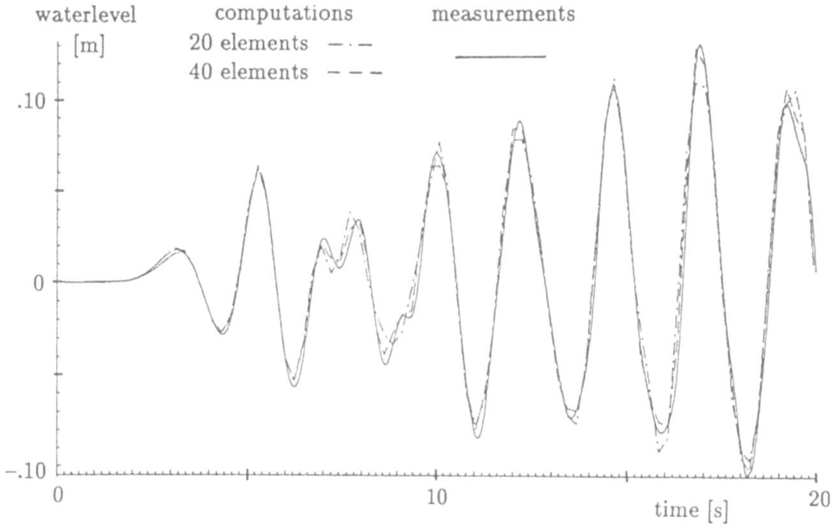

Figure 8: Transient wave elevations in a wave channel: comparison between computed and measured results

The numerical simulations were carried out using 20 and 40 cubic spline elements for the free surface representation. Again the bottom was discretized into three polygonial elements of order four, the flap and the other vertical boundary into two. A constant step size of $\Delta t = 0.06$ seconds was chosen for the time integration . Figure 8 depicts the comparison of measured (straight lines) and calculated (dashed lines) results for the free surface elevation at the chosen point starting from the resting position over a period of 20 seconds. The computations match the measurements with acceptable accuracy. As expected slightly better results are achieved by finer discretization.

CONCLUDING REMARKS

Numerical simulations of nonlinear deep and shallow water waves were carried out with a combined boundary element and time stepping technique. The computations of regular periodic as well as overturning waves showed the improved numerical stability of this method. This was achieved by the implementation of a weighted least squares method, double-node representation of corners, and the use of semi-analytical integration schemes for singular integral kernels. The extension of this procedure towards a numerical wave channel was obtained and resulted in satisfactory agreement with measured data of steady state and transient wave elevations. The computational costs of the improved approach were accepted and can be significantly reduced by taking advantage of modern computer architecture concepts (e.g. vector or parallel processing).

REFERENCES

[1] Betts, P.L. and Mohamad, T.T. *Water waves: A time-varying unlinearized boundary element approach*, in: Finite Element Flow Analysis (Ed. Kawai, T.), University of Tokyo Press, pp. 923–929, 1982.

[2] Rienecker, M.M. and Fenton, J.D. *A Fourier approximation for steady water waves*, J. Fluid Mech., Vol. 104, pp. 119–137, 1981.

[3] Gravert, P. *Numerische Simulation extremer Schwerewellen im Zeitbereich mit direkter Randelementmethode und Zeitschrittverfahren*, Fortschrittberichte VDI, Reihe 7, Nr. 132, VDI-Verlag, Düsseldorf, 1987 (in german).

[4] Grilli, S.T. and Svendsen, I.A. *Corner problems and global accuracy in the boundary element solution of nonlinear wave flow*, Engineering Analysis with Boundary Elements, Vol. 7, pp. 178–195, 1990.

[5] Haugel, A. *Adaption of tidal numerical models to shallow water wave problems*, Proc. 17th Coastal Engineering Conf., Sidney, Australia, 1980.

[6] Longuet-Higgins, M.S., Cokelet, E.D. The deformation of steep surface waves on water. *A numerical method for computation*, Proc. Roy. Soc. London, A 350, pp. 1–26, 1976.

[7] New, A.L., Mc Iver, P. and Peregrine, D.H. *Computation of overturning waves*, J. Fluid Mech., Vol. 150, pp. 233–251, 1985.

[8] Schaper, H. *Ein Beitrag zur numerischen Berechnung von kurzen Flachwasserwellen mit verbesserten Differenzenverfahren*, Berichte des Instituts für Strömungsmechanik und elektronisches Rechnen im Bauwesen, Hannover, 1985 (in german).

[9] Vinje, T. and Brevig, P. *Numerical simulation of breaking waves*, Proc. of the third International Conference Finite Elements in Water Resources, pp. 5.196–5.211, Lafayette/Miss., 1980.

[10] Wendland, W. *Bemerkungen zu Randelementmethoden und ihren mathematischen und numerischen Aspekten*, in: Mitteilungen der GAMM, Heft 2, pp. 3–27, 1986 (in german).

[11] Williams, J.M. *Limiting gravity waves in water of finite depth*, Phil. Trans. Roy. Soc. London, Ser. A 302, pp. 139–188, 1981.

Time-Domain Simulation of Second-Order Wave Diffraction Around Structures

M. Isaacson, K.F. Cheung

Department of Civil Engineering, University of British Columbia, Vancouver, B.C., V6T 1W5, Canada

ABSTRACT

A time-domain second-order method developed recently to study the nonlinear wave diffraction around a surface-piercing body of arbitrary shape in three dimensions is summarized. The free surface boundary conditions and the radiation condition are satisfied to second order by a numerical integration in time, and the field solution at each time step is obtained by an integral equation method based on Green's theorem. The solution is separated into a known incident potential and a scattered potential to be evaluated. The variation with time of the free surface elevation to first and second order is discussed, and a video film is developed to assist visualization of the predicted wave field.

INTRODUCTION

The accurate prediction of steep wave interactions with large offshore structures generally requires the solution to a nonlinear, moving-boundary wave diffraction problem. In general, two approaches to this problem have been developed. One is based on a full nonlinear solution using a time-stepping procedure, in which the free surface boundary conditions are applied at the instantaneous free surface and a new system of simultaneous equations based on Green's theorem is generated and solved at each time step as the free surface moves to a new position (e.g. Vinje and Brevig [13], Isaacson [6], Dommermuth and Yue [3], and Grosenbaugh and Yeung [5]). The other is based on a perturbation procedure in the frequency-domain which provides a second-order solution defined over a time-independent boundary including the still water surface (e.g. Molin [10], Eatock Taylor and Hung [4], Vada [12], Abul-Azm and Williams [1], Kim and Yue [9], and Wu and Eatock Taylor [14]).

An alternative approach developed recently by Isaacson and Cheung [7, 8] applies a perturbation procedure to the time-stepping method in order to provide a time-domain second order solution for the two- and three-dimensional problems respectively. Because of the application of the perturbation procedure, the full nonlinear problem is reduced to a second-order problem defined on a time-independent boundary, so that a solution to the system of equations obtained through the discretization of Green's theorem is required only once rather than at each time step. Their method has been found to be computationally efficient, and numerically very stable. The computed wave forces and runup to second order, and the corresponding components at first and second order have been validated through extensive comparisons with previous theoretical and experimental results and are not repeated here. In the present paper, emphases are placed on the analysis of the free surface elevation to first and second order and on the development of a video film which provides a convenient visualization of the predicted wave field.

THEORETICAL FORMULATION

Full nonlinear problem

The full nonlinear boundary value problem defining the fluid motion is first considered. With reference to figure 1, the three-dimensional problem is defined with a right-handed Cartesian coordinate system (x,y,z), in which x and y are measured horizontally and z is measured vertically upwards from the still water level. The body located at the centre of the domain is rigid and fixed in space. The incident waves are two-dimensional in the x-z plane and are progressive in the positive x direction.

Let t denote time and η the free surface elevation above the still water level. The seabed is assumed impermeable and horizontal along the plane $z = -d$. With the fluid assumed incompressible and inviscid, and the flow irrotational, the fluid motion can be described by a velocity potential ϕ which satisfies the Laplace equation within the fluid domain $\mathscr{D}$.

$$\nabla^2\phi = 0 \qquad \text{in } \mathscr{D} \tag{1}$$

The potential ϕ is subject to boundary conditions on the seabed, the instantaneous wetted body surface S_w and the free surface S_f, given respectively as

$$\frac{\partial\phi}{\partial z} = 0 \qquad \text{at } z = -d \tag{2}$$

$$\frac{\partial\phi}{\partial n} = 0 \qquad \text{on } S_w \tag{3}$$

$$\frac{\partial\phi}{\partial z} - \frac{\partial\eta}{\partial t} - \frac{\partial\phi}{\partial x}\frac{\partial\eta}{\partial x} - \frac{\partial\phi}{\partial y}\frac{\partial\eta}{\partial y} = 0 \qquad \text{on } S_f \tag{4}$$

$$\frac{\partial\phi}{\partial t} + g\eta + \frac{1}{2}|\nabla\phi|^2 = 0 \qquad \text{on } S_f \tag{5}$$

Here n denotes distance in the direction of the unit normal vector **n** directed outward from the fluid region and g is the acceleration due to gravity. Equations (2) and (3) correspond respectively to the kinematic boundary conditions on the impermeable seabed and body surface where the normal fluid velocity component vanishes. Equation (4) is the kinematic free surface boundary condition requiring that the fluid particle velocity normal to the free surface is equal to the velocity of the free surface itself in that direction, while equation (5) is the dynamic free surface boundary condition stating that the pressure at the free surface, expressed in terms of the Bernoulli equation, is equal to zero, thereby neglecting the effects of surface tension. In addition, a suitable radiation condition is to be applied on a control surface S_c some distance from the body as indicated in figure 1. The major difficulties associated with this problem arise primarily from the poorly defined radiation condition involving transient waves and from the two nonlinear free surface boundary conditions applied on the instantaneous free surface, which itself is unknown *a priori.*

Second-order expansion

When the water depth is not small compared with a typical wavelength and the wave height is moderate, it is possible to reduce the two nonlinear free surface boundary conditions originally derived on the instantaneous free surface to conditions evaluated at the still water level by the Stokes second-order expansion procedure. As the location of the free surface is unknown *a priori*, the full nonlinear free surface boundary conditions, equations (4) and (5), are first expanded about the still water level with a Taylor series expansion:

$$\left(\frac{\partial\phi}{\partial z} - \frac{\partial\eta}{\partial t} - \frac{\partial\phi}{\partial x}\frac{\partial\eta}{\partial x} - \frac{\partial\phi}{\partial y}\frac{\partial\eta}{\partial y}\right)_{z=0} + \eta\frac{\partial}{\partial z}\left(\frac{\partial\phi}{\partial z} - \frac{\partial\eta}{\partial t} - \frac{\partial\phi}{\partial x}\frac{\partial\eta}{\partial x} - \frac{\partial\phi}{\partial y}\frac{\partial\eta}{\partial y}\right)_{z=0} + \cdots = 0 \qquad (6)$$

$$\left(\frac{\partial\phi}{\partial t} + g\eta + \frac{1}{2}|\nabla\phi|^2\right)_{z=0} + \eta\frac{\partial}{\partial z}\left(\frac{\partial\phi}{\partial t} + g\eta + \frac{1}{2}|\nabla\phi|^2\right)_{z=0} + \cdots = 0 \qquad (7)$$

The problem may now be cast in a time-independent geometry which includes the still water surface S_o in place of the instantaneous free surface S_f. To proceed with the Stokes second-order expansion, components at first and second order are separated by introducing perturbation expansions for ϕ and η:

$$\phi = \varepsilon(\phi_1^w + \phi_1^s) + \varepsilon^2(\phi_2^w + \phi_2^s) + \cdots \qquad (8)$$

$$\eta = \varepsilon(\eta_1^w + \eta_1^s) + \varepsilon^2(\eta_2^w + \eta_2^s) + \cdots \qquad (9)$$

where ε is a perturbation parameter which is small. The superscripts w and s indicate respectively the incident and scattered components, while the subscripts 1 and 2 indicate respectively components at first and second order. Substitution of the power series representation for ϕ and η into the Laplace equation and boundary conditions, including the expanded free surface boundary conditions, gives rise to separate boundary value problems for each of the ε and ε^2 terms in the power series.

Solution to second order

The problem is now applied to a time-independent domain D bounded by the seabed, the body surface below the still water level S_b, the control surface S_c and the still water surface S_o. In the k-th order problem, the scattered potential satisfies the Laplace equation in D

$$\nabla^2\phi_k^s = 0 \qquad \text{in D} \qquad (10)$$

and is subject to the boundary conditions applied on the seabed, the body surface and the still water surface, given respectively as

$$\frac{\partial\phi_k^s}{\partial z} = 0 \qquad \text{at } z = 0 \qquad (11)$$

$$\frac{\partial\phi_k^s}{\partial n} = -\frac{\partial\phi_k^w}{\partial n} \qquad \text{on } S_b \qquad (12)$$

$$\frac{\partial\phi_k^s}{\partial z} - \frac{\partial\eta_k^s}{\partial t} = f_k' \qquad \text{on } S_o \qquad (13)$$

$$\frac{\partial\phi_k^s}{\partial t} + g\eta_k^s = f_k'' \qquad \text{on } S_o \qquad (14)$$

At first order, the forcing terms f_1' and f_1'' are zero and the boundary value problem is homogeneous, while at second order f_2' and f_2'' are given respectively by

$$f_2' = -\left(\frac{\partial\phi_2^w}{\partial z} - \frac{\partial\eta_2^w}{\partial t}\right) + \frac{\partial\phi_1}{\partial x}\frac{\partial\eta_1}{\partial x} + \frac{\partial\phi_1}{\partial y}\frac{\partial\eta_1}{\partial y} - \eta_1\frac{\partial^2\phi_1}{\partial z^2} \qquad (15)$$

$$f_2'' = -\left(\frac{\partial \phi_2^s}{\partial t} + g\eta_2^w\right) - \frac{1}{2}|\nabla\phi_1|^2 - \eta_1\frac{\partial^2\phi_1}{\partial z\partial t} \tag{16}$$

The second-order forcing terms are composed of the known incident wave components at second order and quadratic terms which can be evaluated from the first-order solution. The quadratic terms in f_2'' physically represent an oscillatory pressure field at double the incident wave frequency applied to the still water surface which would generate second-order forced waves subject to the geometric constraint defined by the quadratic terms of f_2'. Since the solution at first order includes the sum of the incident and scattered waves, the second-order forced waves generated by the quadratic terms are composed of three components:

(a) a plane wave component associated with the self-interaction of the first-order incident waves;
(b) a scattered wave component associated with the self-interaction of the first-order scattered waves; and
(c) a third component associated with the cross-interactions between the first-order incident and scattered waves.

The plane wave component (a) is part of the incident wave field and is excluded from the scattered potential. Its effects on the solution at second order is then accounted for by the body surface boundary condition. These second-order forced waves are phase-locked with the first-order wave system and do not satisfy any simple dispersion relation.

The effective pressure fields imposed on the still water surface by the forced wave components (b) and (c) are non-uniform in space and are concentrated near the body. The motion of these two components of forced waves on the still water surface would generate second-order free waves at double the incident wave frequency. Imposing the body surface boundary condition, the diffraction of all three components of forced waves gives rise to additional second-order free waves. These second-order free waves propagate at the corresponding celerity independent of the first-order wave system and satisfy the linear dispersion relation in the form of

$$(2\omega)^2 = gk_2 \tanh(k_2 d) \tag{17}$$

where k_2 is the wave number of the second-order free waves and ω is the angular frequency of the incident waves. With the first-order wave number denoted by k, in deep water k_2 approaches 4k while in shallow water k_2 approaches 2k. The scattered potential at second order is then composed of the forced wave components (b) and (c) and free waves generated by all three components of forced waves. Mathematically these forced and free waves represent respectively the inhomogeneous and homogeneous solutions of the boundary value problem.

With the problem now defined in terms of scattered components, a radiation condition applied to the scattered potential can be introduced. Provided the control surface is located sufficiently far from the body, the Sommerfeld radiation condition, originally derived on the basis of spatially and temporally periodic conditions of a potential function, can be extended to account for unsteady wave motions. In the time domain, the radiation condition for the scattered potential at k-th order is given by

$$\frac{\partial \phi_k^s}{\partial t} + c\frac{\partial \phi_k^s}{\partial r} = 0 \qquad \text{on } S_c \tag{18}$$

where r is the direction of scattered wave propagation at the control surface and c is the celerity of the scattered waves at the control surface in that direction. Based on Orkanski [11], the Sommerfeld radiation condition is first applied to estimate the time-dependent celerity on the boundary of the flow which in turn is used to calculate the boundary condition on the control surface for an advanced time.

With the problem now defined for a time-independent domain, the potential at k-th order on the surface of the domain is given by the an integral equation involving a Green's function G:

$$\phi_k^s(x) = \frac{1}{2\pi}\int_{S'}\left[G(x,\xi)\frac{\partial\phi_k^s}{\partial n}(\xi) - \phi_k^s(\xi)\frac{\partial G}{\partial n}(x,\xi)\right] dS \qquad (19)$$

Here ξ represents a point on the surface S' over which the integration is performed and n is measured from the point ξ. In the present context S' would comprise of S_b, S_c, S_o and the seabed. However, with the body and subsequently the wave field symmetric about the x axis, and the seabed horizontal, it is more efficient to simulate one half of the total surface with the seabed excluded from S' and to choose a Green's function that accounts for the double symmetry about the x axis and the seabed. This is

$$G(x,\xi) = \sum_{k=1}^{4} \frac{1}{|\xi_k - x|} \qquad (20)$$

The source points ξ_k are defined such that $\xi_1 = (\xi, \psi, \zeta)$ is a point on S'; $\xi_2 = (\xi, -\psi, \zeta)$ is the image of ξ_1 about the x axis; $\xi_3 = (\xi, -\psi, -(\zeta+2d))$ is the image of ξ_2 about the seabed; and $\xi_4 = (\xi, \psi, -(\zeta+2d))$ is the image of ξ_1 about the seabed.

With either the potential or its normal derivative on each portion of the boundary S' evaluated from the corresponding boundary conditions, the solutions to the integral equation can be obtained by a numerical procedure from which the remaining unknowns can be evaluated.

NUMERICAL FORMULATION

Field solutions at first and second order

The integral equation is solved by a numerical procedure in which the boundaries S_b, S_c and S_o are discretized into finite numbers of planar quadrilateral facets. The corresponding values of ϕ_k^s and $\partial\phi_k^s/\partial n$ are taken as constant over each facet and applied at the facet centroid. Equation (19) is thereby reduced to the following set of simultaneous equations:

$$\sum_{j=1}^{N_b} A_{ij}\phi_{kj}^s + \sum_{j=N_b+1}^{N} B_{ij}\left(\frac{\partial\phi_k^s}{\partial n}\right)_j = -\sum_{j=1}^{N_b} B_{ij}\left(\frac{\partial\phi_k^s}{\partial n}\right)_j - \sum_{j=N_b+1}^{N} A_{ij}\,\phi_{kj}^s \quad (i = 1, \ldots, N) \qquad (21)$$

in which N is the total number of facets in the formulation and N_b is the number of facets on S_b. It is understood that the surfaces are treated in the order of S_b, S_c and S_o as the index j increases from 1 to N. The coefficients B_{ij} and A_{ij} correspond to integrals of the Green's function and its normal derivative over area of the j-th facet respectively. When $i = j$, the coefficients are evaluated by a closed form integration, and when $i \neq j$ the integration is performed by a Gaussian numerical integration.

It is important to note that the coefficients in equation (21) are invariant in time. Thus, the solution to the system of equations is required only once rather than at each time step, compared to a more conventional nonlinear time-domain solution in which a new set of simultaneous equations must be generated and solved at each time step. Furthermore, the coefficients are functions of geometry only and consequently the same set of equations can be applied to different incident wave conditions with modifications only made to the input vector on the right-hand side. This is relatively efficient in comparison to a frequency domain analysis in which a new set of equations must be generated and solved for each individual incident wave condition. The input vector on the right-hand side consists of the normal derivative of the scattered potential on the body surface which is known from the incident potential, and the potential on the control surface and the still water surface which is to be evaluated by an explicit time-integration procedure applied to the corresponding boundary conditions.

Time-integration of radiation condition

In the present method, the radiation condition is integrated numerically in time to obtain the required scattered potential on the control surface. The radiation condition, equation (18), is applied to the component normal to the control surface. The potential on the control surface at time (t+Δt) can be expressed in terms of the solutions obtained at previous time-steps by expanding the condition with a central difference in time and a leap-frog difference in space:

$$\phi_k^s(\mathbf{x},t+\Delta t) = \frac{1-c\Delta t/\Delta n}{1+c\Delta t/\Delta n}\,\phi_k^s(\mathbf{x},t-\Delta t) + \frac{2c\Delta t/\Delta n}{1+c\Delta t/\Delta n}\,\phi_k^s(\mathbf{x}-\mathbf{n}\Delta n,t) \tag{22}$$

where Δt is a time-step size; Δn is a small distance in the order of a characteristic facet diameter; **x** is a point on the control surface; **n** is the normal vector at **x**; and c is now defined as the apparent celerity of the scattered waves in the direction of the normal vector on the control surface. Literally, the boundary value is extrapolated from the values of ϕ_k^s near the boundary inside the domain at previous time-steps through a knowledge of the celerity c on the control surface.

The celerity on the control surface at a time t is approximated by its value at the previous time step (t–Δt) at locations (**x**–**n**Δn) which are just within the control surface. This is consistent with the flow of information from within the fluid domain to the boundary. From equation (18), the numerical value of c at (t–Δt) and (**x**–**n**Δn) is

$$c = -\left[\frac{\Delta n}{\Delta t}\right]\left[\frac{\phi_k^s(\mathbf{x}-\mathbf{n}\Delta n,t) - \phi_k^s(\mathbf{x}-\mathbf{n}\Delta n,t-2\Delta t)}{\phi_k^s(\mathbf{x}-\mathbf{n}\Delta n,t) + \phi_k^s(\mathbf{x}-\mathbf{n}\Delta n,t-2\Delta t) - 2\phi_k^s(\mathbf{x}-2\mathbf{n}\Delta n,t-\Delta t)}\right] \tag{23}$$

Unfortunately, there is a numerical difficulty in using equation (23) in that the celerity is undefined when the denominator is zero. In order to circumvent this numerical difficulty, the following numerical scheme is carried out. When the absolute values of both $\partial\phi_k^s/\partial t$ and $\partial\phi_k^s/\partial n$ are less than certain prescribed values, the scattered potential at the advanced time $\phi_k^s(\mathbf{x},t+\Delta t)$ in equation (22) is simply given by $\phi_k^s(\mathbf{x},t)$. This is equivalent to taking c = 0, and is used because ϕ_k^s changes very little in one time step on account of $\partial\phi_k^s/\partial t$ being small. As the Sommerfeld radiation condition requires the scattered waves at large distance from the body to be outgoing, only the outward celerity (i.e. c ≥ 0) is used in describing the boundary condition. Finally, for stability reasons, the maximum value of the celerity calculated by equation (23) is limited to have a maximum value of Δn/Δt. Thus, the numerical values of c are given overall by:

$$c = \begin{cases} 0 & \text{for } c' \le 0, \text{ or when } \partial\phi_k^s/\partial t,\ \partial\phi_k^s/\partial n \text{ are small} \\ c' & \text{for } 0 < c' < \Delta n/\Delta t \\ \Delta n/\Delta t & \text{for } c' \ge \Delta n/\Delta t \end{cases} \tag{24}$$

where c' is the initially calculated value of the celerity from equation (23). The validity of this numerical radiation condition in the application of second-order wave diffraction has been confirmed by Cheung [2].

Time-integration of free surface boundary conditions

The free surface boundary conditions are used in an explicit time-stepping procedure to provide the free surface elevation and the velocity potential on the still water surface at each time step. Initially, the free surface elevation at a new time $(t+\Delta t)$ is first evaluated in terms of the known solution up to time t by the first order Adam-Bashforth equation as

$$\eta_k^s(t+\Delta t) = \eta_k^s(t) + \frac{\Delta t}{2}\left[3\left(\frac{\partial \eta_k^s}{\partial t}\right)_t - \left(\frac{\partial \eta_k^s}{\partial t}\right)_{t-\Delta t}\right] \tag{25}$$

The dynamic free surface boundary condition, equation (14), can now be used to provide $\partial\phi_k^s/\partial t$ at $(t+\Delta t)$. In turn the potential ϕ_k^s may then be obtained by the first order Adam-Moulton equation:

$$\phi_k^s(t+\Delta t) = \phi_k^s(t) + \frac{\Delta t}{2}\left[\left(\frac{\partial \phi_k^s}{\partial t}\right)_t + \left(\frac{\partial \phi_k^s}{\partial t}\right)_{t+\Delta t}\right] \tag{26}$$

With the normal derivative of the velocity potential on S_b known from the incident wave potential and the velocity potential on S_c and S_o evaluated from the time-stepping equations, the flow in the domain at the advanced time $(t+\Delta t)$ can now be solved from the discretized boundary integral equation (21). The output vector provides $\partial\phi_k^s/\partial n$ on S_o at $(t+\Delta t)$. The kinematic free surface boundary condition, equation (13), can now be used to provide the corresponding values of $\partial\eta_k^s/\partial t$ at the advanced time $(t+\Delta t)$. This completes the calculation of all the boundary parameters at time $(t+\Delta t)$ and the computation can then proceed to the next time step.

At $t = 0$, the scattered potential is identically zero and the body has no effects on the flow. The initial conditions are known and correspond to an undisturbed Stokes second-order wave field in the domain. With the imposition of the body surface boundary condition for $t > 0$, scattered waves at first and second order are developed and propagate away from the body. To avoid an abrupt initial condition and allow a gradual development of the scattered potential, the right-hand side of equation (12), which corresponds to the imposition of the body surface boundary condition, is multiplied by a modulation function such that the body imposes its existence gradually over the first wave cycle (see Isaacson & Cheung [7]).

Free surface elevation

With the velocity potential known, the free surface elevation to second order is given explicitly by the dynamic free surface boundary condition. For regular wave diffraction to second order, the free surface elevation η may be expressed in the usual way as a sum of three components η_1, η_0 and η_2 which are respectively the first-order oscillatory component at the incident wave frequency, the second-order steady component and the second-order oscillatory component at the twice the frequency. The expressions for η_1, η_0 and η_2 are given respectively as:

$$\eta_1 = -\frac{1}{g}\left(\frac{\partial \phi_1}{\partial t}\right) \qquad \text{on } S_o \tag{27}$$

$$\eta_0 = -\frac{1}{g}\left\langle\frac{1}{2}|\nabla\phi_1|^2 + \eta_1\frac{\partial^2\phi_1}{\partial z\partial t}\right\rangle \qquad \text{on } S_o \tag{28}$$

$$\eta_2 = -\frac{1}{g}\langle\frac{\partial\phi_2}{\partial t} + \frac{1}{2}|\nabla\phi_1|^2 + \eta_1\frac{\partial^2\phi_1}{\partial z\partial t}\rangle - \eta_0 \quad \text{on } S_o \tag{29}$$

where < > denotes a time average. The runup and rundown to second order can be obtained respectively from the maximum and minimum values of the free surface elevation η.

RESULTS AND DISCUSSION

Discretization scheme

In order to illustrate and examine the present method, the special case of regular wave diffraction around a bottom-mounted surface-piercing circular cylinder is considered in this paper. The numerical model corresponds to that of a rectangular domain with a circular cylinder located at the centre. A FORTRAN program GEOM3D has been developed to discretize the body surface, the control surface and the still water surface into finite numbers of planar quadrilateral facets, and to generate and solve the system of simultaneous equations. Depending on the geometry and incident wave frequencies of interest, the number of facets used in the discretization may vary from 1500 to 2300. The coefficients A_{ij} and B_{ij} are evaluated by a 16-point Gaussian integration, and double precision is used throughout.

Figure 2 shows an example of the surface discretization scheme. The seabed and one half of the total surface are excluded due to the choice of the Green's function that accounts for double symmetry. Smaller facets are used on the body surface and on the nearby still water surface in order to describe the more rapid variation of the potential near the body and to speed up the convergence of runup and wave force calculations. The number of facets in the z direction is determined such that the facets nearest to the still water surface are square and the facets nearest to the seabed have an aspect ratio of approximately 1.5. A second FORTRAN program WAVE3D has been developed to time step the development of the flow and to calculate the free surface elevation and wave forces at each time step.

Flow visualization

For the numerical model shown in figure 2, approximately 100 kilobytes of numerical results of the free surface elevation are generated at each time step. The most effective way to present such a large volume of data is through a computer animation of the free surface movement. Not only does this provide a convenient visualization and analysis of the predicted wave field, but it also provides a visual assessment of the correctness of the numerical solution.

For particular conditions corresponding to ka = 2, kd = 1 and H/L = 0.08, the variation with time of the free surface elevation to first and second order has been transformed to a colour film as an aid to viewing the development of the flow. The symbols a, H and L denote respectively the radius of the cylinder, the wave height and the wavelength. The domain is 4 wavelengths long and 4 wavelengths wide. The vertical scale of the free surface elevation, except as noted otherwise, is exaggerated 8 times in order to make the interactions more pronounced. Although numerical results were obtained over one side of the x axis, the free surface elevation has been plotted over the entire domain in order to show a complete picture of the flow development.

Numerical results were obtained using the IBM 3090/150S computer at the University of British Columbia and were transferred to a Macintosh IIx computer for post-processing. A spreadsheet program WINGZ Version 1.1 and an image analysis program IMAGE Version 1.33g were used to generate colour-contoured three-dimensional and plan views of the free surface at 40 equal time-steps in sequence spanning a single wave period and shown on a 60 x 60 grid. These images were stored in disk and cycled through a loop in order to generate a moving picture extending over many wave periods. The computer animation was then transferred to videotape using the RasterOps ColorBoard 364 and the RasterOps Video Expander that enable the conversion of a Red-Green-Blue video signal to a National Television Standards Committee interlaced colour signal without significant loss of resolution and colour detail.

For convenience, in the results described here reference is made to the attached figures which show views at only two instants, half a cycle apart; although a better assessment of the results include a review of the corresponding video film.

First-order free surface

By studying the first-order solution, some insight may be gained into the nature of the quadratic forcing terms in equations (15) and (16), which in turn dictate the second-order wave field. Figure 3 shows views of the first-order free surface which are 180° out of phase. In the figure, the incident waves propagate from left to right in the positive x direction, while the scattered waves propagate concentrically from the cylinder and their amplitudes decay with distance away from the cylinder. In the upwave region, the scattered waves and the incident waves propagate in the opposite direction forming a system of partial standing waves in front of the cylinder and giving rise to a strong second-order forcing. Along the sides of the cylinder, the first-order scattered waves propagate perpendicularly to the incident waves producing a fairly complicated and intensive forcing near the cylinder. In the downwave region, the scattered waves with decaying amplitude propagate in approximately the same direction as the incident waves and the second-order forcing is relatively small.

Since figures 3a and 3b correspond to instants half a cycle apart and the wave field is harmonic in time, the free surface elevation in the two plots are identical in amplitude but have opposite signs. Consequently, both cases correspond to the same quadratic forcing which acts at twice the incident wave frequency. The amplitude of the quadratic forcing is proportional to the square of amplitude of the first-order free surface elevation. In order to examine the intensity of the quadratic forcing, the first-order wave crest envelope is shown in figure 4. This has a maximum on the upwave side of the cylinder, and its amplitude decreases gradually along the side of the cylinder to a minimum near the lee quarter of the cylinder. In the upwave region, the wave crest envelope is significantly modulated, indicating the presence of a partial standing wave system which extends to as far as the lee quarter. Its amplitude decreases slowly with distance away from the cylinder and with the azimuthal angle measured from the negative x axis. In the downwave region, the wave crest envelope is not modulated and its amplitude decreases rapidly to that of the first-order incident waves at the far-field.

Second-order free surface

At second order, two distinct components of the free surface can be identified. One is a spatially-varying, corrugated mean component due to the non-zero mean of the second-order forcing terms in the free surface boundary conditions, while the other is the second-order oscillatory component which itself is composed of second-order forced and free waves.

The mean water surface elevation with the vertical scale exaggerated 40 times is shown in figure 5. On the upwave side, regions of setup and setdown in mean water level are observed to locate respectively at the same locations as the high and low regions of the wave crest envelope in figure 4. It is also observed that the mean water surface modulation occurs mainly in the region near the negative x axis and extends to a distance relatively far away from the cylinder. Due to the large tangential fluid velocity at first order that occurs along the side of the cylinder, the mean water level is always negative and has a minimum in this region. In the downwave region, the mean water elevation decreases rapidly and monotonically from a small setup at the lee of the cylinder to zero at the far-field away from the cylinder.

Views of the oscillatory second-order free surface which correspond to the same instants as in figures 3a and 3b are shown in figures 6a and 6b respectively. It is noted that both instants give rise to the same results, since they are half a cycle apart and are subject to the same quadratic forcing. This also indicates the effectiveness of the radiation condition which avoids reflection of scattered waves from the control surface so as to produce a steady state second-order solution whose frequency is twice the wave frequency.

In the upwave region, the wave field is dominated by second-order forced waves phase-locked with the first-order partial standing wave system. The second-order interactions in the downwave region are mainly free waves satisfying the linear dispersion relation in the form of equation (17) and propagating away from the cylinder at the corresponding celerity independent of the first-order wave system. From figure 6 and the associated video film, these second-order free waves can easily be identified to originate at the lee and near the lee quarter of the cylinder in which three-dimensional wave pulses are emitted alternatingly. Along the side of the cylinder both forced and free waves are dominant and wave field is more complicated. Second-order free waves of smaller amplitudes are also generated along the upwave surface of the cylinder and propagate away from the cylinder concentrically.

Free surface to second order

The free surface to second order is obtained by a superposition of the results in figures 3, 5 and 6, and the corresponding views of the free surface at the same two instants half a wave cycle apart are shown in figure 7. Figure 7a corresponds to the instant when the surface elevation at the upwave side of the cylinder is a maximum, and figure 7b corresponds to the instant half a cycle later. Although the second-order oscillatory components at the two phase angles are identical, their effect on the solution to second order is unique at each phase angle.

The runup to second order and the amplitudes of the free surface elevation at first and second order around the cylinder periphery are functions of azimuthal angle as well as the incident wave frequency. Due to the redirection of incident wave momentum, the maxima of the first-order free surface elevation and the second-order steady setup always occur at the upwave side of the cylinder. The second-order forced waves are phased-locked with the first-order profile to produce an overall wave profile whose crests are steeper and troughs flatter, as may be clearly observed. Taken together with the second-order steady setup, the maximum of the runup to second order also occurs at upwave side. For the example shown, the maximum runup to second order is found to be about 50% higher than the first-order solution, and the free surface profiles clearly indicate the importance of second-order effects.

CONCLUSIONS

A time-domain second-order diffraction method which treats nonlinear interactions of ocean waves with large fixed structures of arbitrary shape in three dimensions is summarized. By the application of a perturbation procedure, the full nonlinear problem defined on a moving boundary is reduced to a second-order problem defined over a time-independent boundary, so that a solution to the resulting system of equations is required only once rather than at each time step.

The present method is applied to a study of the nonlinear diffraction of regular waves around a bottom-mounted surface-piercing circular cylinder. A video film has been developed to provide a convenient visualization of the predicted wave field. Various aspects of the free surface to second order and the components at first and second are illustrated. The importance of second-order effects in the evaluation the total free surface elevation is also highlighted.

REFERENCES

1. Abul-Azm, A.G. & Williams, A.N. Second-order diffraction loads on truncated cylinders. Journal of Waterway, Port, Coastal & Ocean Engineering, ASCE, Vol. 114, No. 4, pp. 436-454, 1988.
2. Cheung, K.F. Time-domain solution for second-order wave diffraction. Ph.D. Thesis, University of British Columbia, Vancouver, B.C., 1991.
3. Dommermuth, D.G. & Yue, D.K.P. Numerical simulations of nonlinear axisymmetric flows with a free surface. Journal of Fluid Mechanics, Vol. 178, pp. 195-219, 1987.
4. Eatock Taylor, R. and Hung, S.M. Second order diffraction forces on a vertical cylinder in regular waves. Applied Ocean Research, Vol. 9, No. 1, pp. 19-30, 1987.
5. Grosenbaugh, M.A. & Yeung, R.W. Nonlinear free-surface flow at a two-dimensional bow. Journal of Fluid Mechanics, Vol. 209, pp. 57-75, 1989.
6. Isaacson, M. Nonlinear wave effects on fixed and floating bodies. Journal of Fluid Mechanics, Vol. 120, pp. 267-281, 1982.
7. Isaacson, M. and Cheung, K.F. Second order wave diffraction around two-dimensional bodies by time-domain method. Applied Ocean Research, in press.
8. Isaacson, M. and Cheung, K.F. Time-domain solution of second-order runup and wave forces. Proceedings of the First International Conference of Offshore and Polar Engineering, in press.
9. Kim, M.H. and Yue, D.K.P. The complete second-order diffraction solution for an axisymmetric body. Part 1. Monochromatic incident waves. Journal of Fluid Mechanics, Vol. 200, pp. 235-264, 1989.
10. Molin, B. Second-order diffraction loads upon three-dimensional bodies. Applied Ocean Research, Vol. 1, No. 4, pp. 197-202, 1979.
11. Orlanski, I. A simple boundary condition for unbounded hyperbolic flows. Journal of Computational Physics,Vol. 21, pp. 251-269, 1976.
12. Vada, T. A numerical solution of the second-order wave-diffraction problem for a submerged cylinder of arbitrary shape. Journal of Fluid Mechanics, Vol. 174, pp. 23-37, 1987.
13. Vinje, T. and Brevig, P. Numerical calculations of forces from breaking waves. Proceedings of the International Symposium on Hydrodynamics in Ocean Engineering, Trondheim, Norway, pp. 547-566, 1981.
14. Wu, G.X. & Eatock Taylor, R. The second order diffraction force on a horizontal cylinder in finite water depth. Applied Ocean Research, Vol. 12, No. 3, pp. 106-111, 1990.

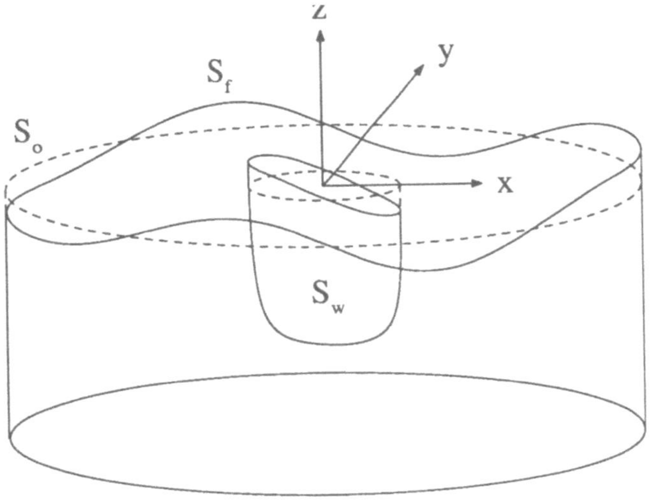

Figure 1. Definition sketch.

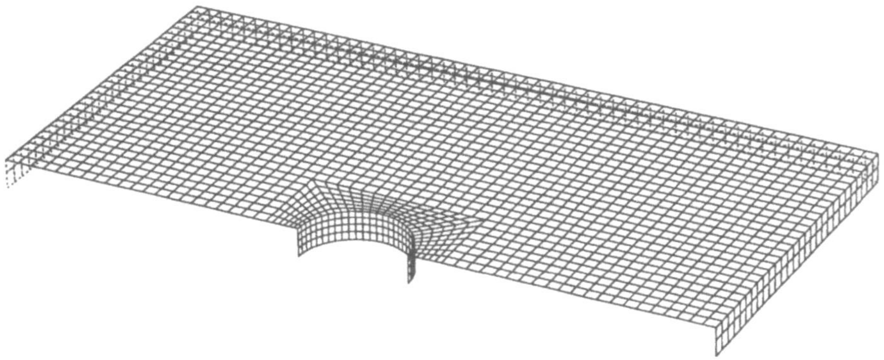

Figure 2. Example of surface discretization with $N_b = 112$ and $N = 1974$.

(a)

(b)

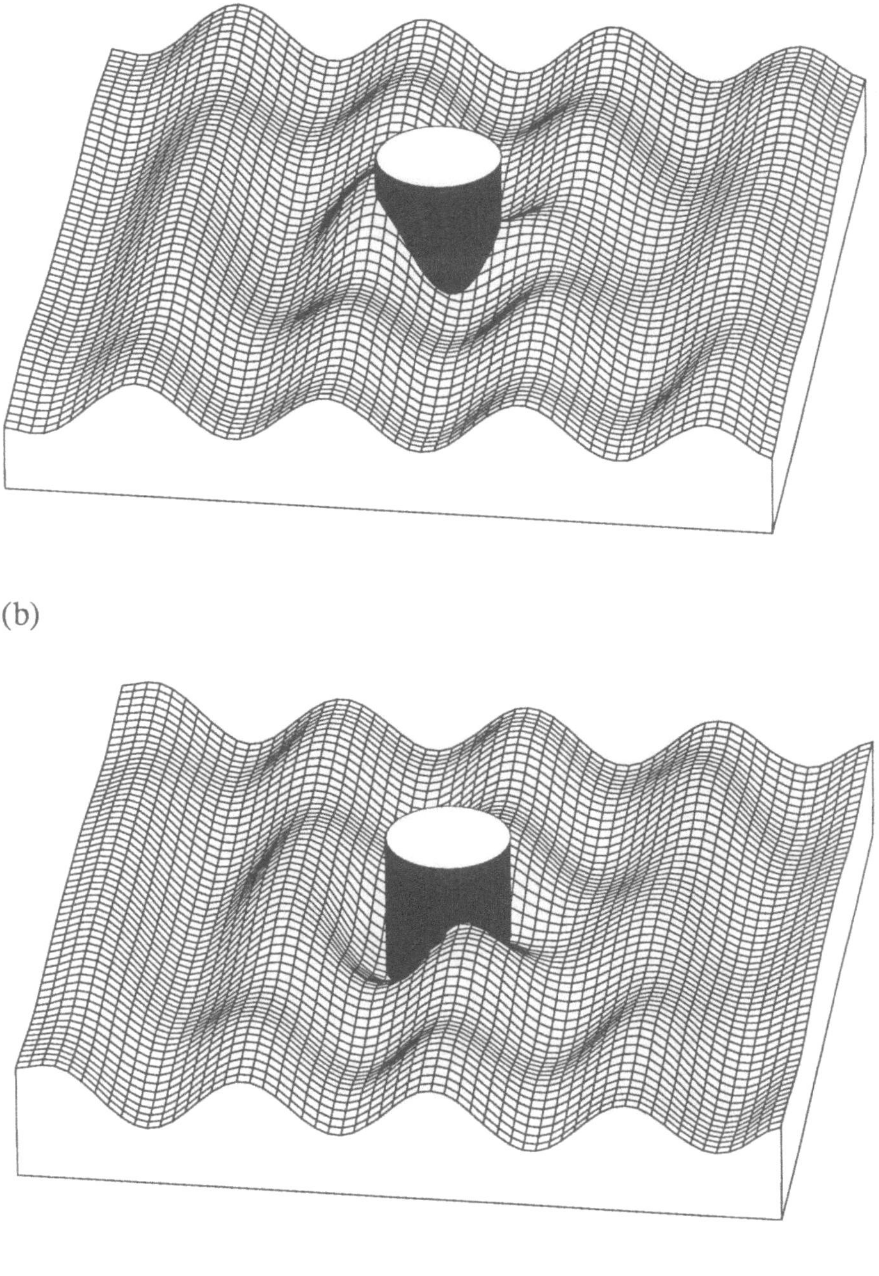

Figure 3. First-order free surface elevation for ka = 2, kd = 1 and H/L = 0.08. (a) $\theta/\pi = 0.75$. (b) $\theta/\pi = 0.25$.

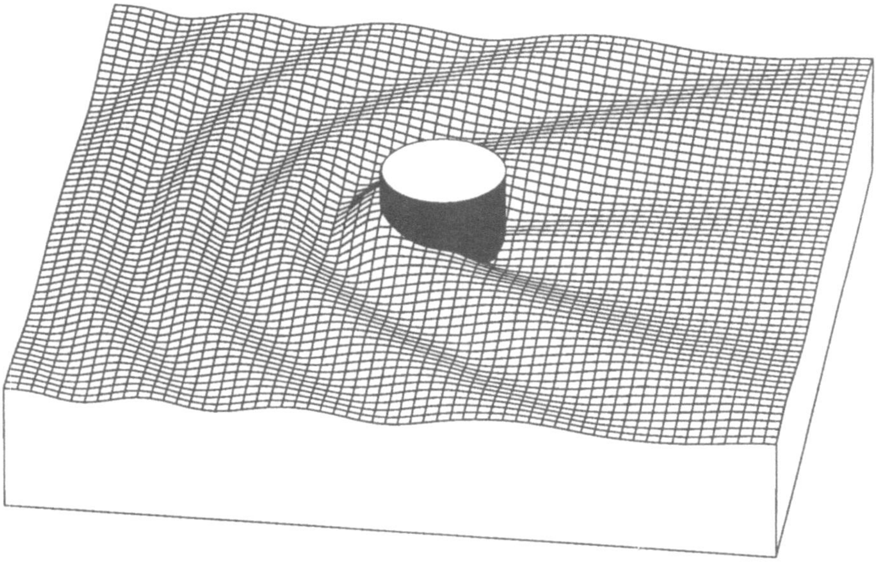

Figure 4. First-order wave crest envelope for ka = 2, kd = 1 and H/L = 0.08.

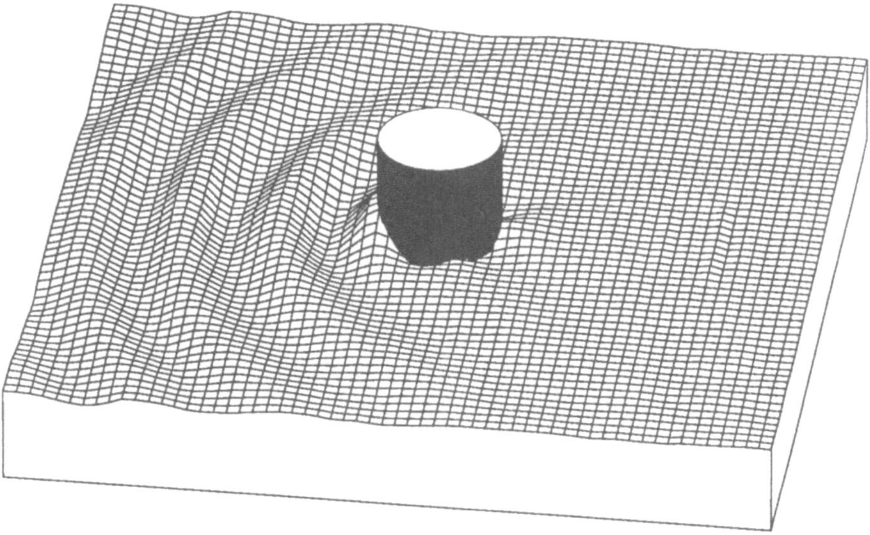

Figure 5. Second-order mean surface elevation for ka = 2, kd = 1 and H/L = 0.08.

(a)

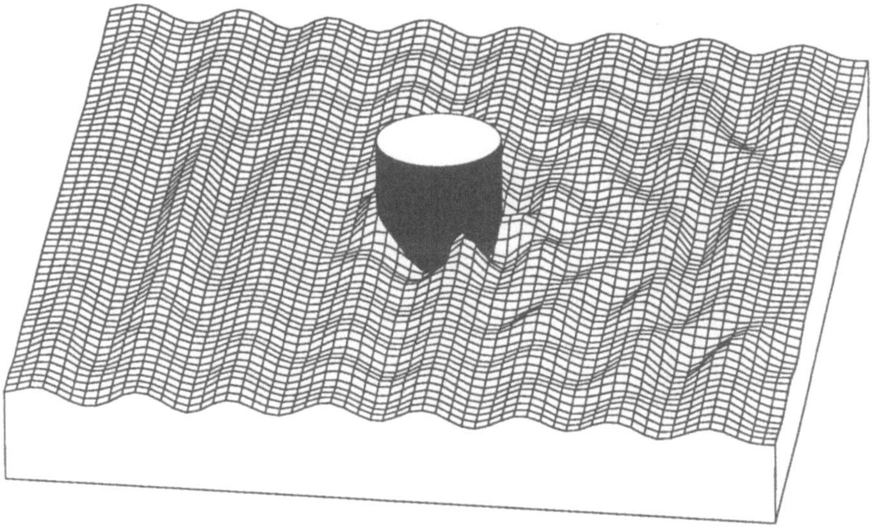

(b)

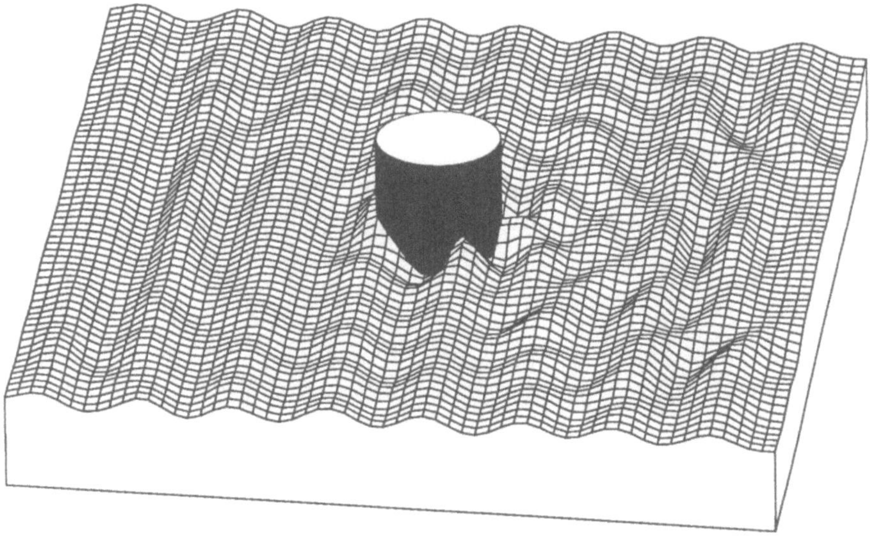

Figure 6. Oscillatory second-order free surface elevation for ka = 2, kd = 1 and H/L = 0.08. (a) $\theta/\pi = 0.75$. (b) $\theta/\pi = 0.25$.

(a)

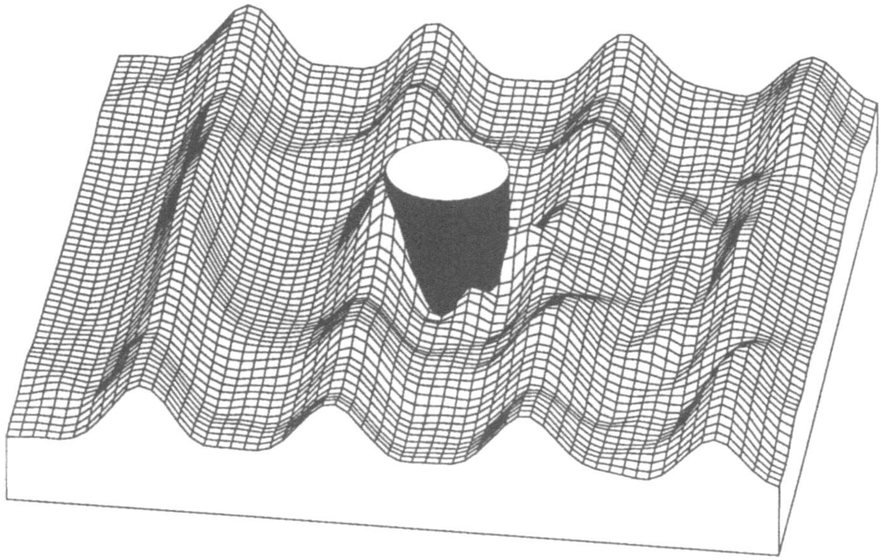

(b)

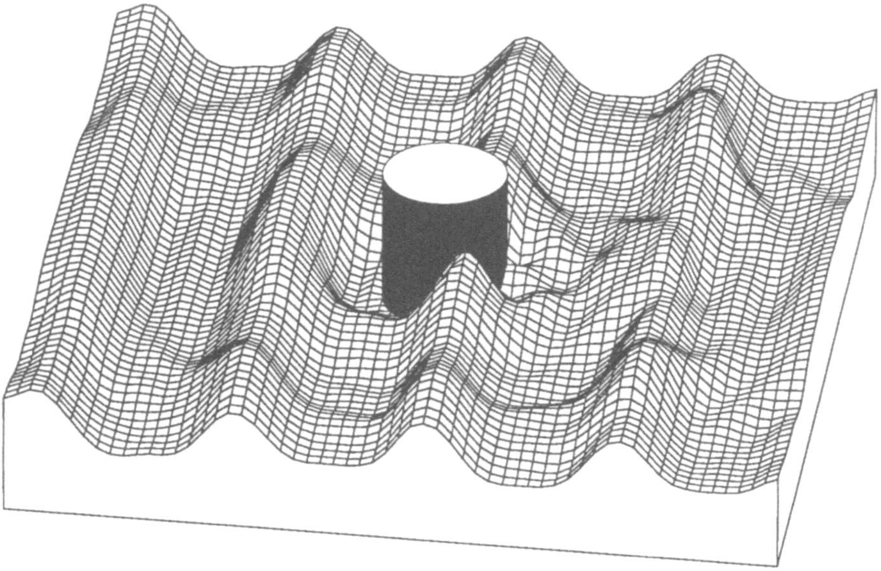

Figure 7. Free surface elevation to second order for $ka = 2$, $kd = 1$ and $H/L = 0.08$. (a) $\theta/\pi = 0.75$. (b) $\theta/\pi = 0.25$.

Shallow Water Waves

R.B. Guenther (*), J.A. Crow (*), F.H. Helfenstein (**)
(*) Department of Mathematics, Oregon State University, Corvallis, Oregon 97731-4605, USA
(**) Department of Mathematics, Central Oregon Community College, Bend, Oregon 97701-5998, USA

1. INTRODUCTION

Surface water waves with wave lengths greater than the depths of the water on which they propagate are called shallow water waves. These waves have been the subject of intensive investigations and many models have been proposed to describe them under various conditions. See Lamb [4] and more recently Peregrine [5]. The most widely used models at present are those based upon inviscid, potential flows. In 1971, Gallagher [1] noticed that it was necessary to introduce an empirical damping term in order to fit his data.

In this paper, we give a derivation of the model equation based upon the Navier-Stokes' equations. The equations are vertically integrated and the interaction with the bottom is taken into account. The damping Gallagher reports on is accounted for in a natrual way. The numerical results are obtained using standard techniques. However, these considerations raise another question which we partially address. The development of the flow is profoundly affected by the bottom topography. Due to erosion processes, the shape of the bottom changes and ideally it should be determined simultaneously with the solution to the flow problem. Thus, we should treat the problem depicted schematically in Figure 1, where $h(x,y,t)$ gives the free surface elevation of the water and $h_0(x,y,t)$ describes the bottom topography, and solve a coupled system for the flow and the free surfaces h and h_0. In this paper, we make use of the fact that erosion processes occur much more slowly than flow processes. For a given $h_0(x,y,t)$, we can determine $h(x,y,t)$ and the accompanying flow over a certain period of time, then we determine $h_0(x,y,t)$ and return to the problem of determining the free surface again. Thus, the problem essentially uncouples.

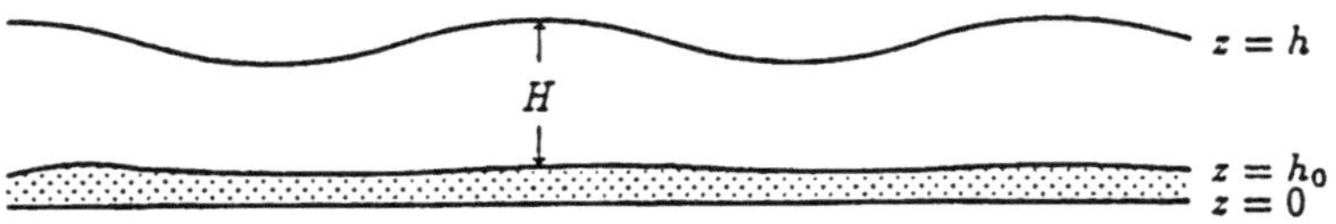

Figure 1.

2. THE FLUID SURFACE

In view of the above remarks, we model the fluid surface independently of the motion of the fluid bed. Refer to Figure 1 for the notation. In [2], certain assumptions were made on the flow which enabled us to derive the partial differential equation

$$\frac{\partial^2 h}{\partial t^2} + \frac{\nu}{H_o \delta} \frac{\partial h}{\partial t} = g \operatorname{div}(H \nabla h). \tag{2.1}$$

Here ν is the kinematic viscosity, H_o the still water depth, δ a representative length characterizing the interaction of the surface and the bottom, and g is the magnitude of the acceleration due to gravity. If there is a steady underlying flow, say an offshore current, V, then the equation must be modified to read

$$\frac{\partial^2 h}{\partial t^2} + \left(\frac{\nu}{H_o \delta} + V \cdot \nabla\right)\frac{\partial h}{\partial t} = g \operatorname{div}(H \nabla h). \tag{2.2}$$

Equation (2.1) contains several equations which are of interest. For example, if the effects of the viscosity are insignificant and the water is of constant depth, then $\nu = 0$, $h_o = 0$ and $H = h$ so that (2.1) becomes

$$\frac{\partial h^2}{\partial t^2} = \frac{1}{2} g \Delta h^2 .$$

Let us still assume the bottom to be horizontal and we define η by

$$h = H_o + \eta .$$

Then (2.1) becomes

$$\frac{\partial^2 \eta}{\partial t^2} + \alpha \frac{\partial \eta}{\partial t} = g \operatorname{div}((H_o + \eta) \nabla \eta) .$$

Now scale the lengths by H_o and the time by $\sqrt{H_o/g}$ to obtain the scaled equation

$$u_{tt} + \mu u_t = \operatorname{div}((1 + u) \cdot \nabla u) . \qquad (2.3)$$

Observe also in equation (2.1) that the local speed of propagation of the wave is $\sqrt{gH}$, which indicates that the crests of the wave have a tendency to travel faster than the troughs and so the waves will tend to break. We can also give an estimate when this breaking will occur.

The analytical and numerical treatment is based upon the following existence and uniqueness theorem.

Theorem (Crow). Let $T > 0$ be given. ϕ and ψ have compact support in R^2 and the initial energy (measured in a Sobolev norm) is sufficiently small, then there exists precisely one solution $u \in C^2(R^2 \times [0,T])$ to (2.3) which assumes the initial data

$$u(x,t,0) = \phi(x,y), \qquad u_t(x,y,0) = \psi(x,y).$$

Moreover, the solution depends continuously on the initial conditions.

The cross section of a typical incoming wave which is periodically forced has the form shown below. At $t = 10$ s , breaking starts.

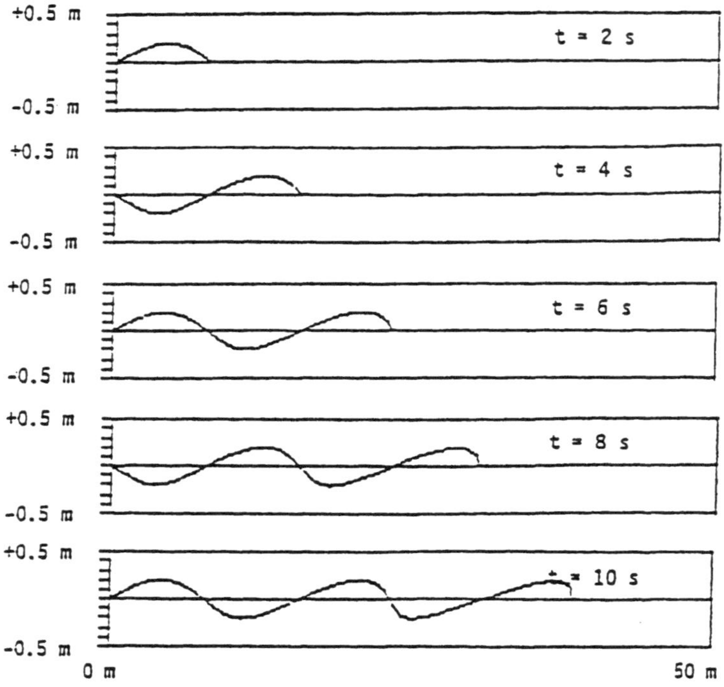

Figure 2.

3. THE EROSION PROBLEM

Flow over a bed of sand causes the bed to undergo a distinct evolution. At very slow flows, the bed remains unchanged. With increased flow velocities, the bed passes through a set of stages, the most dramatic occurring when the fluid is in a state of highly turbulent motion. Due to the mathematical difficulties associated with turbulence, we shall restrict ourselves to slower flows where the transport of the soil particles is caused by rolling and saltation and takes place in a thin layer near the bed.

The physical description of the erosion process is obtained by balancing sediment mass, energy and momentum. The erosion producing flow is governed by the Navier-Stokes' equations for open channel flow, although away from the bottom, we could assume potential flow. We assume the sediment transport occurs in a narrow region near the bed and the "law of the wall" (see Schlichting [6] or Gust and Southard [3]) applies. The vorticity of the flow near the streambed is a fundamental factor for the erosion process.

The complete mathematical model taking these assumptions into account is rather involved. However, a finite element model for the two dimensional case has been constructed. The following figures show the results of calculating the typical dune migrations which are observed under the above assumptions.

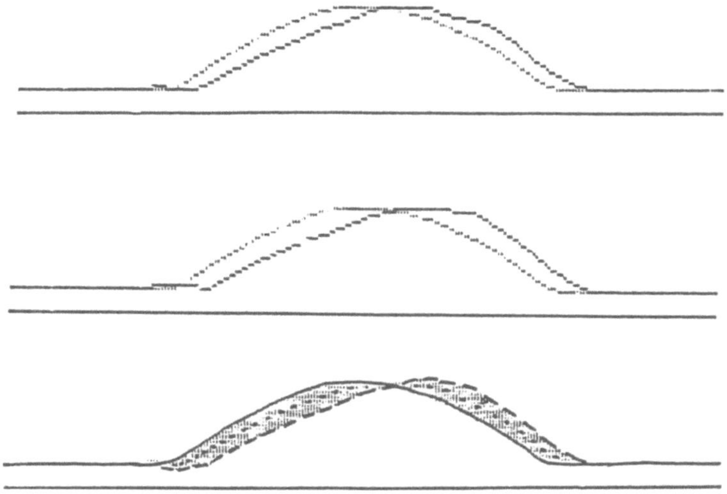

Dune Migration --- Law of the Wall

Figure 3.

BIBLIOGRAPHY

1. B. Gallagher. Generation of surf beat by nonlinear wave interaction, J. Fluid Mech., 49(1971), 1-20.

2. R. B. Guenther and J. A. Crow. Damping of shallow water waves, in Free Boundary Problems, K.-H. Hoffmann and J. Sprekels (ed), International Series of Numerical Mathematics, Vol. 95, Birkhäuser Verlag, Basel, 1990.

3. G. Gust and J. B. Southard. Effects of weak bed load on the universal law of the wall, J. Geo. Res., 88(1983), 5939-5952.

4. H. Lamb. Hydrodynamics, Dover, New York, 1945.

5. D. H. Peregrine. Equations for water waves and the approximations behind them, in Waves on beaches and resulting sediment transport, R. E. Meyer (ed), Academic Press, New York, 1972.

6. H. Schlichting. Boundary-Layer theory, 7th edition, McGraw-Hill, New York, 1979.

Numerical Calculation of Surface Waves

V.R. Kogan, V.V. Kuznetsov

Institute of Applied Physics, Academy of Science, N. Novgorod, USSR

ABSTRACT

In this paper we present a new method for calculation of nonstationary periodic surface waves. From the point of view of calculation, this method is based on both the technique of rapid operator reference, corresponding to Dirichlet problem, and the analytical theory of boundary value problem. New physical results on dynamics of nonstationary gravity-capillary waves were obtained using numerical calculation.

INTRODUCTION

Progressive waves of infinitely small amplitudes were considered in classical works by G. Stokes, T. Levi-Civita, J. Wilton as well as in the paper by S.I. Secerge-Zencovich, U.P. Krasovsky. These approximate theories are for steady, symmetric waves and describe the surface wave in a form of series. The region of convergence is very small ($2a/\lambda=10^{-4}$). Powerful computers allow to study waves of finite amplitude. The main mathematical difficulty of the problem arises essentially from the need to satisfy the Bernoulli's condition and singular Gilbert equation. It was realized in works by B. Chen, P.G. Saffman, L.W. Schwartz, J.M. Vanden-Broeck and D.H. Peregrine. K.I. Babenko (1987) obtained new operational equation for gravity waves, which helps to investigate wave bifurcations. M.S. Longuet-Higgins and E.D. Cokelet (1976) significantly forwarded numerical simulation of nonstationary gravity waves. Using this method V.R Kogan obtained the results of numerical simulation of gravity-capillary waves. This problem is related to a class of singularly-perturbed problems and appears to be more complicated than simulation of purely gravity waves because of member with a small parameter and a leading derivative (capillary effect) emerging in Cauchy-Lagrange condition

We used calculation aspects of analytical function theory (i.e. a method of rapid calculation of a singular Gilbert integral) and variation theory of conformal mapping [1] as well as duality of Dirichlet problem solution and conformal mapping of the fluid region onto a unit circle or a ring.

The above mentioned approach can be treated as a generalization of classical works by Stokes, Levi-Civita on progressive waves onto non-stationary case. For steady waves, this problem was reduced to deriving one conformal mapping. Here we have one-to-one correspondence of points of the fluid surface and those on the boundary S^1 of a unit circle, the velocity of points on S^1 being time-independent. In case of nonstationary potential waves the situation is drastically different. The conformal mapping $G(z,t)$ here doesn't depend on time and all information on nonstationary surface waves (dynamics, stability, bifurcations) can be described in terms of some dynamics system on S^1. Thus, numerical simulation is reduced to solution of a system of ordinary differential equations on S^1 and calculation of wave profile deformation using formulas of variation of theory and Gilbert integral.

Taking into account viscosity in the near-surface layer algorithm modification is given and is the generalization for nonpotential fluid. Using the above algorithm they numerically simulated nonstationary gravity-capillary waves and obtained new physical results on their dynamics. We also obtain the results on water of finite depth and parameters of application the shallow-water theory.

STATEMENT OF THE PROBLEM

Let (x,y) denote rectangular coordination with the x-axis horizontal and the y-axis vertical. The motion is assumed to be periodic in the x-direction. The fluid is incompressible, constant in density, inviscid and moving due to gravity and surface tension. The motion may be assumed to be started from rest by conservative forces, so that it is irrotational at all time. At very fixed t the velocity potential $\phi(t,x,y)$ satisfy Laplaci's equation ($\nabla^2\phi = 0$) inside the region of fluid D_z and dynamic condition derived from Bernoulli's equation [2], [3].

Using conformal mapping

$$re^{i\vartheta}=\zeta=e^{-iz} \quad (z = x+iy) \tag{1}$$

are transformed D_z into D_ζ inside the contour C which corresponds to the fluid surface. (r,ϑ) are polar coordinates in the ζ-plane. Let (s,n) be tangential and normal coordinates at a typical point in the C. A system of equation for ϕ, r and ϑ determining a motion of points on the free surface has a form [2], [3]

$$\frac{D\vartheta}{Dt} = r^2 \frac{\partial\vartheta}{\partial s}\frac{\partial\phi}{\partial s} - r\frac{\partial r}{\partial s}\frac{\partial\phi}{\partial n}, \quad \frac{Dr}{Dt} = r^2 \frac{\partial r}{\partial s}\frac{\partial\phi}{\partial s} + r^3 \frac{\partial\vartheta}{\partial s}\frac{\partial\phi}{\partial n},$$

$$\frac{D\phi}{Dt} = -P_s - \ln r + \frac{1}{2} r^2\left[\left[\frac{\partial\phi}{\partial s}\right]^2 + \left[\frac{\partial\phi}{\partial n}\right]^2 \right], \tag{2}$$

$$P_s = P_a - W_e K, \quad K=r\left[r\left[\frac{\partial^2 r}{\partial s^2}\frac{\partial\vartheta}{\partial s} - \frac{\partial r}{\partial s}\frac{\partial\vartheta^2}{\partial s^2}\right] - \left[\frac{\partial r}{\partial s}\right]^2 \frac{\partial\vartheta}{\partial s}\right]$$

where K is local curvature of C and P_a is atmospheric pressure (further taken as $P_a=0$). The system (2) is written in a dimensionless form and W_e is Weber number.

The system is to be closed by an equation for determining a normal derivative $\partial\phi/\partial n$ on C. It is obtained by a limiting transformation (e.g. from Sokhotsky-Plemel's formula) and has a form

$$\int_c \ln R \frac{\partial\phi}{\partial n}\, ds = \mathrm{P} \int_c \phi \, d\alpha + \pi\cdot\phi \tag{3}$$

(R,α) be the polar coordinates of P with respect to arbitrary point Q, lying on the $C(t)$.

METHOD OF NUMERICAL CALCULATION OF THE SYSTEM (2) AND SCHWARTZ OPERATOR

The values of ϑ, r, ϕ are to be evaluated at a finite number N of points on the boundary which correspond to fixed particles. We label these with the suffix j, where $j = 1,2,\ldots N$. The corresponding values of r,ϑ,ϕ are similarly labelled r_j, ϑ_j, ϕ_j. The function $r_j, \vartheta_j,$

ϕ_j were approximated using cubic splines. More specifically, a 5-point formula is used for the second derivatives. The local errors are of order $(\Delta s)^4$, where Δs is the maximum arclength between two adjacent points s_j. The singular integral equation (3) was regulated similar to [2]. It results in a system of algebraic equation

$$A_{ij} \left[\frac{\partial \phi}{\partial n} \right]_j = F_i(\phi) \tag{4}$$

where A_{ij} is matrix whose elements depend only on the form of the contour $C(t)$. $F(\phi)$ is determinated by the value of the potential on $C(t)$. In [2] the system was solved by a Gaussian method i.e. it directly led to

$$R_{ij} \left[\frac{\partial \phi}{\partial n} \right] = Q_{ij}\, F_j(\phi) \tag{5}$$

It follows from Green function that having determined $\partial\phi/\partial n$ we thus solve Dirichlet problem in D_ζ. It can be used for constructing a conformal mapping D_ζ onto a unit circle and include it into calculation.

Let us consider the analytic function $g = g_1 + ig_2$ in the region D_ζ such as

$$g_1 \Big|_{c(t)} = - \ln|\zeta - \zeta_o|, \tag{6}$$

where $\zeta \in C(t)$ and ζ_o is the image of points atinfinite depth in the (x,y) plane. Substituting (6) into a right-hand side (5), let us reversely find $\partial g_1/\partial n$. From Cauchy-Remain equation we have

$$g_2(s) = \int_o^s \frac{\partial g_1}{\partial n} ds \tag{7}$$

Thus we know the boundary value of the analytic function g onto $C(t)$ and, subsequently, the boundary value of the function

$$G(\zeta,\zeta_o) = (\zeta - \zeta_o)\, e^{g+i\beta} \tag{8}$$

(β is real constant). $G(\zeta,\zeta_o)$ extended analytically inside D_ζ.

Thus the given algorithm is a constructive way for conformal mapping of D_ζ onto a unit circle; G is Schwartz operator (we used the terms of [4]). As the matrix R, Q depend only on $C(t)$ it is necessary to calculate a reverse Gaussian method to construct conformal mapping of G. At N=60 the computation of $\partial\phi/\partial n$ an G for each time step tares ≈ 10s using PC IT-386.

MODERNISATION OF COMPUTATION ALGORITHM USING SCHWARTZ OPERATOR

Let ζ_o=0 in formula (6) and let G map D_ζ onto a unit circle $D_{\bar\zeta}$ in the $\bar\zeta$-plane. (ρ,ϑ) are polar coordinates in the $\bar\zeta$-plane, and we have

$$\bar\vartheta = \vartheta + g_2 + \beta \tag{9}$$

Due to conformability of G

$$\left.\frac{\partial\phi}{\partial n}\right|_c = |G'|\left.\frac{\partial\phi}{\partial\rho}\right|_{\bar c} \tag{10}$$

where $\bar C(t)$ is the image of $C(t)$ for mapping G. While $\partial\phi/\partial\bar\vartheta$ and $\partial\phi/\partial\rho$ are real and imaginary parts of the analytical function $dW/d\bar\zeta$ in the circle $|\rho|\le 1$, then on the $\bar C(t)$ they are linked with Gilbert transformation.

$$\frac{\partial\phi}{\partial\rho} = \frac{1}{2\pi}\int_0^{2\pi}\frac{\partial\phi}{\partial\bar\alpha}\,\mathrm{ctg}\frac{\bar\alpha-\bar\vartheta}{2}\,d\bar\alpha \tag{11}$$

Let

$$\frac{\partial\phi}{\partial\bar\alpha} = \sum_{k=1}^{N} a_k\cos(k\bar\alpha) + b_k\sin(k\bar\alpha) \tag{12}$$

then [5]

$$\frac{\partial\phi}{\partial\bar{\rho}} = \sum_{k=1}^{N} a_k \sin(k\bar{\vartheta}) - b_k \cos(k\bar{\vartheta}) \tag{13}$$

i.e.

$$\frac{\partial\phi}{\partial n} = |G'| \sum_{k=1}^{N} a_k \sin(k\bar{\vartheta}) - b_k \cos(k\bar{\vartheta}) \tag{14}$$

As $g_2|_{c(t)}=-\ln r$, it is easy to get $|G'|_{c(t)}=\partial\vartheta/\partial s+\partial g_2/\partial s$, or, using Couchet-Riemann equation, $|G'|_{c(t)}=\partial\vartheta/\partial s+ \partial g_1/\partial n$.

The formula (14) was used to calculate a normal derivative $\partial\phi/\partial n|_{c(t)}$. Due to harmonic nature of ϕ in the region $D_{\bar{\zeta}}$, the function $\partial\phi/\partial\bar{\alpha}$ belongs to Holder class H(μ) at μ=1. Then (c.f. [5]) $\partial\phi/\partial\rho$ belongs to the class H(1-ε) where ε is unlimitedly small positive constant, i.e. Gilbert transformation is smoothing. A modernized algorithm turned out to be more stable and precise as compared to that given in [2]. The calculation time here was increased by 5%.

Table 1

time	algorithm of the work [2]		
	A	$\bar{K}$	E
0.0	$-0.18\cdot10^{-8}$	$0.37\cdot10^{-8}$	0.1094701
1.2	$-0.81\cdot10^{-4}$	$0.69\cdot10^{-8}$	0.1094682
2.4	$-0.42\cdot10^{-4}$	$0.89\cdot10^{-8}$	0.1094650
3.6	$0.25\cdot10^{-4}$	$0.34\cdot10^{-7}$	0.1094649
4.8	$0.40\cdot10^{-4}$	$0.31\cdot10^{-6}$	0.1094678
6.0	$0.21\cdot10^{-3}$	$0.20\cdot10^{-5}$	0.1094649

Table 2

time	modernisational algorithm		
	A	$\bar{K}$	E
0.0	$-0.23 \cdot 10^{-6}$	$0.18 \cdot 10^{-8}$	0.1094698
1.2	$-0.15 \cdot 10^{-6}$	$0.20 \cdot 10^{-8}$	0.1094692
2.4	$-0.42 \cdot 10^{-6}$	$0.33 \cdot 10^{-8}$	0.1094689
3.6	$0.12 \cdot 10^{-6}$	$0.30 \cdot 10^{-8}$	0.1094686
4.8	$0.37 \cdot 10^{-6}$	$0.12 \cdot 10^{-7}$	0.1094688
6.0	$0.12 \cdot 10^{-5}$	$0.82 \cdot 10^{-7}$	0.1094680

The tables 1,2 gives mean curvatures of profile $\bar{K}$, total energy E and A-integral from a normal potential derivative for contour $C(t)$ calculated using different algorithms. Here $\bar{K}, E$, and A are invariants of the potential flow:

$$\bar{K} = \frac{1}{2\pi} \int_{o}^{2\pi} K(\vartheta) d\vartheta = 0, \qquad E = Q + V + P \tag{15}$$

where the kinetic and potential wave energes have the forms

$$Q = \frac{1}{2} \int_{o}^{2\pi} \int_{-\infty}^{y_s} (\nabla\phi)^2 dxdy = \frac{1}{2} \int_{c(t)} \phi \frac{\partial\phi}{\partial n} ds \tag{16}$$

$$V = -\frac{1}{2} \int_{o}^{2\pi} \ln^2 r d\vartheta , \tag{17}$$

respectively, and the energy due to surface tension has a form

$$P = W_e \left[\int_0^{S_n} \frac{ds}{r(s)} - 2\pi \right] ; \qquad (18)$$

$$A = \int_{c(t)} \frac{\partial \phi}{\partial n} ds = 0. \qquad (19)$$

It is evident from the table that the invariants (15), (19) in the modernized algorithm are preserved better than those in the algorithm given in [2].

EQUATION FOR MOTION OF GRAVITY-CAPILLARY WAVES ON THE A UNITE CIRCLE

Let $C(t_1)$, $C(t_2)$ be the forms of the fluid wave in the ζ-plane which correspond to time moments t_1, t_2. $C(t_2)$ being obtained from $C(t_1)$ by a shift along the coordinates of the system.

$$\frac{D\zeta}{Dt} = \left[\frac{\partial w}{\partial \zeta} \right]^* \qquad (20)$$

where $w(\zeta, t) = \phi(\zeta, t) + i\psi(\zeta, t)$ is time-dependent complex potential. (Here (*) is complex conjugation). Denote

$$T_t^c : C(t_1) \longrightarrow C(t_2) \qquad (21)$$

The transformation T_t^c and Schwartz operator lead to the transformation of the circumference $C(t)$ according to the formula

$$\bar{T}_t^c = G_2 \, T_t^c \, G_1^{-1} \qquad (22)$$

As $W(\zeta, t)$ is the analytic function at each t, then T_t^c (and, subsequently, $\bar{T}_t^c$) are nonanalytic.

Lagrange derivative in (r, ϑ) coordinates has a form

$$\frac{D}{Dt} = \frac{\partial}{\partial t} + \frac{\partial\phi}{\partial\vartheta}\frac{\partial}{\partial\vartheta} + r^2\frac{\partial\phi}{\partial r}\frac{\partial}{\partial r} \tag{23}$$

We can easily check the relations

$$\frac{\partial}{\partial r} = \frac{\partial r}{\partial s}\frac{\partial}{\partial s} + r\frac{\partial\vartheta}{\partial s}\frac{\partial}{\partial n}\,, \quad \frac{\partial}{\partial\vartheta} = r^2\frac{\partial\vartheta}{\partial s}\frac{\partial}{\partial s} - r\frac{\partial r}{\partial s}\frac{\partial}{\partial n}\,, \tag{24}$$

Using conformability of G, as well as (20), (23) we get an equation for Lagrange derivative in $(\rho,\bar{\vartheta})$ coordinates:

$$\frac{D}{Dt} = \frac{\partial}{\partial t} + r^2|G'|^2\left[\frac{\partial\phi}{\partial\bar{\vartheta}}\frac{\partial}{\partial\bar{\vartheta}} + \frac{\partial\phi}{\partial\rho}\frac{\partial}{\partial\rho}\right] \tag{25}$$

From (20) we find

$$\frac{D\bar{\zeta}}{Dt} = |\zeta|^2|G'|^2\left[\frac{\partial w}{\partial\bar{\zeta}}\right]^* \tag{26}$$

For the time Δt we have from (26)

$$\Delta\rho = r^2|G'|^2\frac{\partial\phi}{\partial\rho}\Delta t + o(\Delta t^2) \tag{27}$$

This relation determines radial deformation of the unit circle $C(t)$ during time Δt.

Let us conformally map the deformation contour onto a unit circle. It is known from the theory of variational conformal mapping that small $\Delta\rho$ lead to shifts of points on the unit circle

$$\Delta\ddot{\vartheta} = \frac{1}{2\pi}\int_0^{2\pi}\Delta\rho(\xi)\,\mathrm{ctg}\frac{\xi - \bar{\vartheta}}{2}\,d\xi \equiv g(\Delta\rho) \tag{28}$$

is Gilbert transformation with density $\Delta\rho$. Finally, we get a system of equation on the unit circle $C(t)$ which fully describes dynamics of gravity-capillary waves

$$\frac{D\vartheta}{Dt} = r^2 \; |G'|^2 \left[\frac{\partial\vartheta}{\partial\bar{\vartheta}} \frac{\partial\phi}{\partial\bar{\vartheta}} - \frac{1}{r} \frac{\partial r}{\partial\bar{\vartheta}} \, g\left[\frac{\partial\phi}{\partial\bar{\vartheta}} \right]\right]$$

$$\frac{Dr}{Dt} = r^3 \; |G'|^2 \left[\frac{1}{r} \frac{\partial r}{\partial\bar{\vartheta}} \frac{\partial\phi}{\partial\bar{\vartheta}} + \frac{\partial\vartheta}{\partial\bar{\vartheta}} \, g\left[\frac{\partial\phi}{\partial\bar{\vartheta}} \right]\right] \tag{29}$$

$$\frac{D\phi}{Dt} = -P_s - \ln r + \frac{1}{2} r^2 |G'|^2 \left[\left[\frac{\partial\phi}{\partial\bar{\vartheta}} \right]^2 + \left[g\left[\frac{\partial\phi}{\partial\bar{\vartheta}} \right]\right]^2 \right]$$

$$\frac{D\bar{\vartheta}}{Dt} = r^2 \; |G'|^2 \; \frac{\partial\phi}{\partial\bar{\vartheta}} + g\left[r^2 |G'|^2 \frac{\partial\phi}{\partial\bar{\rho}} \right]$$

The above method proved to be more stable and twice as quick as the one, described in [3].

NUMERICAL CALCULATION OF VISCOUS LIQUID

The paper [6] derives a model equation for Couchet-Lagrange integral with viscosity forces in quasi-potential approximation taken into account

$$\frac{\partial\phi}{\partial t} = -P_s - y - \frac{1}{2} (\nabla\phi)^2 + 4\nu \frac{\partial^2\phi}{\partial s^2} \tag{30}$$

where ν is coefficient of kinematic viscosity. A total time derivation from the potential (with all above taken into account) has a form

$$\frac{D\phi}{Dt} = -P_s - \ln r + \frac{1}{2} r^2 |G'|^2 \left[\left[\frac{\partial\phi}{\partial\bar{\vartheta}} \right]^2 + \left[g\left[\frac{\partial\phi}{\partial\bar{\vartheta}} \right]\right]^2 \right] +$$

$$4\nu \left[\; |G'|^2 \frac{\partial^2\phi}{\partial\bar{\vartheta}^2} + |G'| \frac{\partial\phi}{\partial\bar{\vartheta}} \frac{\partial |G'|}{\partial\bar{\vartheta}^2} \right] \tag{31}$$

The method was generalized for nonpotential flows using Green formula for solving Puasson equation as well as for waves on the surface waves of finite depth. In particular, we proved the Korteweg-De Vries equation to be valid for the parameters $a/\lambda<0.1$, $h/\lambda<0.1$.

RESULTS OF NUMERICAL CALCULATION

Using a computer, we numerically calculated dynamics of gravity-capillary waves with initial curvatures $2a/\lambda\in[0.03,\ 0.16]$. The initial profile and potential were assigned as

$$y_o = -\frac{2a\pi}{\lambda}\cos x, \qquad \phi_o = -\frac{2a\pi}{\lambda}\exp(y_o)\sin x \qquad (32)$$

In a result of numerical calculation ripples were found to appear not only on a wave crust (that was assumed earlier, e.g. [7]), but on a front slope near the angle and its appearance is associated with the fact that the profile symmetry is violated. At the moment of ripple emergence, higher space Harmonics in the profile wave rapidly increase. Fig. 1a,b gives wave profiles at $t=0.0$ and $t=6.6$ at $2a/\lambda=0.06$.

The results of solving the system (29) at $2a/\lambda=0.06$ are given in Fig.2 which shows a circumference $\rho=1$ using by a vector field (28). For the sake of visualization, a small vector field $D\rho/Dt=\varepsilon$ was added to the vector field (28). Figure 2a (time $t=0.0$,no ripples) and Fig.2b (time $t=6.6$, capillary ripples on the front slope) are drastically different. When ripples appear, an unstable point divides into two unstable points and one stable one. Further the situation is repeated: unstable points bifurcate, the character of bifurcations the depending on the relation between the curvature derivative and steepness. Note, that two specific points of the vector field $D\bar{\vartheta}/Dt$ are present in the gravity wave dynamics up to the moment of wave collapse, these points being instable on the front slope and stable on the back one.

When we take viscosity into account, the profile wave dynamics varies. Capillary ripples do not propagate along the profile. Figures 3a,b gives a curvature of wave with and without viscosity taken into account.

REFERENCES

1. Longuet-Higgins,M.S. and Cokelet,E.D. The deformation of steep surface waves on water. Part 1. A numerical method of computation. Proc. Roy. Soc. A. Vol.350, pp. 1-26, 1976.

2. Kogan,V.R. and Churilov, V.A. Numerical simulation nonstationary gravity-capillary waves on water of infinite depth. Izvestiya AN USSR, Mexanica Zhidcosti and Gaza Vol.22, n. 2, pp. 124-129, 1987.

3. Kogan,V.R. and Kuznetsov,V.V. The method of numerical calculation nonstationary gravity-capillary waves of finite amplitude. Izvestiya AN USSR, Zhurnal Vichislitelnoy Matematiki and Matematicheskoy Fiziki , Vol.29, n. 6, pp. 844-852, 1989.

4. Lavrentev,M.A. and Shabat,B.V. The method of theory function of complex value. M. Science, 1965.

5. Muschelishvili,N.I. The singular integrial equations. M. Science, 1968.

6. Ruvinsky,K.D. and Freiman,G.I. About of generetion gravity-capillary waves by step gravitary waves. Izvestiya AN USSR, Fizika Atmosfer and Okeana, Vol.17, n. 7, pp. 746-753, 1981.

7. Longuet-Higgins,M.S. The deformation of capillary waves by step gravity waves. J Flu ed Mech Vol.16, n. 1, pp. 138-159, 1963.

8. Babenko,K.I. Some remarks to theory surface waves of finite amplitude. Doklad Akademy Nauk, Vol.249, n. 5, pp. 1033-1037, 1987.

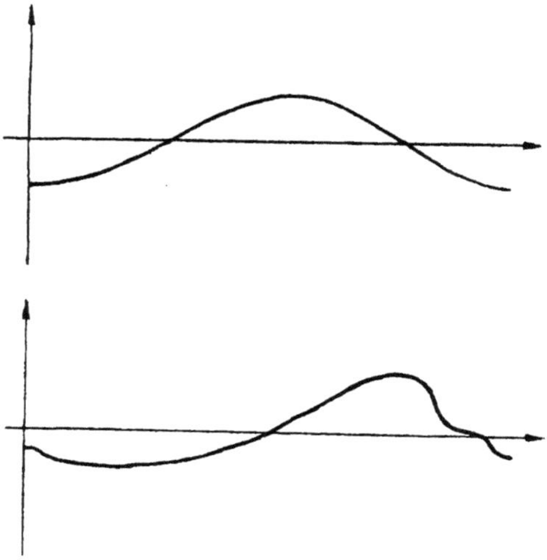

Fig.1 a, b. Profile of gravity-capillary waves at t = 0.0 and t = 6.0

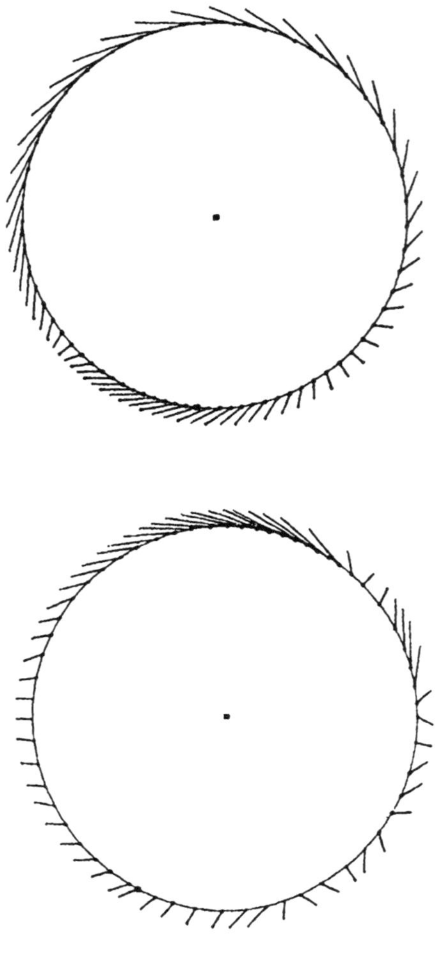

Fig.2 a, b. The motion of points onto the unite circle at t = 0.0 and t = 6.0

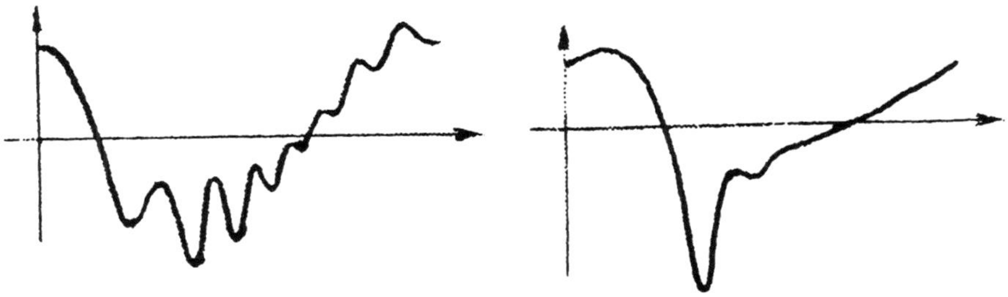

Fig.3 a,b. Curvature of gravity-capillary waves with viscosity at t = 33.0 and without viscosity t = 16.0.

SECTION 3: CAVITATIONAL FLOW

A Numerical Approach for Modelling Cavitating Flows

M.S. Ingber (*), C.E. Hailey (**)

(*) Department of Mechanical Engineering, University of New Mexico, Albuquerque, New Mexico 87131, USA

(**) Aerodynamics Department, Sandia National Laboratories, Albuquerque, New Mexico 87185, USA

ABSTRACT

The flow about a submerged partially- or supercavitating axisymmetric body is considered in this paper both at zero and non-zero angle of attack. The direct boundary element method is used to solve for the potential flow field and to determine the cavity shape. This is accomplished by performing a Newton-Raphson iteration in which the cavity shape is perturbed in an effort to satisfy the nonlinear dynamic boundary condition along the surface of the cavity. For problems with non-zero angle of attack, the flow field is three-dimensional and the cavity is not symmetric. In these cases, the convergence of the Newton-Raphson iteration is sensitive to the initial estimate of the cavity shape. Methods of generating these initial estimates are discussed which have been successful in leading to convergent solutions. Also efficient methods of parameterizing the cavity are developed to reduce the computational effort.

INTRODUCTION

The effects of cavitation are of primary concern in the design of submerged bodies such as projectiles and propellers. The shape and extent of the cavity will significantly influence the resultant forces and moments acting on a submerged body. In some cases, cavities are desirable because of the resultant drag reduction. In other cases, cavities are not desirable because of noise, erosion, or an uncontrollable resultant moment. There is presently no unified theory for modelling cavitating flows over bodies because of the complexities of the flow field and the lack of adequate experimental data. In a design setting, water tunnel testing of a model is time-consuming and expensive. Hence, timely and inexpensive estimates of cavity shapes are desirable.

The flow field about a submerged partially- or super-cavitated three-dimensional body poses several difficult problems. These problems include determining the point at which the dividing streamline which defines the cavity detaches from the body, the angle at which the dividing streamline leaves the body, the shape of the aft portion of the cavity in the region of turbulent two-phase flow, and the length of the cavity. Several closure models have been proposed for approximating the location of the point of detachment [1-4], the angle at which the dividing streamline leaves the body [1-4], and the shape of the aft portion of the cavity [5-9]. Apparently, however, no closure model has been proposed for determining the length of the cavity. This is surprising because none of these closure problems are independent.

Several numerical methods have been developed to predict cavity shapes on two-dimensional and axisymmetric bodies [4,10-15]. However, common to many of these methods, the length of the cavity has been assumed to be known a priori. In practice, the length of the cavity is not known a priori, and therefore, some type of closure model is necessary in order to predict this length. In this paper, we develop a boundary element method (BEM) capable of determining a constant pressure streamline representing the cavity surface and couple the BEM with a closure model capable of predicting the length of the cavity. In order to validate the closure model, we compare our results to water tunnel data.

An integral boundary-layer solver is implemented with the computer code in order to calculate the drag on a cavitating body. The drag force is composed of three components, one due to the pressure distribution over the body, one due to the laminar and/or turbulent skin friction, and one due to the base pressure. For blunt bodies, the component of the force due to the skin friction is negligible. However, for slender bodies, the skin friction component of the force can be significant. In the results, we compare our drag, normal force and pitching moment calculations to water tunnel measurements. We also compare measured pressure distributions with our calculations for axisymmetric bodies at zero and non-zero angle of attack.

BOUNDARY ELEMENT FORMULATION

The governing differential equation, assuming incompressible, irrotational, and inviscid flow, about a body in a uniform stream in terms of the potential function ϕ is given by Laplace's equation. The potential function can be broken into two parts; one part due to the freestream, ϕ_{fs}, and the other part due to the perturbation caused by the body, ϕ_p. The governing system can be written as follows:

$$\nabla^2\phi = \nabla^2\phi_{fs} + \nabla^2\phi_p = 0, \tag{1}$$

where ∇^2 is the three-dimensional Laplacian operator. On the solid boundaries, the no-penetration condition is given by

$$\phi' = \phi'_{fs} + \phi'_p = 0, \tag{2}$$

where the ′ represents the derivative in the direction of the outward normal to the surface. On the cavity surface, the kinematic and dynamic boundary conditions are given by

$$\phi' = \phi'_{fs} + \phi'_p = 0, \tag{3}$$

$$\left(\frac{\partial \phi_p}{\partial x}\right)^2 + \left(\frac{\partial \phi_p}{\partial y}\right)^2 + V_\infty^2 = C^2, \tag{4}$$

where V_∞ is the free stream velocity. The constant C represents the liquid velocity along the cavity and is related to the cavitation number, K, through the equation

$$K = \frac{P_\infty - P_v}{1/2\rho V_\infty^2} = C^2 - 1, \tag{5}$$

where P_∞ is the free stream pressure, P_v is the vapor pressure, and ρ is the liquid density. It is convenient to formulate the problem in terms of the perturbed potential. Since ϕ_{fs} is harmonic, the governing differential equation for the perturbed potential, ϕ_p, is also given by the Laplace equation.

This problem is highly nonlinear not only because the dynamic boundary condition, Eq. 4, represents a nonlinear boundary condition but also because the position of the cavity surface is unknown a priori. Since this position is of primary concern, the boundary element method is well suited for this problem. Following standard formulation procedures for the direct boundary element method (see, *e.g.*, Brebbia, Telles, and Wrobel [16]), ϕ_p can be represented in terms of a boundary integral as follows:

$$\eta(\xi)\phi_p(\xi) = \int_\Gamma [G'(\xi,x)\phi_p(x) - G(\xi,x)\phi'_p(x)]d\Gamma(x), \tag{6}$$

where Γ represents the wetted surface of the body and the cavity surface, $G(\xi,x)$ is the free-space Green's function given by $G(\xi,x) = 1/\mid \xi - x \mid$, and the coefficient $\eta(\xi)$ can be calculated from the formula

$$\eta(\xi) = \int_\Gamma G'(\xi,x)d\Gamma(x). \tag{7}$$

The integral equation (6) is discretized by approximating the boundary Γ by 9-node quadrilateral and 6-node triangular quadratic, isoparametric boundary elements. Collocating Eq. 6 at each of the N nodes within the boundary elements leads to a system of N equations

$$[G_{ij}]\{\phi_j\} = [H_{ij}]\{\phi'_j\}, \tag{8}$$

where ϕ_j and ϕ'_j represent the values of the perturbed potential function and its normal derivative, respectively, at the collocation nodes within the boundary elements.

In traditional boundary element methods, the position of the boundary Γ is known and either the potential ϕ_p, the flux ϕ_p', or a linear combination of the two are specified at each collocation node, and Eq. 8 is rearranged (assembled) in order to solve for the remaining unknowns. In the present analysis, the portion of the boundary given by the streamline defining the cavity is unknown a priori and only one of the two boundary conditions (Eqs. 3 and 4) can be prescribed along the cavity surface. In general, if the kinematic condition is prescribed (Eq. 3) then the dynamic condition (Eq. 4) will not be satisfied, and vice versa. In the next section, an iterative numerical procedure is outlined to determine the location of the cavity surface and to satisfy the kinematic and dynamic boundary conditions along this surface simultaneously.

CAVITY MODELLING

We assume that the detachment zone and the angle that the leading edge of the cavity surface makes with the attached flow is known. In particular in this study, we limit our attention to bodies without laminar separation bubbles for which the smooth separation condition holds [1,2]. The smooth separation condition states that the leading edge of the cavity is at the location where the local pressure drops below the cavitation pressure on the fully wetted body under noncavitating conditions. Further, the smooth separation condition demands that the cavity surface be tangent to the attached flow at the point of detachment. Along each axial ray, we assume a length for the cavity. The initial estimate for the cavity shape can be parameterized in several ways. For axisymmetric flow fields (axisymmetric bodies at zero angle of attack), the cavity shape can be defined by the radii at the various body stations along the axis of symmetry. For nonaxisymmetric flow fields (axisymmetric bodies at non-zero angle of attack), the parameterization of the cavity surface is significantly more complicated. In order to minimize the number of parameters necessary to describe the cavity, the discretization of the cavity is accomplished with rays in the direction of the axis of the body and transverse rings (Fig. 1). The thickness of the cavity is specified at three evenly spaced body stations along each ray and the shape of the cavity between the body stations is determined by an iterpolation scheme. A further complication to modelling nonaxisymmetric cavities is due to the fact that the detachment and reattachment zones are fully three-dimensional.

In order to model the cavity just in front of the reattachment zone, we use a displacement thickness model [15]. If the boundary element node just upstream of the reattachment node is a midside node, the displacement thickness is chosen as 75% of the cavity thickness at the next upstream node under the cavity. If the boundary element node just upstream of the reattachment node is an edge node, the displacement thickness is chosen as

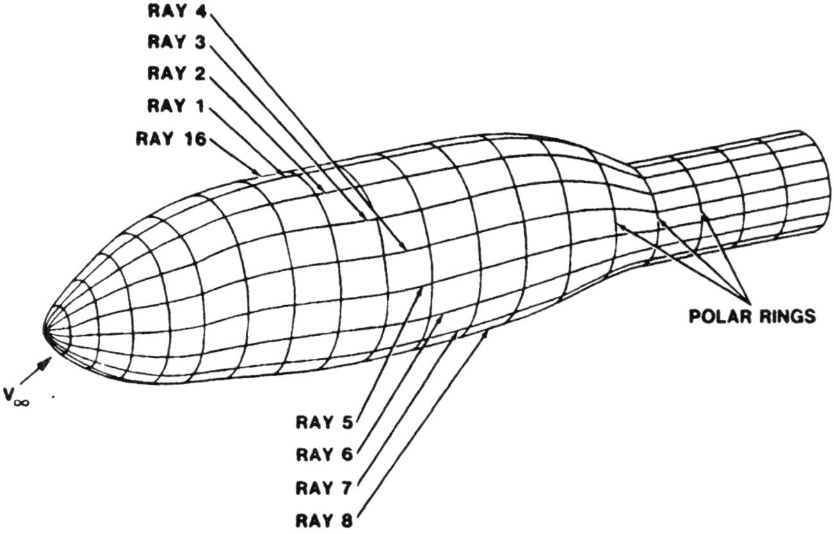

Figure 1: A typical boundary element discretization for a cavitating axisymmetric body at non-zero angle of attack.

25% of the cavity thickness at the next upstream node under the cavity. Using this displacement model, the pressure at the last two nodes under the cavity are not given by the cavitation pressure. Nevertheless, the flow is two-phase bubbly and turbulent in this region, thus making a cavity termination model necessary.

Using the initial estimate for the cavity shape described by its parameterization, the no-penetration condition on the solid surface and the analogous kinematic boundary condition on the cavity surface are inserted into the discretized boundary element equations (Eq. 8). The solution of these equations then yields the values of the perturbed potential function at the nodes of the boundary elements. However, as mentioned above, the condition of constant velocity (Eq. 4) along the surface of the cavity will not in general be satisfied. The velocities along the outer edge of the cavity are a function of the parameterization of the cavity. Therefore, the condition of constant velocity along the outer edge of the cavity can be written symbolically as

$$\mathbf{F}(p_1, p_2, ..., p_s) = 0, \tag{9}$$

where the p_i's are the parameters defining the cavity. The nonlinear implicit vector equations represented by Eq. 9 are solved by iteration using a Newton-Raphson scheme. The partial derivatives $\partial\mathbf{F}/\partial p_i$ are determined by perturbing the p_i's individually and determining the potential solution for the perturbed cavity shape.

A solution for the cavity shape that satisfies Eqs. 1-4 except at the leading and trailing edge of the cavity can be generated through the Newton-

Raphson iteration. We have already mentioned that, at the trailing edge, we are using a cavity termination model (the displacement thickness model), and therefore, we do not require that the local pressure be equal to the cavitation pressure in this region. At the leading edge of the cavity, we do require the local pressure to be equal to the cavitation pressure but this condition will not in general be satisfied because the cavity thickness at the leading edge is prescribed to be zero and is not a parameter of the cavity modelling. However, the requirement that the cavity surface must be tangent to the attached flow has not yet been enforced. This requirement can be enforced by adjusting the cavity length. Through numerical experiment, a qualitative relationship was determined between the assumed cavity length, the cavity thickness, and the angle at which the cavity leaves the body. As the cavity length is increased, the cavity becomes thicker and the angle at which the cavity leaves the body increases. This qualitative relationship is used to suggest a new closure model for cavity modelling. The cavity length is chosen so that the dividing streamline leaves the body at the proper angle; in particular, the dividing streamline must be tangent to the attached flow. When this condition is met the dynamic boundary condition (Eq. 4) will be satisfied at the leading edge of the cavity.

The closure model is easily implemented for axisymmetric bodies through an iteration in which the cavity length is adjusted until the dividing streamline makes the proper angle with the attached flow at the point of separation since this iteration involves only one parameter. The closure model is more difficult to implement for nonaxisymmetric flow fields (axisymmetric bodies at non-zero angle of attack) for two reasons. First, the parameterization of the cavity is more complicated resulting in a much larger number of parameters necessary to describe the cavity. Hence, determining the locations of the constant pressure streamlines assuming given lengths along each axial ray is far more time-consuming. Second, the zones of separation and reattachment of the dividing streamlines are fully three dimensional. Small errors in determining the length of the cavity along the various rays can cause significant errors in the cavity thickness. A manifestation of these errors is an oscillation in the cavity thickness in the polar direction about a ring. In fact, in several cases, our initial estimate of the cavity shape was so poor that the Newton-Raphson iteration failed to converge. Therefore, it was necessary to develop a method of generating the initial estimates. This was done by extending the method developed by Wolfe, Hailey, and Oberkampf [17] for axisymmetric cavities. Along each ray, the point of detachment and cavity length are specified. Then the initial cavity shape along that ray is approximated by an arc of a circle which is tangent to the body at the point of detachment and has the prescribed length. With this initial estimate, a cavity shape is determined and then the lengths of the cavity along each ray are adjusted until the cavity surface is parallel to the attached flow everywhere along the zone of detachment.

BOUNDARY LAYER FORMULATION

Once the converged cavity solution has been obtained from the potential solution, the boundary-layer calculations are performed. The familiar von Karman momentum integral equation is the starting point,

$$\frac{C_f}{2} = \theta\left(\frac{2+H}{U}\frac{dU}{ds}\right) + \frac{d\theta}{ds} + \frac{\theta}{r}\frac{dr}{ds}. \tag{10}$$

where $U = U(s)$ is the outer flow velocity determined from the BEM calculations as a function of the parameter of arclength s along Γ. The skin-friction coefficient, C_f, is given by

$$C_f = \frac{\tau_w}{\frac{1}{2}\rho U^2}, \tag{11}$$

where τ_w is the wall shear stress. The momentum thickness, θ, is given by

$$\theta = \int_0^\infty u(U-u)dn, \tag{12}$$

and the displacement thickness, δ^*, is given by

$$\delta^* = \int_0^\infty (U-u)dn. \tag{13}$$

The variable H is called the shape factor, $H = \delta^*/\theta$. The Karman momentum integral (Eq. 10) is valid for both laminar and turbulent boundary layers.

For laminar flow, the solution of Eq. 10 is obtained from Thwaites' one-parameter scheme [18]. The fundamental assumption of a one-parameter scheme is that C_f, H, and θ are functions of the freestream velocity $U(s)$, and a parameter λ, where $\lambda = \theta^2/\nu(dU/ds)$. In Thwaites' scheme,

$$\theta^2 \approx \frac{0.45\mu}{r^2U^6\rho}\int_0^s r^2U^5 ds. \tag{14}$$

Once the momentum thickness, θ, is obtained from Eq. 14, the remaining laminar boundary layer parameters (H and C_f) are obtained from a tabulated correlation that depends on λ [19]. The values of H and C_f predicted by Thwaites' method have been found to be within $\pm 5\%$ for favorable pressure gradients.

The point of transition is calculated in the present work by the approximate method of Schlichting [19]. The degree of surface roughness will influence the separation point. The method described by Wolfe and Oberkampf [20] for including distributed surface roughness in the transition model was employed in the current model. However, in all cases discussed in this paper, the Reynolds numbers were low enough so that transition to turbulence did not take place, and hence, we omit the discussion of the turbulent boundary layer. The boundary-layer calculation is performed on the wetted-flow portions of the body. The skin-friction drag is calculated for the wetted portion of the body only, based on the assumption that the cavity-covered section of the body does not contribute to the skin friction drag.

RESULTS

Axisymmetric Cavities

We first consider cavities on axisymmetric bodies at zero degree angle of attack. The closure model is easily implemented in these cases since the iteration in which the cavity length is adjusted until the dividing streamline is tangent to the attached flow involves only one parameter. As an example, we consider a body comprised of a 45° conical nose and a cylindrical mid-body with the cavitation number given by $K = 0.30$. The total length of the body is 5.207 calibers. For this body with the abrupt change in normal direction at the juncture between the conical nose and the cylindrical mid-body, the smooth separation condition requires that the cavity detachment is at the body juncture and that the cavity leaves the body at a 22.5° angle. Solutions for three assumed cavity lengths are shown in Fig. 2. The cavity in Fig. 2a is seen to leave the body at an angle much smaller than 22.5° and the cavity in Fig. 2c is seen to leave the body at an angle much greater than 22.5°. We use the term cavity in a loose sense here since the pressure is below the cavitation pressure at the leading edge for the cavity in Fig. 2a and the pressure is above the cavitation pressure at the leading edge for the cavity in Fig. 2c. The cavity in Fig. 2b is seen to be approximately tangent to the attached flow at the point of separation, that is, this cavity leaves the body at approximately the prescribed angle of 22.5°. Hence, using our closure model, the cavity in Fig. 2b is chosen as the one best representing the actual physical cavity. This cavity is compared to the water tunnel photographs and measurements of Rouse and McNown [21] in Fig. 3. The numerical calculations compare quite well with the water tunnel data except in the reattachment zone. However, the flow is two-phase bubbly and turbulent in this region causing the current modelling to be insufficient.

As a second example, we consider a 1-caliber ogive cylinder with the cavitation number given by K=0.24. The length of the body is 4.866 calibers. Using the smooth separation condition, the leading edge of the cavity was calculated to be located at a distance of 0.3660 calibers from the nose of the ogive body. Solutions for three cavity lengths are shown in Fig. 4. The cavity in Fig.4a is seen to penetrate into the ogive surface. The cavity in Fig.4b is seen to be essentially tangent to the ogive surface at its leading edge. The cavity in Fig. 4c is seen to make an oblique angle with the ogive surface. Hence, using our closure model, the cavity shown in Fig. 4b is chosen as the one best representing the actual physical cavity. This cavity

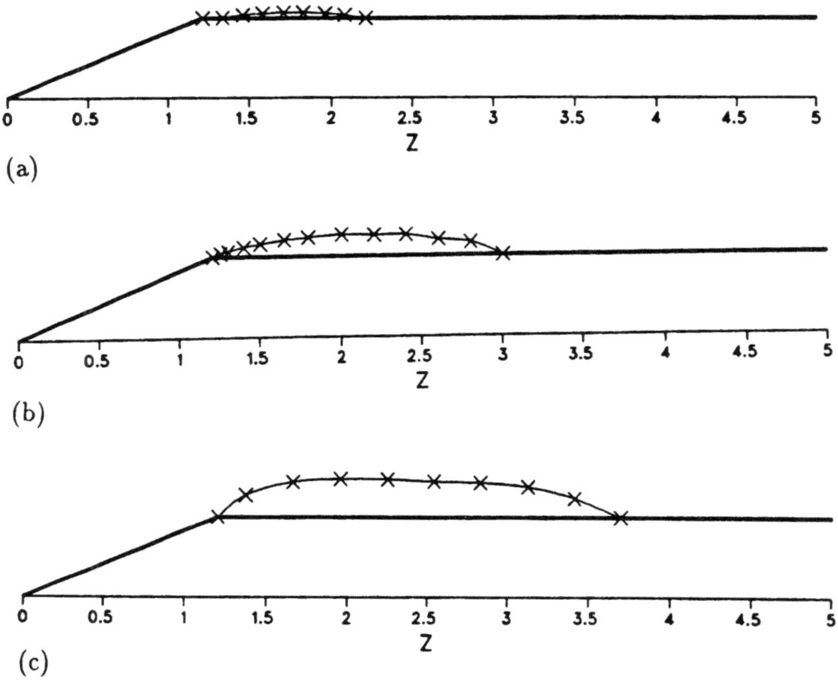

Figure 2: Calculated cavity shapes for the 45° cone-cylinder body with cavitation number $K = 0.3$. (a) Cavity length $L = 1.0$. (b) Cavity length $L = 1.8$. (c) Cavity length $L = 2.6$.

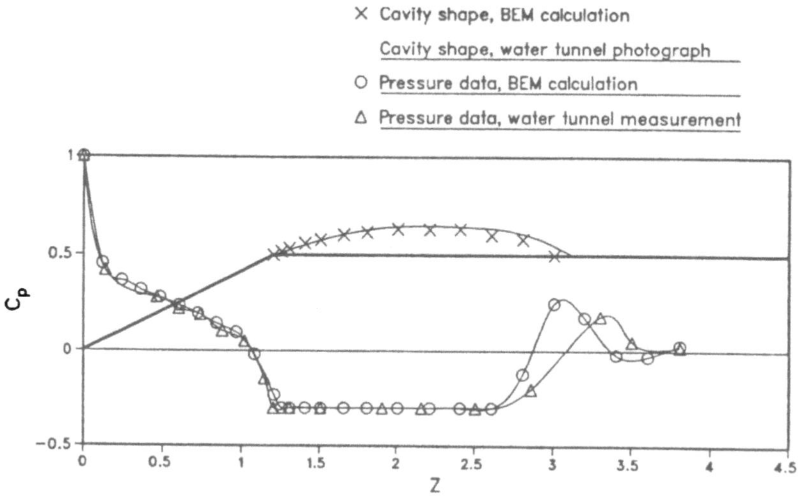

Figure 3: Comparison of BEM and water tunnel results for the 45° cone-cylinder body with cavitation number $K = 0.3$.

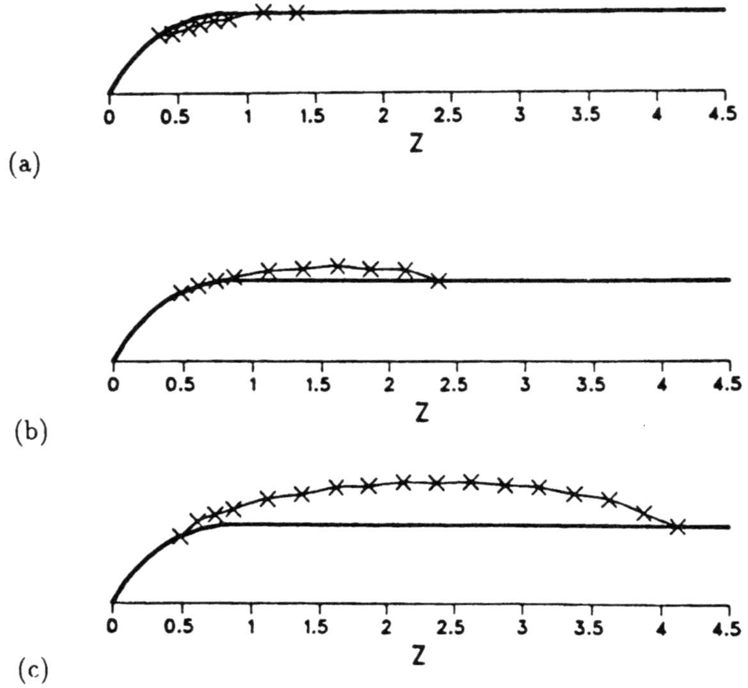

Figure 4: Calculated cavity shapes for the 1-caliber ogive body with cavitation number $K = 0.24$. (a) Cavity length $L = 1.1$. (b) Cavity length $L = 2.2$. (c) Cavity length $L = 3.8$.

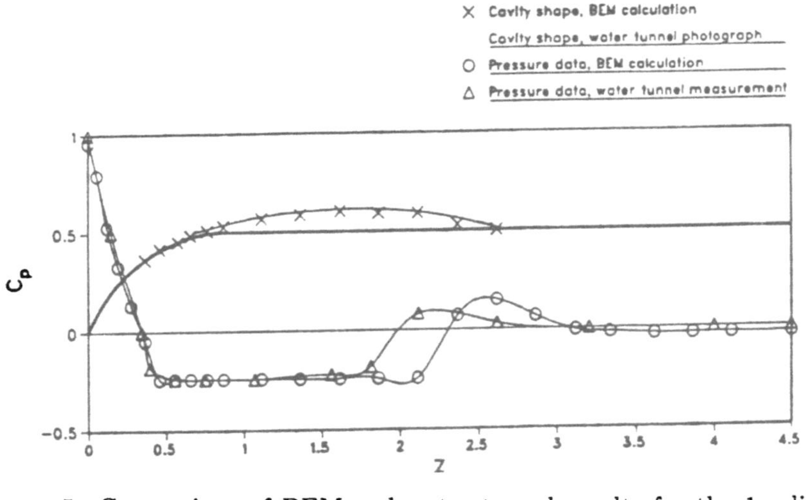

Figure 5: Comparison of BEM and water tunnel results for the 1-caliber ogive body with cavitation number $K = 0.3$.

is compared to the water tunnel data of Rouse and McNown [21] in Fig. 5. Again, the comparison is seen to be quite good.

Drag predictions for 10°, 20°, and 30° cones at various cavitation numbers are shown in Fig. 6. Here, the coefficient of drag, C_D, is defined by

$$C_D = \frac{F_x}{1/2 \rho V_\infty^2 S}$$

where F_x is the drag force and S is the reference area given by $S = \pi D^2/4$. The freestream velocity is assumed to be 20 fps which is a reasonable value for a water tunnel. The cone base diameter, D, was 25.4 mm for all cases. The Reynolds number based on the diameter for these calculations is 1.36×10^5 which results in a laminar boundary layer in all cases. Also plotted in the figure are experimental results from Cox and Maccoll [22] and the numerical predictions of Wolfe *et al.* [17]. The numerical method of Wolfe *et al.* is based on an axial distribution of sources and sinks, and thus, is limited to axisymmetric flow fields and relatively slender bodies. The BEM results show good agreement with both the water tunnel measurements and the numerical predictions of Wolfe *et al.* The largest component of drag due to viscous effects was experienced for the 10° cones. As an example, at a cavitation number of 0.04, approximately 8.5% of the total drag was due to viscous effects. Thus, the viscous drag is not negligible.

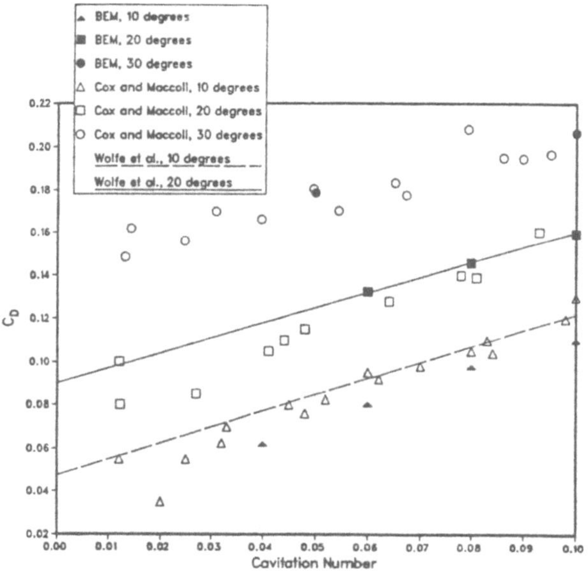

Figure 6: Comparisons of the BEM results with the water tunnel results of Cox and Maccoll and the numerical results of Wolfe *et al.* for the drag on 10°, 20°, and 30° cones.

Nonaxisymmetric Cavities

The first example of a nonaxisymmetric cavity is given by the body comprised of a 45° conical nose and a cylindrical midbody at 6° angle of attack with the cavitation number given by $K = 0.4$. This example was chosen since the results could be compared with the water tunnel data of Rouse [23]. Because of the discontinuous body slope, the points of detachment are located at the juncture between the nose and midbody. A comparison of the BEM results and the water tunnel data for the coefficient of pressure along the leeward and windward sides of the body is shown in Fig. 7. The positive body stations are positioned on the leeward side of the body while the negative body stations are positioned on the windward side of the body. It is seen in Fig. 7 that the cavity length on the leeward side is longer than on the windward side. Further, the pressure is seen to be higher on the windward side of the conical nose. The experimental and numerical cavity lengths and pressures agree reasonably well between the water tunnel data and the BEM results.

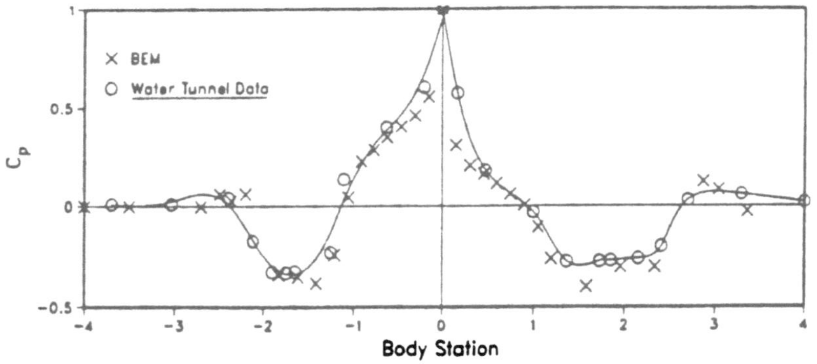

Figure 7: Comparison of the BEM results with the water tunnel results of Rouse for the coefficient of pressure on a body comprised of a 45° conical head and a cylindrical midbody at 6° angle of attack.

The second example of a nonaxisymmetric cavity is given by supercavitation on a 30° cone at -2.5° angle of attack with the cavitation number given by $K = 0.1$. The solution showed some oscillations in cavity thickness along the midside node rays. The forces and moment presented for this case were obtained by integrating the edge node pressures only. This example was chosen because results for the forces and moment could be compared with the water tunnel data of Kiceniuk [24]. The cavity detachment point is located at the base of the cone which is consistent with the experiment where air was introduced into the cone base region to produce a cavity. The cavity shapes on the leeward and windward sides are shown in Fig. 8. It is seen that the cavity is thicker on the leeward side as expected. The normal force coefficient, C_N, and the pitching moment coefficient, C_M, are defined respectively by

$$C_N = \frac{F_z}{1/2\rho V_\infty^2 S} \quad ; \quad C_M = \frac{M_y}{1/2\rho V_\infty^2 S},$$

where F_z is the normal force, M_y is the resultant moment. The moment reference point is the center of the cone base. The BEM predicted results given by $C_D = 0.198$, $C_N = -0.044$, and $C_M = -0.036$. Kiceniuk measured in the water tunnel $C_D = 0.200$, $C_N = -0.040$ and $C_M = -0.045$ in his water tunnel experiments. The normal and drag forces show good agreement between measurement and theory. The pitching moment results show some discrepancy. We are in the process of running more cases and reducing recently obtained water tunnel data in order to determine the reliability of the BEM results.

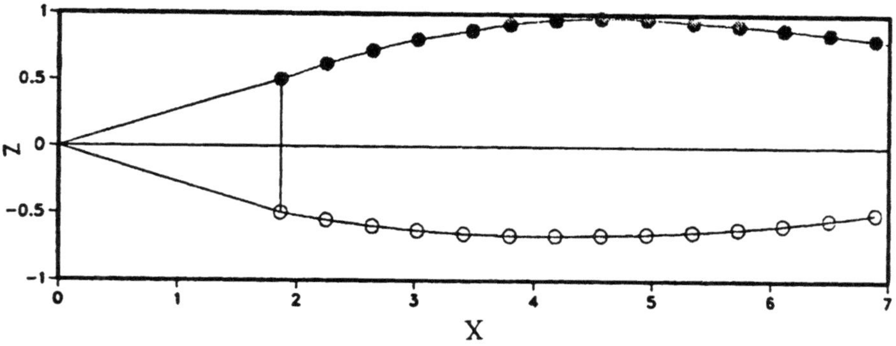

Figure 8: The cavity shape on the leeward (top) and windward (bottom) sides of a 30° conical body at -2.5° angle of attack.

SUMMARY

The boundary element method has been combined with a cavity closure model and integral-boundary layer solver to determine cavity shapes and resultant forces and moments on axisymmetric bodies both at zero and non-zero angle of attack. This approach has been successful in predicting cavity lengths and drag forces for the zero angle of attack cases, that is, for axisymmetric flow fields. We believe our calculations for non-zero angle of attack are among the first to predict nonaxisymmetric cavity shapes. We are in the process of evaluating the accuracy of these calculations.

ACKNOWLEDGMENT

This work was supported by Sandia National Laboratories and the U.S. Department of Energy under contract DE-AC04-76DP00789.

REFERENCES

1. Brillouin, M., *Ann. Chim. Phys.*, **23**: 145-230, 1911.

2. Villat, H., "Sur la validitè des solutions de certains problémes d'hydrodynamique," *J. Math.*, **10**(6): 231-290, 1914.

3. Arakeri, V. H., "Viscous effects on the position of cavitation separation from smooth bodies," *J. Fluid Mech.*, **68**(4): 779-799, 1975.

4. Brennen, C. "A numerical solution of axisymmetric cavity flows," *J. Fluid Mech.*, **37**: 671-688, 1969.

5. Joukowsky, N. E., "A modification of Kirchoff's method of determining a two-dimensional motion of a fluid given a constant velocity along an unknown streamline. II. Determination of the motion of a fluid for any condition given on a streamline," *Rec. Math.*, **25**, 1890.

6. Riabouchinski, D., "On steady fluid motion with free surfaces," *Proc. London Math. Soc.*, **19**: 206-215, 1920.

7. Kreisel, G., "Cavitation with finite cavitation numbers," *Admiralty Res. Lab. Rep. No. r1/h/36*, 1946.

8. Wu, T. Y., "A wake model for free-streamline flow theory: Part 1, fully and partially developed wake flows and cavity flows past an oblique flat plate," *J. Fluid Mech.*, **13**: 161-181, 1962.

9. Tulin, M. P., "Supercavitating flows - Small perturbation theory," *J. Ship Res.*, **7**(3): 16-37, 1964.

10. Mogel, T. R. and Street, R. L., "A numerical method for steady state cavity flows," *J. Ship Res.*, **18**(1), 22-31, 1974.

11. Furuya, O., "Nonlinear calculation of arbitrary shaped supercavitating hydrofoils near a free surface," *J. Fluid Mech.* **68**(1), 21-40, 1975.

12. Pellone, C. and Rowe, A., "Supercavitating hydrofoils in non-linear theory," in *Third Intl. Conf. on Numerical Ship Hydrodynamics, Paris, June 1981*, (ed. J. C. Dern & H. J. Haussling), Bassin d'essais des Carènes, Paris, France, 1981.

13. Dagan, A., "Axially symmetric cavitation flow at small cavitation numbers," *Int. J. Num. Meth. Fluids*, **8**, 943-955, 1988.

14. Aitchison, J. M. and Karageorghis, A., "Numerical solution of a free surface problem by a boundary element method," *Int. J. Num. Meth. Engrg.*, **8**, 91-96, 1988.

15. Lemonnier, H. and Rowe, A., "Another approach in modelling cavitating flows," *J. Fluid Mech.*, **195**, 557-580, 1988.

16. Brebbia, C. A., Telles, J. C. F., and Wrobel, L. C., *Boundary Element Techniques*, Springer Verlag, Berlin, 1984.

17. Wolfe, W. P., Hailey, C. E., and Oberkampf, W. L., "Drag of bodies of revolution in supercavitating flow," *J. Fluids Engrg.*, **111**(3): 300-305, 1989.

18. White, F. M., *Viscous Fluid Flow*, McGraw-Hill, Inc., New York, 1974.

19. Schlichting, H., *Boundary Layer Theory*, McGraw-Hill, Inc., New York, 6th Edition, 1968.

20. Wolfe, W. P., and Oberkampf, W. L., "Drag Prediction for Projectiles and Finned Bodies in Incompressible Flow," AIAA Paper No. 85-0104, 1985.

21. Rouse, H. and McNown, J. S., "Cavitation and pressure distribution. Head forms at zero angle of Yaw," *State University of Iowa, Engrg. Bull.*, No. 32, 1948.

22. Cox, R. N. and Maccoll, J. W., "Recent contributions to basic hydroballistics," *Symposium on Naval Hydroballistics*, NAS-NRC Publication 515, Washington, D.C., Chapt. IX, pp.215-239, 1957.

23. Rouse, H., "Cavitation and pressure distribution. Head forms at angles of yaw," *State University of Iowa, Engrg. Bull.*, No. 42, 1962.

24. Kiceniuk, T., "An experimental study of the hydrodynamic forces acting on a family of cavity-producing conical bodies of revolution inclined to the flow," Report No. E-12.17, Hydrodynamics Laboratory, California Institute of Technology, 1954.

Boundary Element Simulation of Axisymmetric Cavity Flow Problems

L.C. Wrobel

Computational Mechanics Institute, Wessex Institute of Technology, Ashurst Lodge, Ashurst, Southampton SO4 2AA, England

ABSTRACT

This paper presents a boundary element formulation for solution of axisymmetric Riabouchinsky cavity flow problems. An iterative procedure for adjusting the free surface position is developed and shown to be stable and convergent. Numerical results are compared with finite difference and finite element solutions, showing the robustness of the BEM model.

INTRODUCTION

The present paper considers the axially symmetric flow of an incompressible inviscid fluid past a disk, placed perpendicular to the flow direction, in a channel of finite width and infinite length. Immediately behind the disk a cavity is formed containing air at a constant pressure; the primary objective of the study is to find the shape and size of this cavity.

It is well-known from experiments what sort of shape the cavity should have behind the disk. However, it is not clear at all, from experiments or theory, what is the cavity form in the region far from the disk. This phenomenum, known as the closure condition, is still open to discussion, and several models have been presented in the literature [1],[2]. In practice the downstream closure region is marked by considerable turbulence so that no potential flow solution can expect to be accurate in this region. However, most authors agree that, within limits, the choice of closure model has little effect on the flow at the upstream end of the cavity [3].

Herein, the model proposed by Riabouchinsky [4] is considered. The model assumes that an image disk can be placed in the flow at some point downstream from the original disk and that the flow geometry will be symmetric about the line midway between the disks (figure 1). Owing to the axial symmetry of the flow, only one-quarter of the region in figure 1 needs to be considered in the numerical analysis (figure 2).

The determination of the free surface position CD is required as part of

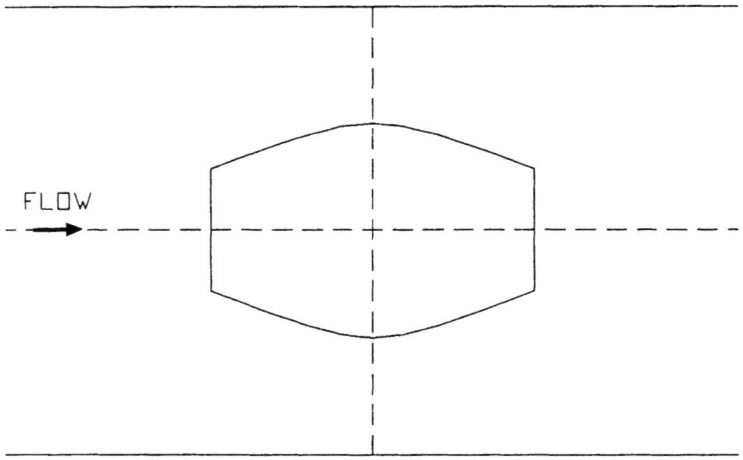

Figure 1: Riabouchinsky cavity model

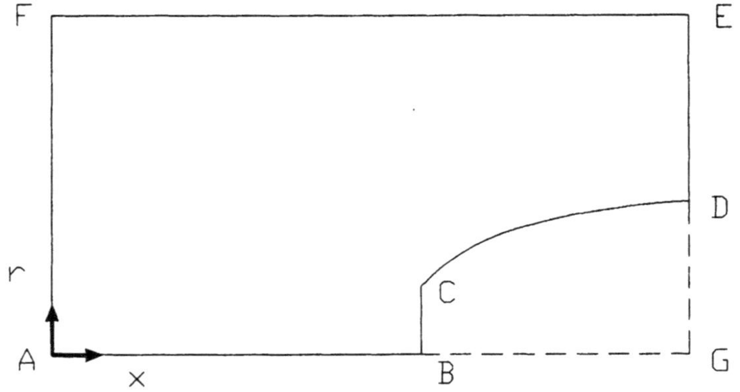

Figure 2: Definition of problem region

the solution of the problem. This generally involves an initial guess of this free streamline and the application of a systematic shifting algorithm to improve its location.

Previous attempts using the finite difference method (FDM) [5] and the finite element method (FEM) [6] presented many difficulties due to the need to mesh the entire flow domain. The presence of a singularity at the separation point C requires the mesh to be very refined near this point and coarser elsewhere, *i.e.* a variable grid spacing.

The most useful information from a solution of a cavity flow problem are the spatial location of the free surface and a knowledge of velocities and/or pressures along the boundaries. This realization leads one to feel that domain-type techniques are computationally inefficient, and that the problem is ideally suited to a boundary element (BEM) solution. This paper presents a BEM procedure in which linear elements are employed and the free surface position is determined iteratively. The iteration algorithm is similar to the one employed by Wrobel [7] for plane cavity flows.

FORMULATION OF THE PROBLEM

A sketch of the Riabouchinsky cavity model adopted in this work can be seen in figure 2. The computational region is truncated at a certain distance upstream, where the flow is assumed to be parallel to the channel walls.

For the steady, irrotational, axisymmetric flow of an incompressible, inviscid fluid the following equations apply using cylindrical coordinates (x, r):

$$\frac{\partial u}{\partial x} + \frac{1}{r} v + \frac{\partial v}{\partial r} = 0 \quad \text{(continuity)} \tag{1}$$

$$\frac{\partial u}{\partial r} - \frac{\partial v}{\partial x} = 0 \quad \text{(irrotationality)} \tag{2}$$

$$p + \frac{1}{2}\rho q^2 + \rho g r = \beta \quad \text{(Bernoulli)} \tag{3}$$

in which u, v are the velocity components in directions x and r, respectively, $q = (u^2 + v^2)^{1/2}$ is the velocity magnitude, p is the pressure, ρ is the fluid density, g the acceleration due to gravity and β is Bernoulli's constant.

Introducing the Stokes' stream function ψ such that

$$u = \frac{1}{r}\frac{\partial \psi}{\partial r} \qquad v = -\frac{1}{r}\frac{\partial \psi}{\partial x}$$

and substituting into (1) and (2), one notes that the continuity equation is identically satisfied and the irrotationality condition gives rise to the equation

$$\frac{\partial^2 \psi}{\partial r^2} - \frac{1}{r}\frac{\partial \psi}{\partial r} + \frac{\partial^2 \psi}{\partial x^2} = 0 \tag{4}$$

The boundary conditions of the problem are as follows, refering to figure 2: line $ABCD$ is a streamline and, without loss of generality, its value can be set to $\psi = 0$. The channel wall, line EF, is also a streamline with value $\psi = \psi_H = q_0 H^2/2$, in which q_0 is the onset velocity magnitude and H the channel radius. Line DE is a symmetry axis, so $\partial\psi/\partial x = 0$ can be applied. At the truncating boundary AF the velocity field is parallel to the channel wall; thus, it is possible to apply either $\psi = q_0 r^2/2$ or $\partial\psi/\partial x = 0$.

The problem is solved by assuming an initial guess for the free surface position and using Bernoulli's equation to correct it during an iteration process. It is customary to assume that the hydrostatic pressure is approximately uniform over the region of interest and to neglect the effects of gravity; as a consequence of (3), the fluid velocity is uniform in magnitude on the free streamline. This is an accurate assumption when $gr << q^2$, which holds true in this case [8].

Denoting quantities referenced to the cavity or free streamline by the subscript c, we may write from (3)

$$p_e + \frac{1}{2}\rho q^2 = \frac{1}{2}\rho q_c^2 = \beta \tag{5}$$

in which $p_e = p - p_c$ and the constant value of the pressure p_c has been incorporated into β. This expression permits computing the pressure at any point once the velocity field has been determined.

An important parameter in cavity flows is the Prandtl cavitation number σ, defined by

$$\sigma = \frac{q_c^2}{q_0^2} - 1$$

The problem can be made non-dimensional by putting

$$x^* = x/H \qquad r^* = r/H \qquad \psi^* = \psi/\psi_H \qquad q^* = q/q_0$$

The numerical problem to be solved is then described by the following equations (dropping the *'s for simplicity):

$$\frac{\partial^2\psi}{\partial r^2} - \frac{1}{r}\frac{\partial\psi}{\partial r} + \frac{\partial^2\psi}{\partial x^2} = 0 \quad \text{in } ABCDEFA \tag{6}$$

$$\psi = 0 \quad \text{on } AB, BC, CD \tag{7}$$

$$\frac{\partial\psi}{\partial x} = 0 \quad \text{on } DE \tag{8}$$

$$\psi = 1 \quad \text{on } EF \tag{9}$$

$$\frac{\partial\psi}{\partial x} = 0 \quad \text{on } AF \tag{10}$$

$$\frac{1}{r}\frac{\partial\psi}{\partial n} = -q_c \quad \text{on } CD \tag{11}$$

with the cavitation number given by

$$\sigma = \frac{q_c^2}{4} - 1$$

NUMERICAL SOLUTION

For the numerical solution of the problem, an initial free surface location has to be assumed. The boundaries of the region are then discretized and the problem solved using a standard BEM methodology [9].

The first step in the BEM formulation is to obtain a boundary integral equation equivalent to equation (6). In the present case, this is of the form

$$c(\xi)\,\psi(\xi) + \int_\Gamma q^*(\xi,\chi)\,\psi(\chi)\,d\Gamma(\chi) = \int_\Gamma \psi^*(\xi,\chi)\,q(\chi)\,d\Gamma(\chi) \tag{12}$$

in which ξ and χ are the source and field points, Γ is the contour $BCDEFA$ which excludes the rotational symmetry axis AB, $c(\xi)$ is a geometric coefficient dependent on the location of the source point, $q(\chi) = \frac{1}{r(\chi)}\frac{\partial\psi(\chi)}{\partial n(\chi)}$, $q^*(\xi,\chi) = \frac{1}{r(\chi)}\frac{\partial\psi^*(\xi,\chi)}{\partial n(\chi)}$, $n(\chi)$ is the outward normal at the field point, and ψ^* is the fundamental solution of equation (6), given by [1]

$$\psi^* = \frac{1}{(a+b)^{1/2}}[aK - (a+b)E]$$

where

$$a = r^2(\xi) + r^2(\chi) + [x(\xi) - x(\chi)]^2$$

$$b = 2r(\xi)r(\chi)$$

and $K = K(m)$ and $E = E(m)$ are the complete elliptic integrals of the first and second kind, respectively, with parameter $m = 2b/(a+b)$. The derivative of ψ^* with respect to the normal is of the form

$$\frac{\partial\psi^*}{\partial n} = \frac{x(\chi) - x(\xi)}{(a+b)^{1/2}}\left[K - \frac{a}{a-b}E\right] n_x + \frac{r(\chi)}{(a+b)^{1/2}}\left[K + \frac{2r^2(\xi) - a}{a-b}E\right] n_r$$

Next, the boundary Γ is subdivided into elements and the discretized version of equation (12) applied to the nodal points through a collocation technique. This generates a set of N equations, with N the number of boundary nodes. Boundary conditions are then introduced and the resulting system of equations solved by Gauss elimination. Details of the BEM implementation are given in [10].

The boundary condition applied on the free streamline CD is $\psi = 0$. The BEM solution then produces the value of q_c at each point along the free surface. These values are all equal for the correct free surface position (but of course not for the assumed one), and it is essential that a procedure be devised to properly move the free surface in the steps of the iteration scheme so that the constant-velocity condition be more closely satisfied on the moved streamline.

With the objective of obtaining a relation between increments of q_c and r on the free surface points, we define $\overline{q}$ as the average velocity in a vertical section of the flow. Thus, the flow rate Q can be expressed as

$$Q = \overline{q}\,\pi\,(H^2 - r^2) \tag{13}$$

or, in non-dimensional form,

$$\overline{q}\,(1 - r^2) = 1 \tag{14}$$

The next step is to derive an approximate relation between $\overline{q}$ and q_c. It is noted that, in the region behind the disk, $\overline{q}(x)$ should increase with x and be lower than the free surface velocity q_c. The following relation was then adopted

$$\overline{q} = q_c \frac{r}{1 - r^2} \tag{15}$$

which complies with the previous requirements. Substituting in expression (14), we obtain

$$q_c\, r = 1 \quad \text{or} \quad q_c = \frac{1}{r} \tag{16}$$

The iteration procedure will then be as follows:

- i) Assume an initial free surface location;
- ii) BEM solution provides q_c along the free surface;
- iii) Calculate β at each free surface node i, *i.e.*

$$\beta_i = \frac{1}{2}\rho q_{ci}^2 \tag{17}$$

- iv) Calculate $\Delta\beta_i = \beta_c - \beta_i$ in which β_c is the value of β at point C, the separation point. This value was selected because point C, the disk tip, is a fixed free surface point;
- v) Calculate Δr_i by substituting (16) into (17) to obtain

$$\beta_i = \frac{\rho}{2r_i^2} \tag{18}$$

$$\frac{\Delta\beta}{\Delta r} \approx \frac{d\beta}{dr} = -\frac{\rho}{r_i^3} \tag{19}$$

$$\Delta r_i = -\frac{\Delta\beta_i\, r_i^3}{\rho} \tag{20}$$

- vi) Compute the relative norm of increments

$$\sum_{i=1}^{M} \frac{1}{M} \mid \frac{r_i^{k+1} - r_i^k}{r_i^k} \mid \tag{21}$$

where M is the number of free surface nodes and $r_i^{k+1} = r_i^k + \omega \Delta r_i^{k+1}$, ω being a relaxation coefficient. If the norm is smaller than a specified tolerance ϵ the process has converged and is terminated; otherwise, the free surface is moved to its new location and the process returns to step ii.

RESULTS OF ANALYSES

Some results are now presented concerning the problem depicted in figure 2. For simplicity, we call $\overline{BC} = d$, $\overline{BG} = L$, $\overline{DG} = b$ and $\overline{AB} = a$.

For specified values of a and d, the problem contains 3 free parameters, *i.e.* L, b and q_c, and one more needs to be prescribed to fully define the problem. As pointed out in [6], the natural choice from the physical point of view is to fix q_c, which is equivalent to specifying the pressure far upstream. However, small changes in the pressure lead to large changes in the cavity length, and the problem becomes more difficult from the mathematical point of view. Thus, it is preferable to fix the length L and compute b and q_c as part of the solution.

Results are presented in table 1 for $d = 0.25$, $L = 1.0$. These results were obtained using two different initial guesses for the free surface: the straight line $y = d$ and a quarter of an ellipse with centre along the line $\overline{EG}$ at $y = d$, and semi-axes L and 0.15. Also presented are the FEM solution of Aitchison [6] and the FDM one of Brennen [5].

The results of Brennen [5] were given in graphical form, and the values in table 1 have been extrapolated from these graphs. The FDM formulation employed a transformed (ϕ, ψ)-plane (ϕ being a velocity potential) where the flow region becomes rectangular. The problem solution required a graded mesh, point-by-point Gauss-Seidel iteration, over-relaxation, a special treatment of the singularity and the optimization of some parameters to obtain a stable and convergent scheme. Brennen [5] also performed experimental tests in a water tunnel which confirmed the accuracy of his numerical results. The FEM procedure employed a refined moving mesh of linear triangular elements, computing the free surface position through a discrete minimization problem using a quasi-Newton method. This was solved repeatedly for different assumed values of q_c until the average pressure in all elements were positive.

The BEM discretization employed only 40 linear elements, with 9 elements along the disk (line BC), 15 along the free surface (line CD), 12 along the channel wall (line EF) and 4 along the truncation boundary (line FA); the symmetry axis DE was taken into account by reflection and condensation as described in [10]. The discretization was graded due to the singularity at the separation point C such that, along the free surface, the elements length vary from around 0.0125 near point C to 0.2 near point D.

Table 1: Results for $d = 0.25$, $L = 1.0$

Method	b	σ
BEM	0.455	0.690
FEM	0.412	0.548
FDM	0.44	0.66

Table 2: Results for $d = 0.25$, $L = 0.5$

Method	b	σ
BEM	0.398	0.794
FEM	0.358	0.637
FDM	0.40	0.73

The BEM solution required a large number of iterations for a tolerance $\epsilon = 10^{-3}$, and a small value of the relaxation parameter $\omega = 0.05$ to control oscillations. The results, however, were insensitive to the initial free surface position. The convergence pattern for the two different assumed positions is depicted in figures 3 and 4. It can be seen that both cases converge to the same solution, and that the initial oscillations in figure 3 were quickly dampened.

Tables 2 to 4 depict the results obtained by the three methods for the cases of $d = 0.25$, $L = 0.5$; $d = 0.125, L = 1.0$ and $d = 0.125, L = 0.5$. Also included in table 3 is the FDM result for σ obtained by Fox and Sankar [11] using the Regula-Falsi method, with no special considerations for the singularity at the separation point. All tables show the same pattern of results, with good agreement between the present BEM and FDM values interpolated or extrapolated from Brennen's graphs [5], and with the FEM consistently predicting lower values.

It is interesting to point out that in all cases the truncation distance a was assumed equal to 2. The correctness of this assumption was verified from the calculated variation of ψ along the line FA which agreed very well with the exact values (see table 5).

Table 3: Results for $d = 0.125$, $L = 1.0$

Method	b	σ
BEM	0.296	0.288
FEM	0.250	0.209
FDM	0.29-0.30	0.28-0.29
R-F	-	0.113

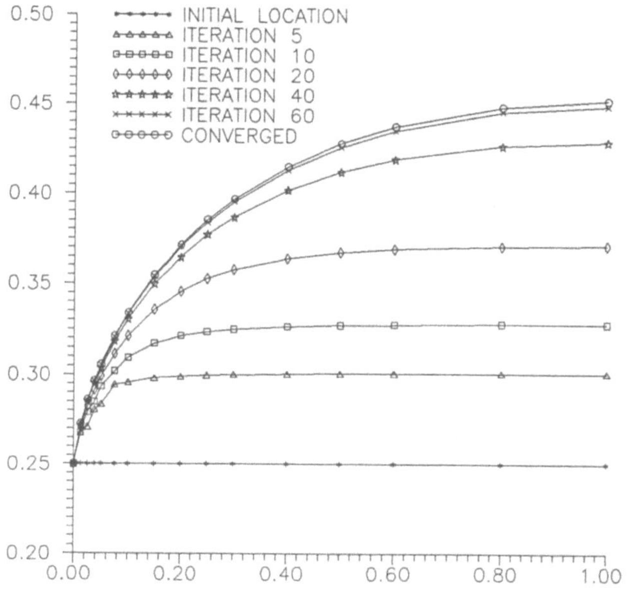

Figure 3: Convergence pattern for straight line as initial guess (distorted scale)

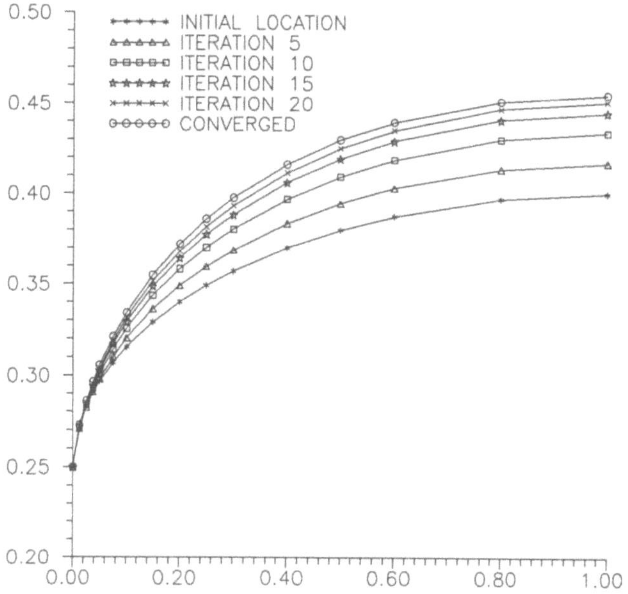

Figure 4: Convergence pattern for ellipse as initial guess (distorted scale)

Table 4: Results for $d = 0.125$, $L = 0.5$

Method	b	σ
BEM	0.245	0.375
FEM	0.203	0.275
FDM	0.24-0.25	0.34-0.35

Table 5: Values of ψ along truncation boundary FA for $d = 0.25, L = 1.0, q_0 = 2$

r	BEM	exact
0.25	0.0623	0.0625
0.50	0.2495	0.25
0.75	0.5616	0.5625

CONCLUSIONS

The BEM procedure for solving axisymmetric cavity flow problems described in the present paper has shown to work well for a series of geometric configurations. The results obtained compare very favourably with the FDM ones of Brennen [5], while the data preparation effort and computer time for the BEM are much smaller.

References

[1] Birkhoff, G. and Zarantonello, E.H., *Jets, Wakes and Cavities*, Academic Press, New York, 1957.

[2] Gilbarg, D., Jets and Cavities, in *Handbuch der Physik* (Ed. Flugge, S.), Vol. IX, Springer-Verlag, Berlin, 1960.

[3] Woods, L.C., *The Theory of Subsonic Plane Flow*, Cambridge University Press, Cambridge, 1961.

[4] Riabouchinsky, D., On steady fluid motion with free surfaces, *Proc. London Mathematical Society*, Vol. 19, pp. 206-215, 1920.

[5] Brennen, C., A numerical solution of axisymmetric cavity flows, *Journal of Fluid Mechanics*, Vol. 37, pp. 671-688, 1969.

[6] Aitchison, J.M., The numerical solution of planar and axisymmetric cavitational flow problems, *Computers and Fluids*, Vol. 12, pp. 55-65, 1984.

[7] Wrobel, L.C., A simple and efficient BEM algorithm for planar cavity flows, submitted for publication in *International Journal for Numerical Methods in Fluids*, 1991.

[8] Batchelor, G.K., *An Introduction to Fluid Dynamics*, Cambridge University Press, Cambridge, 1967.

[9] Brebbia, C.A., Telles, J.C.F. and Wrobel, L.C., *Boundary Element Techniques*, Springer-Verlag, Berlin, 1984.

[10] Wrobel, L.C., A numerical solution of axisymmetric cavity flows using an integral equation method, submitted for publication in *Journal of Fluid Mechanics*, 1991.

[11] Fox, L. and Sankar, R., The Regula-Falsi method for free boundary problems, *J. Inst. Maths Applics*, Vol. 12, pp. 49-54, 1973.

Cavitation Effects in Journal Bearing having a Big Radial Clearance

K. Kryniski
Department of Mechanical Engineering, James Cook University of North Queensland, Qld. 4811, Australia

INTRODUCTION

Due to their reliability and low maintance costs over an extended service time, the journal bearings, also known as fluid-film bearings, are commonly incorporated in the super-critical rotor systems. Together with proven balancing methods, they allow a rotating machine to pass smoothly through the various critical speeds, both during start-ups and shut-downs. However, journal bearings need to be designed very carefully, as at some operating conditions (speed and load), they may introduce the undesired effects, such as unstable operations or sub-harmonic resonances. The standard procedure leading to fluid-film bearing design is based on the bearing capacity, determined by the Sommerfield number [1][2]. When Sommerfield number is determined, all design parameters, such as viscosity, radial clearance, diameter and rotation speed, etc. are matched to satisfy specified engineering requirements. The procedure is considered to be completely reliable and is commonly used in turbo-machinery and high-speed compressor designs. However, the significant divergences between theory and practice were observed with the increase of a bearing radial clearance [3].

The purpose of this paper is to present the approach, which attempts to explain on the background of cavitational flow the most unexpected phenomena, such as multi-equilibrium positions, amplitude jumps, significant reduction of bearing capacity, etc. which have been observed on the actual rotor-bearing system. The paper shows both the advantages and limitations of the standard journal bearing design procedures.

NOTATIONS

c	eccentricity ratio
α	attitude angle
$a=2r/D$	radial clearance ratio
r	radial clearance
D	bearing diameter
$H=h/r$	normalized oil film thickness
$c_t=\partial c/\partial t$	radial velocity ratio
$\alpha_t=\partial \alpha/\partial t$	
$p=p^* a^2/\eta\,\Omega$	normalized pressure
η	viscosity
L/D	length to diameter ratio
f	non-dimensional force
S	Sommerfield Number

FLUID-FILM BEARING MODEL

As a result of a fluid flow induced by the rotating shaft, hydrodynamic pressure is developed in a gap between the journal and bearing. In consequence a hydrodynamic force is generated. The pressure in a gap of the thickness H is described by the Reynold's equation (1):

$$\frac{\partial}{\partial\phi}\left(H^3\frac{\partial p}{\partial\phi}\right)+\left(\frac{D}{L}\right)^2\frac{\partial}{\partial z}\left(H^3\frac{\partial p}{\partial z}\right)=6\left(\frac{\partial H}{\partial\phi}+\frac{2}{\Omega}\frac{\partial H}{\partial t}\right) \qquad (1)$$

The journal motion, relatively to the bearing, is given by the journal centre position and velocity, these are defined with reference to Fig.1 by:

- eccentricity ratio c, which determines a radial journal centre position,
- attitude angle α, defining angular position with reference to the beginning of the sleeve,
- radial velocity ratio c_t,
- tangential velocity ratio $c\alpha_t$ of the journal centre.

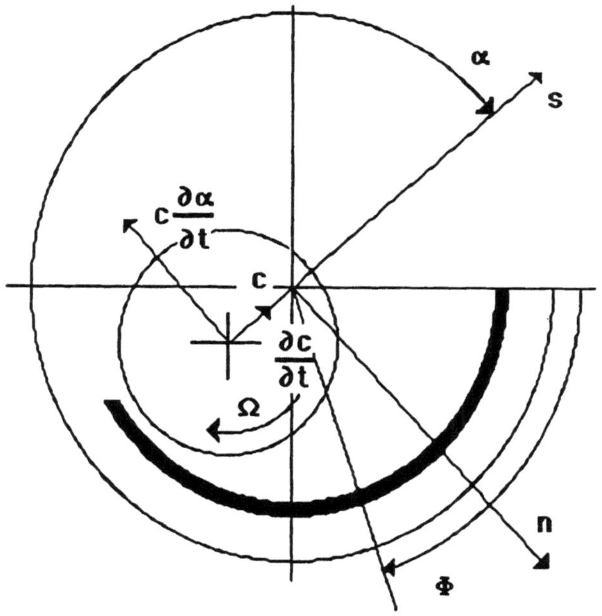

Fig. 1

In consequence, the pressure distribution in circumferential and transversal directions becomes a function of such chosen parameters of motion, i.e.:

$$p(f,z) = p(f,z,c,\alpha,c_t,\alpha_t) \tag{2}$$

The hydrodynamic force components, obtained by integration of the pressure over the area of the sleeve:

$$f_s = -0.5 \int_0^1 \int_0^{\Phi} p \cos(\phi-\alpha) \, d\phi dz$$

$$f_n = -0.5 \int_0^1 \int_0^{\Phi} p \sin(\phi-\alpha) \, d\phi dz \qquad (3)$$

become also function of these motion parameters.

The pressure distribution is obtained by the integration of the Reynold's equation with boundary conditions adjusted to the pressure conditions met at the edge of the fluid-film. However, due to cavitation phenomenon, pressure inside the bearing cannot drop below the critical value, known as a cavitation pressure, which is specific for a chosen lubricant. Therefore, to fulfil such requirements and model a cavitation effect, the procedure integrating the Reynold's equation modifies the solution at each iteration step in such a way, that the pressure, which dropped bellow the cavitation value at some node is replaced immediately by the cavitating value and iteration is repeated. The procedure converges quickly resulting in the self-adjusted boundary conditions, which correspond to the real fluid-film pressure conditions.

BEARING LOAD CHARACTERISTICS

Bearing load capacities can be plotted on the polar diagram, so called hydrodynamic force hodograph [1]. It is very convenient characteristics, indicating whether the bearing can withstand load requirements, allowing for optimum design of bearing orientation for maximum loading capacity and allowing for anticipation of an equilibrium position of the journal centre under a specific loading conditions.

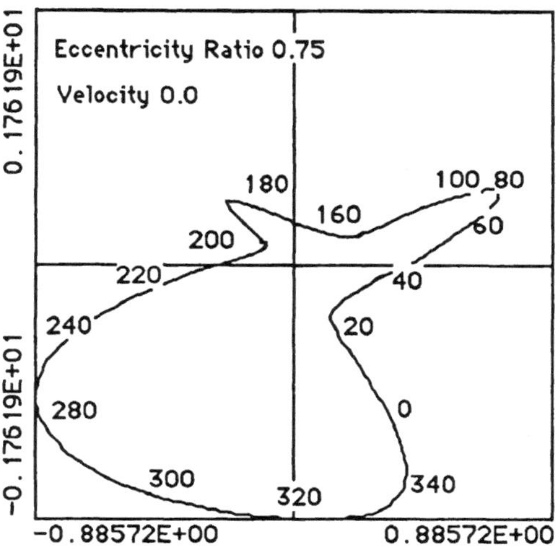

Fig.2

An example of a hydrodynamic force characteristics of a turbo-generator bearing are shown in Fig.2. The polar diagrams were constructed as the locus of the end points of the normalized (non-dimensional) hydrodynamic force vectors, all drawn from the centre of the sleeve for a given eccentricity ratio c, and successive values of the attitude angle α and with the velocity parameters set up to zero. Each of the diagrams consists of several closed curves, which correspond to the appropriate value of the eccentricity.

CAVITATION MODEL

The most common form of cavitation in journal bearings is gaseous cavitation where the continuous liquid phase is disrupted by the presence of gas bubbles. These gas bubbles are due to exsolved gases, coming out of solution when the pressure falls bellow saturation pressure, and to atmospheric gas being "sucked" into the bearing when the pressure falls bellow ambient. Consequently, the minimum possible pressure in the fluid film is the saturation pressure.

In the past a number of different approaches to the cavitation boundary condition problem were suggested [4]. The Gumbel approach also known as "half-Sommerfeld" solution, integrates the Reynold's equation for the continuous case where cavitation does not occur, allowing a full range of positive and negative fluid pressures. The pressure values below cavitation pressure are then made equal to the cavitation pressure. The "Swift-Stieber" solution, satisfies flow continuity, assuming the position of the beginning of the region of cavitation on the bearing sleeve in advance, which, in general is unknown. The boundary conditions applying to the beginning of the cavitation area are:

$$p = p_{cav}$$
$$\frac{\partial p}{\partial \phi} = 0 \qquad (4)$$

The approach used in the developed software corrects the pressure for cavitation effects with every iteration step of the integration of the Reynold's equation. Thus, the approach is akin to the "Swift-Steiber" solution, but with an added advantage in that the position of the cavitation area on the sleeve need not be assumed in advance, it is adjusted by itself automatically.

BOUNDARY CONDITIONS

Two bearing, of the same type, were used in analysis. Both were three lobes, oil lubricated. The bearing geometry is shown in Fig.3. Table 1 specified all relevant parameters.

Diameter [m]	Length [m]	Radial Clearance [m]	Viscosity [Ns/m^2]	Speed [rad/s]
1.0.5000	0.4150	3.375E-4	0.02120	314.00
2.0.0500	0.0415	0.000300	0.02120	314.00

Tab. 1

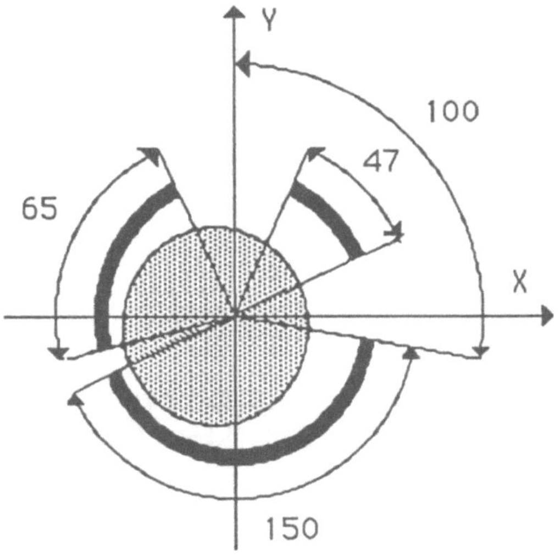

Fig.3

Integrating the Reynold's equation, modified by cavitation effect, the non-dimensional pressure distribution along circumferential direction is obtained. It is presented in Fig.4.

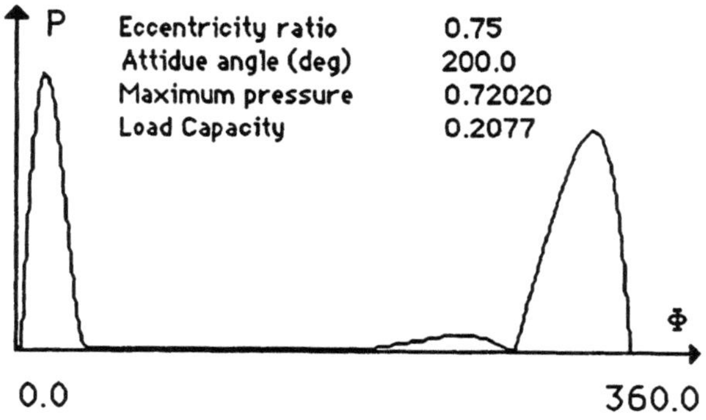

Fig.4

Pressure boundary conditions, as well as the cavitation pressure were set up to zero. This is fully justified by the fact, that pressure in the Reynold's equation is expressed in the non-dimensional form, as a result of the pressure normalisation by the factor involving viscosity, rotation speed and radial clearance ratio:

$$p_{non\text{-}dim} = \frac{p_{dim}\, a^2}{\eta\, \Omega} \qquad (5)$$

Considering inlet or outlet pressure in a range of atmospheric pressure 101.3 kPa, the non-dimensional value of inlet and outlet pressures are equal to 0.0294 for first type of the bearing, a typical design of turbo-generator. This is negligible, as compared with the maximum of non-dimensional pressure developed in the bearing.

The effect of the pressure boundary conditions becomes, however, more significant in a case of a big radial clearance. The normalized atmospheric pressure is 2.323. Even a portion of this is comparable with the pressure building-up by the rotating shaft. Therefore, pressure profile and consequently the bearing load characteristics may be significantly influenced by the inlet, outlet, ambient and cavitating pressures.

The analysis was performed for different boundary conditions. The hydrodynamic force characteristics were predicted for different values of outlet and cavitation pressures. The last one may correspond to different rotation speeds. The results are shown in Fig.5.

EXPERIMENTAL RIG

The experimental installation used consists of a rotor, two bearings, drive system, supporting structure and measuring equipment. The rotor was designed to have critical speeds outside of the speed range of a drive system. It was driven with constant speed in the range of 0-7000 r.p.m. The bearing chosen were two identical three lobe journal bearings which were lubricated with hydraulic oil SHELL

Tellus 46. The oil was supplied to each lobe through a groove to maintain a continuous oil film over each lobe in the bearing. To prevent leakage of oil from the bearing the rubber seals were used on both sides. The stiffness of the seals has been matched to be smaller than that of the oil film, even at low eccentricity ratios.

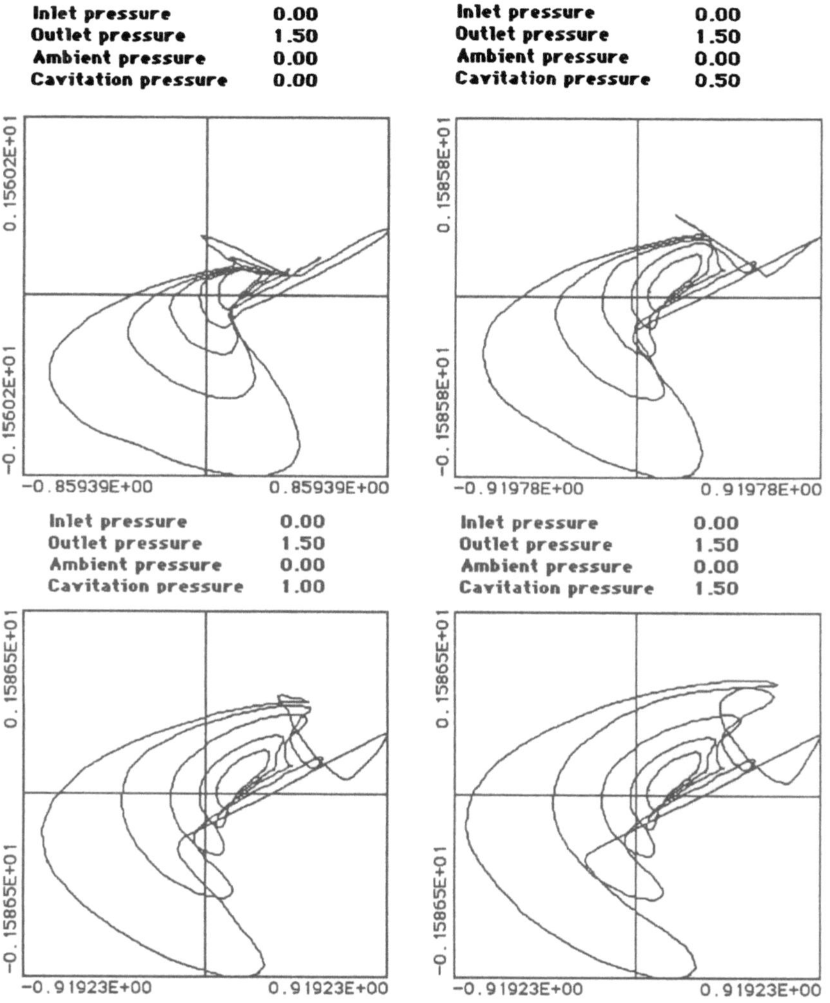

Fig. 5

The displacement of the journal, relative to the bearing, was measured by a BENTLEY-NEVADA 7200 Series proximitor transducer system. Each of the bearings was equipped with two pairs of eddy current transducers. Each pair of transdusers were located outside the bearing at equal spacing from the centre in vertical and horizontal planes. This arrangement allowed measurement of an average journal centre displacement. The signals from the transducers were monitored by the computer based data acquisition system.

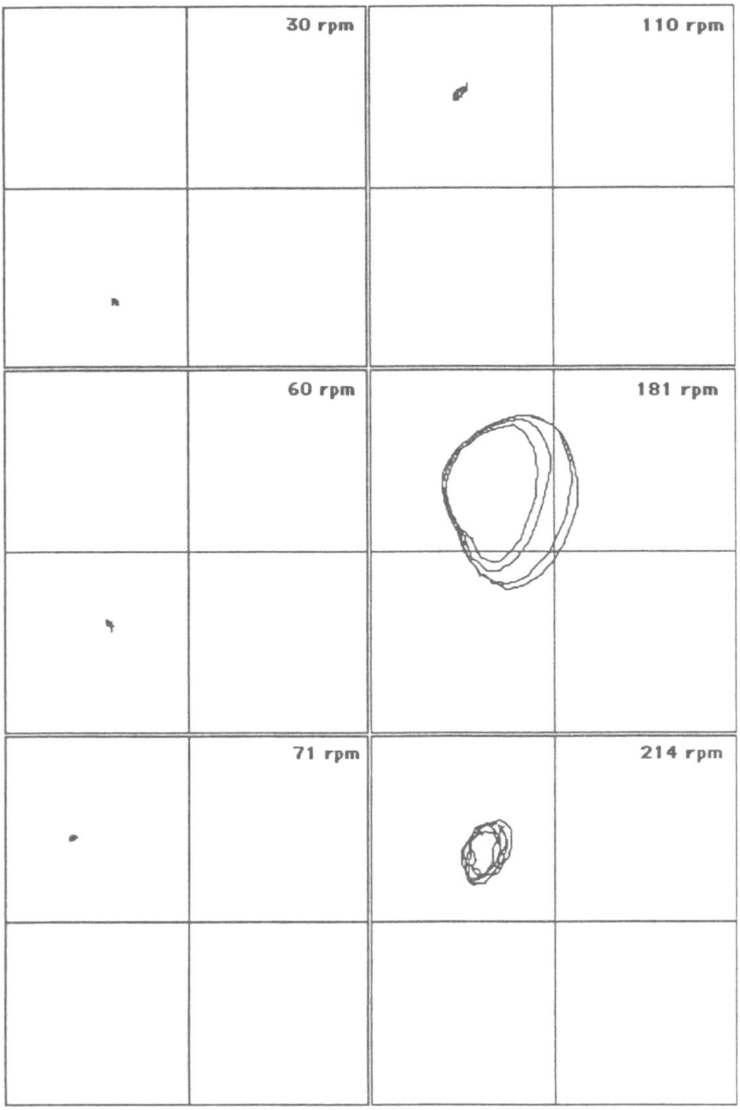

Fig. 6

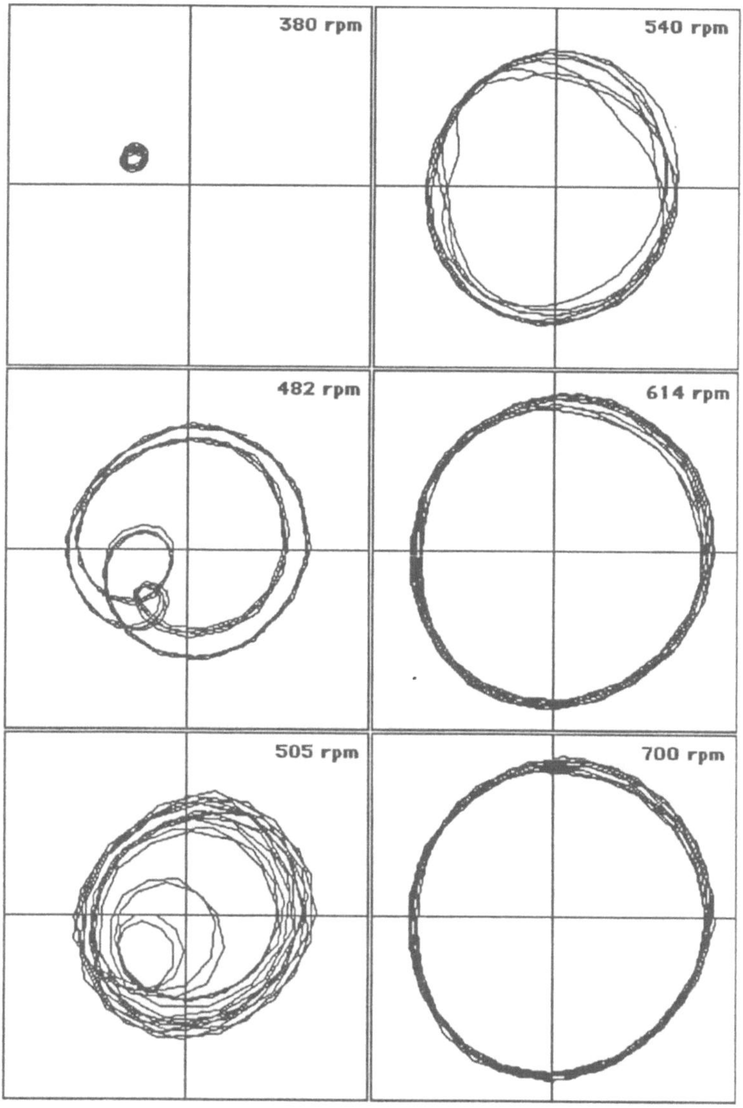

Fig. 7

JUMPING PHENOMENA

Unexpected bearing dynamic performance was recorded on the experimental installation. With a slight increase of rotational speed the journal changed drastically an equilibrium position. Fig. 6 and 7 show the trajectories developed at different speeds.

Due to vibration, the journal, under unchanged load and speed conditions jumped from one position to another, indicating the existence of two equilibrium positions. The phenomenon can be explained on the basis of predicted bearing load characteristics shown on Fig.5.

Each of the force hodographs corresponds to the specific eccentricity ratio. Thus, occurrence of intersecting loops indicate a possibility of two journal positions, where the generated hydrodynamic forces are identical with respect to magnitude and direction. Thus, an external force, e.g. bearing load can be balanced in two different position. As both of these positions are stable, vibrating shaft jumps from one position to the other, as a level of vibration allows for switching of the attraction regions.

REFERENCES

1. Z.A. Parszewski, M. Roszkowski, "Load capacity of multi-sleeve journal bearings", The Archive of Mechanical Engineering, No 4,1967,pp.155.

2. Z.A. Parszewski, M. Roszkowski, "Hydrodynamic theory of journal bearing at steady and unsteady motion", I.Mech.Eng. C87/71, 1971, pp.289.

3. D.X. Li, "Dynamic optimization of multi-bearing rotors in terms of the system configuration parameters", Ph.D. Thesis, 1990, The University of Melbourne.

4 D. Dowson, C.M. Taylor, "Fundamental aspects of cavitation in bearings".

5. Z.A. Parszewski, J.M. Krodkiewski, D.X. Li, "Oil pressure supply outlet influence on bearing characteristics", 3rd International Conference on Rotordynamics, 1990, pp.79.

Numerical Study of Cavitational Flows

A.G. Terentiev (*), K.E. Afanasiev (**)
(*) Chuvash State University, Chebocsary, USSR
(**) Kemerovo State University, Kemerovo, USSR

ABSTRACT

The paper analyses cavitation as a hydrodynamic phenomenon and various theoretical models of the cavitation . It formulates the main boundary problem of cavitation that can be solved as a singular function. Numerical solving of a cavitational flow problem is realized by using Riaboshinsky-model and based on a boundary integral equation for the velocity potential.

INTRODUCTION

Cavitation as a hydrodynamic phenomenon is connected with the appearance of ruptures on continuous liquid. The engineers had met it at the end of the last century in connection with high-speed marine propellers, hydrofoil boats, and low-head pumps and turbines. Approximately from those days they began experimental investigation of cavitation and supercavitating flows. First theoretical works have been published in the 30's. Most intensive theoretical studies of cavitation have been undertaken in the 60's. Up to the present many aspects of cavitation have received quite a thorough interpretation. A review of this subject can be found in the works of Sedov [1], Gurevich [2], Birkhoff and Zarantonello [3], Ivanov [4], Terentiev [5].

The present article deals with supercavitating flows. Cavitation theory is based on an experimentally justified hypothesis about the pressure in a cavity being minimal and constant. The determining non-dimensional parameter of the cavitating flow is cavitation number being defined as

$$K = \frac{P - P_o}{1/2\rho U^2}$$

where P is the ambient liquid pressure, P_o the cavity gas pressure, U the stream speed, and ρ the liquid density.

A simple reasoning shows that for an isolated body in the boundless liquid a stationary cavity with closed boundary mathematically is impossible. Usually this difficulty is overcome by formal presupposition with regards to the character of the flow at the end of the cavern. The analysis of numerical calculations and their comparison with experimental investigations confirm, that all the cavitational schemes at small number K give mutually co-ordinated and also with the experiment results. Thus selection of a certain cavitational scheme is mostly determined by subjective consideration and the possibility of getting comparatively simple solutions. It is multiple number of schemes and subjectivism in the selection that are the most weak points of the non-linear theory of supercavitating flows.

1. THE MAIN PROBLEM OF THE CAVITATION THEORY

For the flat flow it is possible to design uncontradictory wholly determined cavitational model from the main boundary problem by means of extension of a solution class included singularity [5]. If the flow is considered to be steady-state and potential and the fluid is thought to be of negligible gravity, the main problem of cavitating flow is equivalent to the following boundary problem [6]: for a given obstacle a velocity potential is to be found satisfying (1) the Laplace equation $\nabla^2\varphi = 0$ outside the obstacle R and the cavity R_1 (2) the kinematic condition on the wetted boundary C of the obstacle R and on cavity boundary C_1, (3) the dynamic condition $|\nabla\varphi| = V_o =$ const on the boundary C_1, (4) and the condition $|\nabla\varphi| = V < V_o$ outside the R_1.

Note that the first three conditions were formulated as back as 1868 by Helmholtz in connection with the dead zone streamlining of bodies and the last one by Brillouin in 1911. The problem formulated above is valid both for two-and three-dimensional flows. Our discussion is restricted to two-dimensional flows. In this case, the flow can be characterised by the complex potential $w(z) = \varphi + i\psi$, determined on the plane of $z = x + iy$. Since the C_1 boundary is previously unknown, it is desirable to use the function $\omega(w) = \ln(dw/dz) = \ln v - i\theta$. It is shown [5] that according to the Brillouin -

condition the function $\omega(w)$ at the end of a cavity has the singularity of $w^{-1/2}$. This singularity determines the class of functions in which the solution of a boundary value problem is to be found.

As an illustration of the above mentioned let us consider the cavitating flow along the rectangular step bottom (Fig. 1, a). If we accept that the complex potential $w = 0$ at the final point of the cavity, while $w = -a$ at the angle point of the step (Fig. 1, b), then the boundary problem for the function ω (w) has the following conditions: Re $\omega = \ln v_0$ on the interval $(-a, 0)$ and $\mathrm{Im}\omega = 0$ out of the interval. The solution of the mixed boundary problem, limited at the point $\varphi = -a$ and having the indicated singularity at the point $\varphi = 0$, is expressed by the function

$$\omega = A\sqrt{(w+a)/w} + \ln V_0 \qquad (1.1)$$

The parameters A and a are simply determined from the cavitation number K and the altitude of the step h:

$$A = -\ln\sqrt{1+K}, \qquad a = -2hV_0/\pi A\sqrt{1+K} \qquad (1.2)$$

The function $z(w)$ and all the geometrical characteristics of the flow are found from the differential equation :

$$dz = e^{-\omega(w)}\,dw \qquad (1.3)$$

In the framework of this theory the author with his colleagues have studied a wide range of problems on supercavitating flows [5,7,8]. Figs.2-4 represent numerical calculations for a wedge and a plate. Fig.2 shows the dependence of the drag coefficiant

$C_D = 2D / \rho U^2 T$ on the wedge half angle β for various

cavitation numbers. It is obvious that at $K=0$ the curves $C_D(\beta)$ first decrease, then increase with the

wedge angle diminution and obtain the finite values as K for β- 0. Fig 3 has the curves of dependence of the resulting force coefficient C_n on the angle of attack α at various fixed cavitation numbers. The solid lines correspond to the wholly supercavitation case, the dotted ones to the partial cavitation case (the cavity is shorter than the length of the flat plate). It is evident that at small fixed angles of attack for the same cavitation number theoretically there are possible three flow regimes :one with

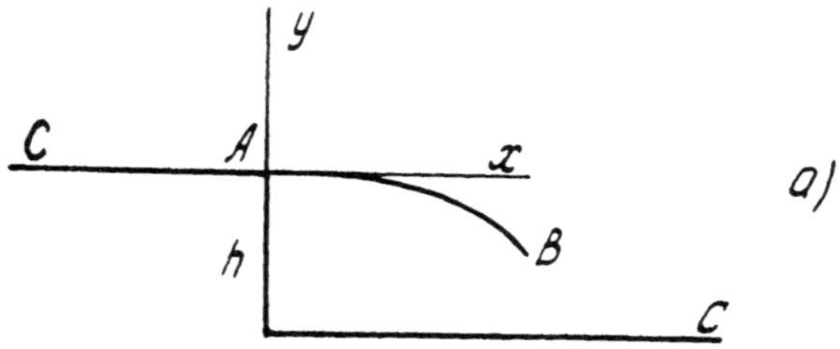

Fig. 1a. Cavity flow along the step bottom

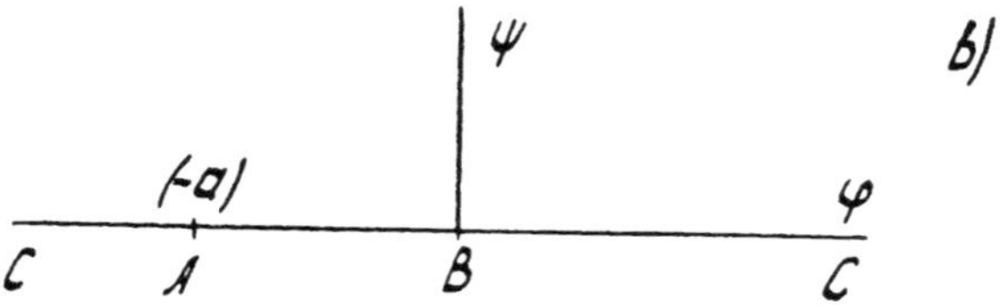

Fig. 1b. Complex potential map

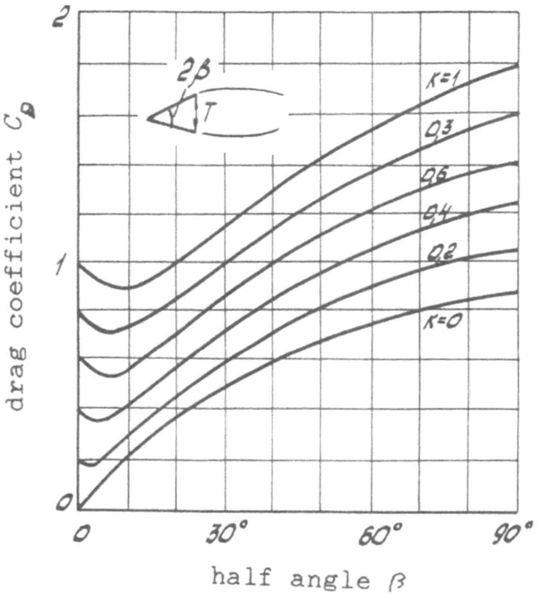

Fig. 2. Dependence of the drag coefficient $C_D = 2D/\rho U^2 T$ on the wedge half angle β and the cavitation number K

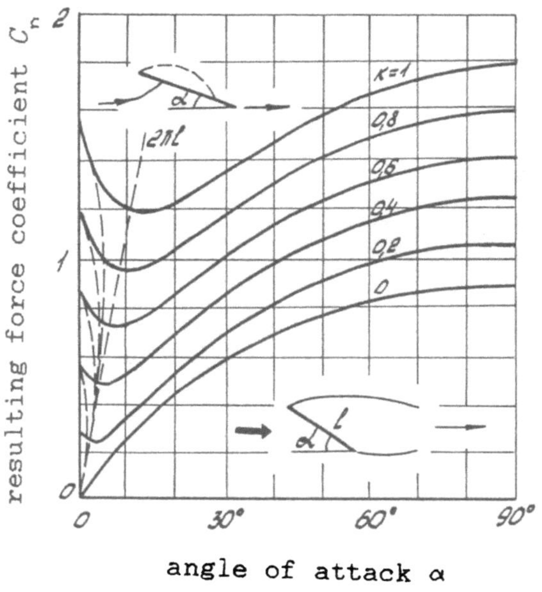

Fig. 3. Dependence of the resulting force coefficient C_n *on the angle of attack* α and the cavitation number *K*.

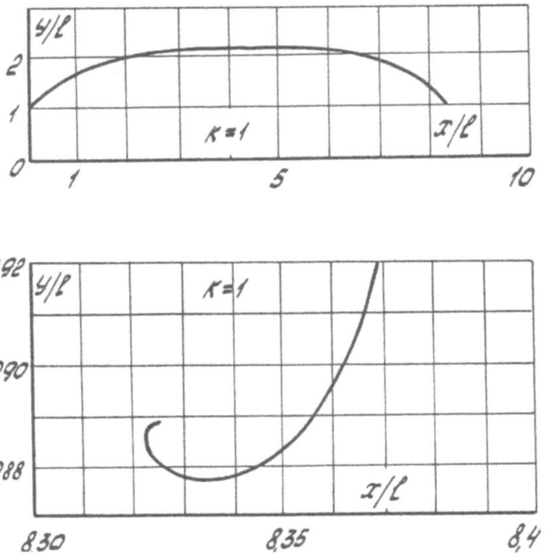

Fig. 4. Configuration of the upper boundary of the cavity.

super- and two with partial cavitation. Physically, only the flow with short cavity seems to be realized which has the lowest lift force value. Fig 4 shows the configuration of the upper boundary of the cavity behind the plate at the angle of attack $\alpha = 90^\circ$ and the cavitation number $K = 1$. In Fig 4b the cavity and the spiral are plotted on a larger scale. Due to the mentioned singularity the free boundaries at the cavity end do not close and form rapidly converging spirals with the flow occurring on a multivalence plane. However, non-single listing of the flow is of local character and it has no influence on integral characteristics of the streamlined body. Thus the results of theoretical calculations in a given scheme conform with the experimental data and could be applied in engineering to design real objects. Unfortunately, due to some singularity of solution at the end of cavity, in the framework of this theory it is difficult to study some concrete problems, specifically, cavitating flows with gravity effects, axially symmetric cavitating flows, and also to apply numerical methods. To these problems it is more convenient to use the cavitational models with artificial closing of the cavity by fictitious firm lines or planes [2]. In the present article it will be used the Riabouchinsky finite-cavity model closing of the cavity by the flat plate or disc.

2. NUMERICAL STUDY OF SUPERCAVITATING FLOWS

Numerical study can be carried out on the basis of integral equations. For the flat flow from Cauchy integral for the potential w in the outward region of the flow we have:

$$\frac{\alpha}{2\pi} w(z) = -\int_C \frac{w(\tau)}{\tau - z} d\tau \qquad (2.1)$$

where α inner angle between tangents at the angle point of a boundary (for the smooth boundary $\alpha = \pi$).

For the investigation of flat flows there could be applied the equation (2.1), the real and imaginary parts being singled out. Numerical solution of the equations for the known contours C can be carried out analogically which are used in the boundary element methods [9].

In cavitational tasks the problem is complicated by the fact that the boundary configuration a priori is not known. Jt must be determined in the process of solution.

Let C_1 be a boundary of a solid body, C_2 a cavity boundary, C_3 solid boundary at the end of the cavity, that is $C = C_1 \cup C_2 \cup C_3$. At the whole boundary C there must be observed following conditions:

$$\psi = 0 \text{ on } C$$
$$\varphi = s \text{ on } C_1 \qquad (2.2)$$

where is the absciss of the cavity boundary. The second condition (2.2) conforms with the dynamic condition $P=P_0$ =const, that is the velocity at the cavity boundary is supposed to be constant and equal to 1. In this case the ambient velocity is a non-dimensional value and expressed through the cavitation number:

$$U = 1/\sqrt{1+K}$$

For numerical calculations it is reasonable instead of the cavitation number K to set the length of the cavity L, the cavitation number being determined after the solution from (2.1). For the determination of the cavity boundary there apply various approaches. For instance, to fix abscisses of the nodes, ordinates being determined from nonlinear system of equations got from one of the conditions (2.2).

Quite an effective approach is one of consequent approximations, based on the correction of the cavity boundary from the condition of maintaining the same quantity of aflowing liquid. Let the boundary of cavity $C_2^{(n)}$ on n-th step be known and determined by the function $\tau(s)$, then the whole boundary $C^{(n)}$ will also be known and an integral equation (2.1) can be studied numerically according to the conditions:

$$\psi = 0 \quad \text{on } C_1 \text{ and } C_3$$
$$\varphi = s \quad \text{on } C_2 \qquad (2.3)$$

There the flow function determined $\psi <> 0$ on $C_2^{(n)}$, that is liquid is flowing through $C_2^{(n)}$. If we approximately take the velocity of liquid between C_2 and $C_2^{(n)}$ equal to 1 then the condition

$$\delta = \psi \qquad (2.4)$$

must be observed in any point of the boundary $C_2^{(n)}$, where δ the distance on the outer normals between boundaries $C_2^{(n+1)}$ and $C_2^{(n)}$. Thus, a new position of the boundary $C_2^{(n)}$ is determined. The process comes to an end at carrying out the condition

$$\max_{s \in C_2} |\delta| < \varepsilon \qquad (2.5)$$

The equation (2.1) can be transformed into [10]

$$\frac{\alpha}{2\pi}w(z)=Uz-\int_c w(s)G_n(z,s)\,ds+\int_c w_n(s)G(z,s)ds \qquad (2.6)$$

where $G(z,s) = \frac{1}{2\pi}\ln\frac{1}{r}$, $r=|z-s|$

For the stationary flow with kinematic conditions $\varphi=0$ ($\psi=0$) from (2.6) follow :

$$\frac{\alpha}{2\pi}\varphi = Ux - \int_c \varphi\, G_n(z,s)ds \qquad (2.7)$$

$$U\,y = \int \frac{d\varphi}{ds} G\,(z,s)ds \qquad (2.8)$$

The first equation (2.7) has been used for calculation of cavitating flows by A.N. Ivanov [4], the second (2.8) by L.G. Guzevsky [11]. The same equations can be composed also for axisymmetric case. The Green's function G and its normal derivative G_n for the axisymmetrical potential φ are known widely [9], but the function G in equation (2.8) for axisymmetric flow is different from the Green's function [11]. The most thoroughly investigated axisymmetric cavitational flow is described by L.G. Guzevsky [11-13].

The aforesaid iteration approach of calculation of the cavity boundary used directly on integral equation (2.6) for the velocity potential φ can also be used for axisymmetric cavitating flows. In this case instead of the condition (2.4) there should be used the condition [14]

$$\delta = \frac{1}{y^{(n)}} \int_0^s y^{(n)}(s')\,\varphi_n\,ds' \qquad (2.9)$$

The integral equation (2.6) has been calculated, employing linear boundary elements [9,10]. Many methods have been approved by us at the Chuvash State University. The program packager enable to calculate the cavitating flows virtually of any configuration.

REFERENCES

1. Sedov, L.I. The flat Problem of Hydrodynamics and Aerodynamics, Nauka, Moskow, 1966 (in Russian).
2. Gurevitch, M.I. The Theory of Ideal Jets, Nauka, Moskow, 1979 (in Russian).
3. Birkhoff, G. and Zarantonello, E.H. Jets, Wakes and Cavities, Academic Press Inc. Publishers, New York, 1957.
4. Ivanov, A.N. Hydrodynamic of Supercavitating

Flows, Sudostrojenie, Leningrad, 1980 (in Russian).
5. Terentiev,A.G. Mathematic Problems of Cavitation, Chuvash State University, Cheboksary, 1981 (in Russian).
6. Birkhoff,G Hydrodynamics, Princeton University Press, Princeton, 1960.
7. Terentiev,A.G. Theoretische Untersuchungen uber die Umstromung von Profilen bei Kavitation, Maschinen-bautechnik, Berlin, Vol. 12, pp 564-567, 1984.
8. Kuznetzov,Yu,V. and Terentiev,A.G. Symmetrical Cavitation Flow around a Wedge, situated between Parallel Walls and Free Boundaries, Book: Jets and Cavitation Flows, Shuvash State University, Cheboksary, pp54-67, 1978 (in Russian).
9. Brebbia,C.A. Telles,J.C.F. and Wrobel,L.C. Boundary Element Techniques, Springer-Verlag, Berlin and New York, 1984.
10. Terentiev,A.G and Afanasiev,K.E. Numerical Methods in Hydrodynamics, Chuvash State University, Cheboksary, 1987 (in Russian).
11. Guzevsky,L.G. Calculation of Axisymmetric Flows with Free Surfaces, Rept. USSR Acad. Sci., Vol. 225, N2, pp 269-271, 1975 (in Russian).
12. Gusevsky,L.G Numerical Analyses of Cavitating Flows, Preprint N40-79, IT Ph USSR Acad. Sci., Novosibirsk, 1979 (in Russian).
14. Krasnov,V.K. and Kuznetzov,Yu.V. Application of Boundary Integral Equation Method for Calculation of Axisymmetric and Flat Cavitating Flows in Tunnel, Book: Actual Problems of Hydrodynamics, Chuvash State University, Cheboksary, pp 71-75, 1989 (in Russian).

SECTION 4: FREE SURFACE FLOW

Free Surface Flow with Surface Tension

S. Räsänen

Turku Institute of Technology, SF-20700 Turku, Finland

ABSTRACT

A numerical model for ideal fluid flow with a free surface and with surface tension is presented. Essential features of the method are a smooth free surface description with piecewise Ferguson curves, the use of velocity components as basic unknowns and a solution procedure where the determination of the position of the surface is treated as an optimum design problem. Numerical results are given for two cases where comparisons with analytical results are possible.

INTRODUCTION

Ideal fluid flow problems have traditionally been solved in terms of the stream function or the velocity potential. This approach still continues although in many cases the direct use of velocity components as basic unknowns would be more advantageous. Free surface flow is especially suited for a velocity treatment because of the complicated boundary conditions.

In this paper a velocity based model for free surface flow with surface tension is described. The model - without surface tension - has earlier been successfully applied to flows over weirs (Räsänen and Salonen[1]), to flows with wakes (Räsänen[2]) and to flows with floating bodies (Räsänen and Salonen[3]).

Accurate description of the form of the free surface is essential in a numerical model for free surface flow. The curvature of the surface has a significant influence on the flow. The inclusion of surface tension in the model provides a test for the capabilities of the surface description because the importance of accurate curvature evaluation obtains an emphasized role.

GOVERNING EQUATIONS

Two-dimensional steady incompressible and irrotational fluid flow with a free surface is considered. The governing equations in terms of the velocity components u and v are the equation of continuity

$$\frac{\partial u}{\partial x} + \frac{\partial v}{\partial y} = 0 \tag{1}$$

and the irrotationality condition

$$\frac{\partial v}{\partial x} - \frac{\partial u}{\partial y} = 0. \tag{2}$$

On the free surface the fluid pressure p must satisfy two conditions. The first one is the Bernoulli equation

$$H = y + \frac{p}{\rho g} + \frac{q^2}{2g}. \tag{3}$$

H is the given total head and y the upward coordinate of the surface, both measured from the same horizontal datum line. ρ is the fluid density, g the acceleration due to gravity and q the flow speed. The second condition is

$$\bar{p} - p = \sigma\kappa \tag{4}$$

where $\bar{p}$ is the atmospheric pressure, σ the surface tension and κ the curvature of the surface, taken positive when the center of curvature is situated above the surface. The two pressure boundary conditions are the main source of difficulties in this kind of a problem. Due to the emerging non-linearity an iterative solution procedure must be devised.

NUMERICAL MODEL

The shape of the free surface is defined through the so called geometry points (see Figure 1). The position vectors of two successive geometry points are $\mathbf{r}_1$ and $\mathbf{r}_2$. Between the points the surface is described by the mapping

$$\mathbf{r}(\xi) = N_1(\xi)\mathbf{r}_1 + N_2(\xi)\mathbf{r}_1' + N_3(\xi)\mathbf{r}_2 + N_4(\xi)\mathbf{r}_2'. \tag{5}$$

$\mathbf{r}'_1$ and $\mathbf{r}'_2$ are the derivatives of the position vectors with respect to the parameter ξ, $0 \leq \xi \leq 1$. The functions N_i are the Hermitean cubic polynomials. The values of the derivatives $\mathbf{r}'_1$ and $\mathbf{r}'_2$ are calculated according to the equations

$$\mathbf{r}'_1 = \frac{d}{\mathbf{t}_1 \cdot \mathbf{e}} \mathbf{t}_1, \quad \mathbf{r}'_2 = \frac{d}{\mathbf{t}_2 \cdot \mathbf{e}} \mathbf{t}_2. \tag{6}$$

where $\mathbf{t}_1$ and $\mathbf{t}_2$ are the unit tangent vectors to the surface at the geometry points, $\mathbf{e}$ is the unit vector in the direction of the chord connecting the geometry points and d is the corresponding chord length. The finite element mesh is made to follow the surface as shown in Figure 1.

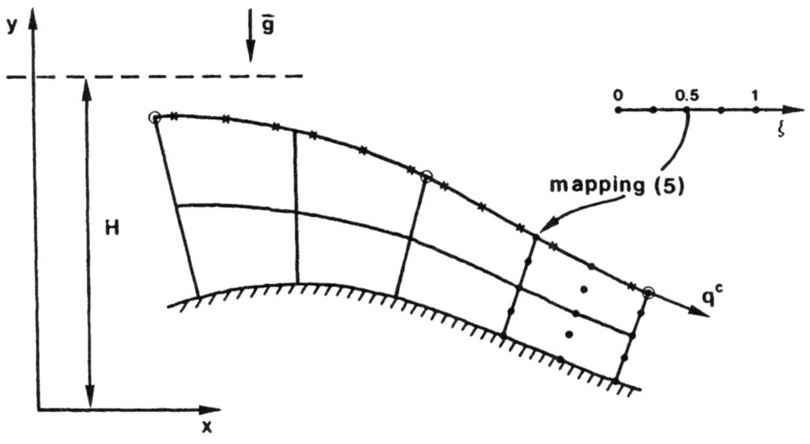

Figure 1. Geometry of the problem and some notations. o geometry point , • node , × integration point.

The smooth free surface description in Equation (5) has been selected on the basis of physical considerations (Räsänen and Salonen[4]). The appearence of sharp corners on the surface should be avoided because the flow speed would attain unrealistic values at the corners. The curvatures calculated from Equation (5) are not in general continuous at a geometry point, but this does not present any problems because in the numerical model the curvatures are evaluated at integration points situated between the geometry points.

The velocity determination is based on the functional

$$\Pi(u,v;\lambda_1,\lambda_2) = $$
$$= \frac{1}{2}\int \left[\left(\frac{\partial u}{\partial x}+\frac{\partial v}{\partial y}\right)^2 + \left(\frac{\partial v}{\partial x}-\frac{\partial u}{\partial y}\right)^2\right] dA + \int \left[\lambda_1\left(\frac{\partial u}{\partial x}+\frac{\partial v}{\partial y}\right)+\lambda_2\left(\frac{\partial v}{\partial x}-\frac{\partial u}{\partial y}\right)\right] dA \qquad (7)$$

where a conventional least square part is supplemented with two Lagrange multiplier terms. The velocity components are approximated with a biquadratic C^0-continuous representation and the Lagrange multipliers λ_1 and λ_2 with a constant C^{-1} -continuous representation. The nodal velocities and the Lagrange multipliers are solved from the system of equations

$$\mathbf{Ka} = \mathbf{P} \qquad (8)$$

obtained by requiring the functional in Equation (7) to be stationary. As boundary conditions to Equation (8) the direction of the velocity is given at the nodes on the boundaries of the domain. On the free surface and on the bottom the velocity is set to be tangential to the boundary. A linear velocity direction distribution is assumed at the inflow and outflow sections. At one point on the free surface - the so called control point - the magnitude of the velocity is given. It is assumed that the fluid pressure fulfills the condition of Equation (4) at the control point. The magnitude of the velocity is then obtained from the Bernoulli equation (3):

$$q^C = \left[2g(H-y^C) - \frac{2}{\rho}(\bar{p}-\sigma\kappa^C)\right]^{1/2}. \qquad (9)$$

The superscript C refers to the control point.

The iterative method for determining the position of the free surface is constructed by minimizing the cost function

$$W = \frac{1}{2}\int (p-\bar{p}+\sigma\kappa)^2 ds \qquad (10)$$

or , in the numerical model, its pointwise counterpart

$$W \approx \frac{1}{2}\sum_{i=1}^{n} w_i(p_i-\bar{p}+\sigma\kappa_i)^2. \tag{11}$$

The summation is over the integration points on the free surface and w_i is the integration weight. The cost function is minimized with respect to the changes of the so called design parameters. The changes are denoted by Δb_j. The design parameters with respect to a current free surface position are selected to be a change in the position of a geometry point in the direction perpendicular to the surface and a change in the direction angle of the tangent to the surface at a geometry point. At the ends of the calculation domain the geometry points are moved in the fixed directions of the inflow and outflow boundary lines.

Changes in design parameters affect the values of pressures and curvatures. Their new values at an integration point due to changes in all the design parameters are

$$\begin{aligned} p_i^{new} &\approx p_i^{old} + \sum_j \frac{\partial p_i}{\partial b_j}\Delta b_j = p_i^{old} + \sum_j c_{ij}\Delta b_j \\ \kappa_i^{new} &\approx \kappa_i^{old} + \sum_j \frac{\partial \kappa_i}{\partial b_j}\Delta b_j = \kappa_i^{old} + \sum_j c_{ij}^{\kappa}\Delta b_j. \end{aligned} \tag{12}$$

Quantities c_{ij} and c_{ij}^{κ} defined in Equation (12) are called pressure and curvature sensitivities, respectively. The minimizing condition of the cost function in Equation (11)

$$\frac{\partial W}{\partial \Delta b_j} = 0 \quad , \quad j = 1, 2, \ldots \tag{13}$$

leads to a linear system of equations for the unknowns Δb_j.

The sensitivities are calculated by making a small predefined change in every design parameter in turn and by employing the approximate relations

$$c_{ij} \approx \frac{\Delta p_i}{\Delta b_j} \quad , \quad c_{ij}^{\kappa} \approx \frac{\Delta \kappa_i}{\Delta b_j}. \tag{14}$$

The curvature change is a mere geometric quantity and it is easily determined from the surface description in Equation (5). The pressure change is obtained

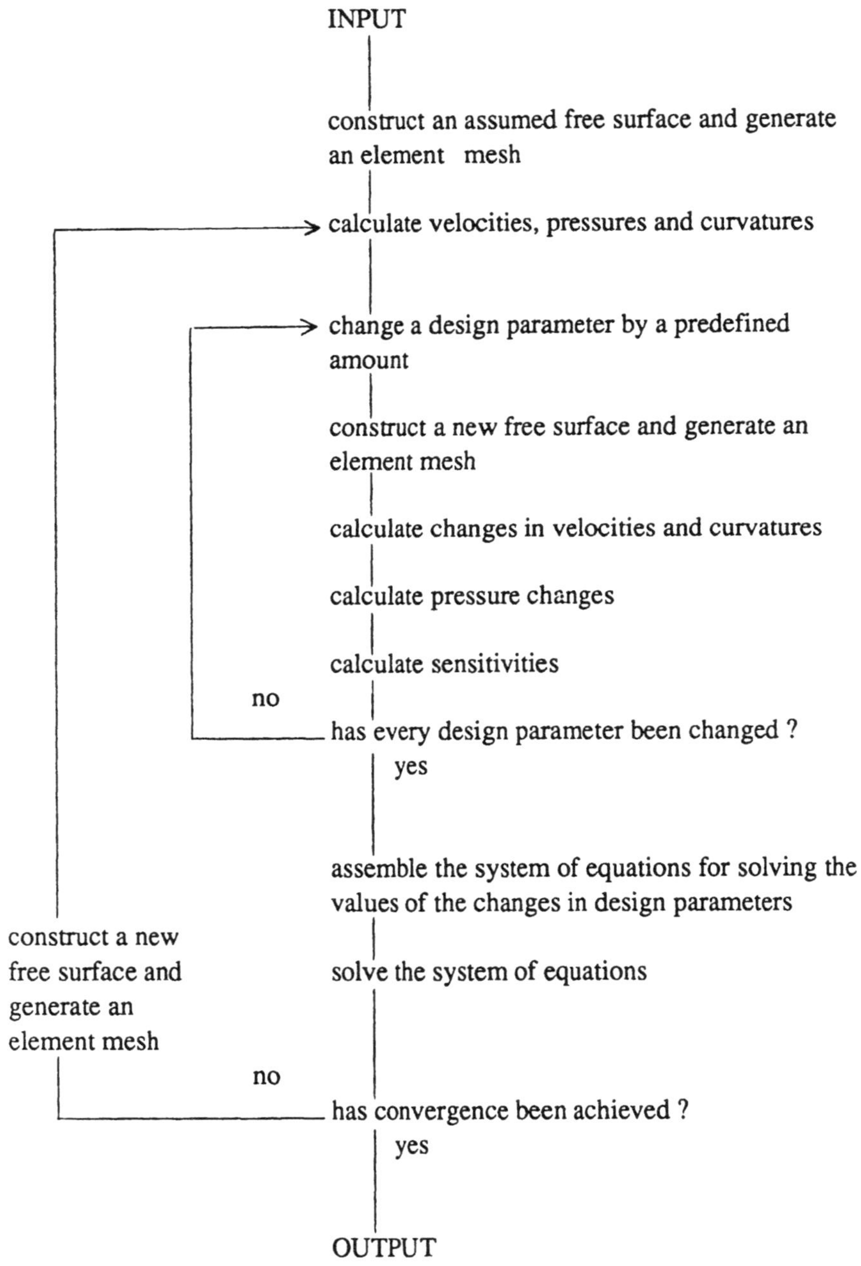

Figure 2. The procedure for solving a free surface problem.

by differentiation from the Bernoulli equation:

$$\Delta p_i \approx -\rho g \Delta y_i - \rho q_i \Delta q_i \,. \tag{15}$$

Changes in speeds needed in Equation (15) are calculated from the approximate equation

$$\mathbf{K}\Delta\mathbf{a} = \Delta\mathbf{P} - \Delta\mathbf{K}\mathbf{a} \tag{16}$$

derived by differentiating equation (8) with respect to a design parameter.The structure of the numerical model is shown schematically in Figure 2.

NUMERICAL EXAMPLES

Nine-noded Lagrangean isoparametric elements have been used in the calculations. Three point Gaussian quadrature has been employed in connection with the cost function in Equation (11). The predefined changes of the design parameters needed for calculating the sensitivities have been selected so that $\Delta b_j/D = 0.01$ for a change in the position of the geometry point (D is the mean depth in a problem) and the change in the direction angle of the tangent is adjusted to produce roughly the same maximum change in the position of the surface. The convergence criterion has normally been $\max|\Delta b_j|/D < 0.001$. The given value of the atmospheric pressure $\bar{p}$ has been zero in all the calculations.

First, the static problem of a cylindrical drop of liquid resting on a plane surface is considered. This extreme case puts heavy demand on the geometric capability of the free surface representation as the difference $\bar{p}$-p will be large and thus the accuracy in the determination of the curvature κ in Equation (4) becomes important. In a static problem the program for calculating the free surface must be modified so that the velocity at the control point is set to zero.

The governing equations have been obtained by modifying the equations for a circular drop presented in Flügge[5] .The notations are shown in Figure 3. The equations are

$$\frac{d \sin\phi}{dx} = \frac{\rho g}{\sigma} z \quad , \quad \frac{dz}{dx} = \tan\phi \,. \tag{17}$$

These are transformed into a dimensionless form by setting

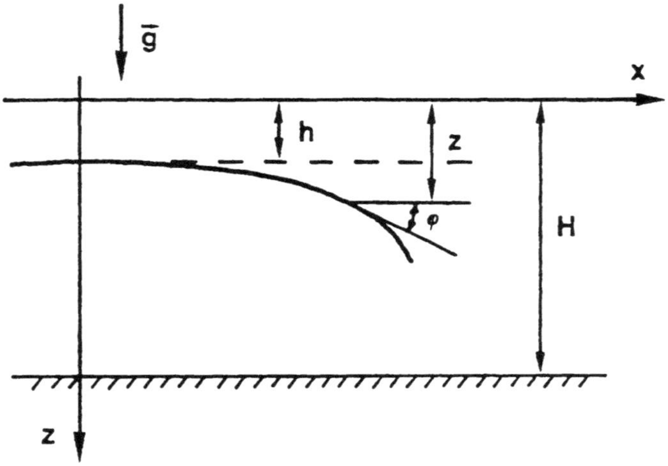

Figure 3. A cylindrical drop on a plane surface. Notations.

$$a^2 = \frac{\sigma}{\rho g} \quad , \quad \eta = \sin\phi \quad , \quad \chi = \frac{x}{a} \quad , \quad \zeta = \frac{z}{a} \tag{18}$$

which leads to equations

$$\frac{d\eta}{d\chi} = \zeta \quad , \quad \frac{d\zeta}{d\chi} = \frac{\eta}{\sqrt{1-\eta^2}} \, . \tag{19}$$

In terms of the dimensionless quantities the boundary conditions are

$$\eta = 0 \quad , \quad \zeta = \frac{h}{a} \quad , \quad \text{when} \quad \chi = 0 \, . \tag{20}$$

Differential equations (19) with the boundary conditions of Equation (20) can be solved numerically. An IMSL-routine[6] based on Runge-Kutta iteration has been used. This very accurate reference solution is called the analytical one in the following.

Calculations have been performed for the values H = 1 m, h = 0.1 m $g = 9.81\ ms^{-2}$ and $\sigma/\rho = 5\ m^3s^{-2}$. The shape of the drop is shown in Figure 4. No differences could be detected between the analytical and calculated curves and thus only the finite element solution is drawn. The result has been

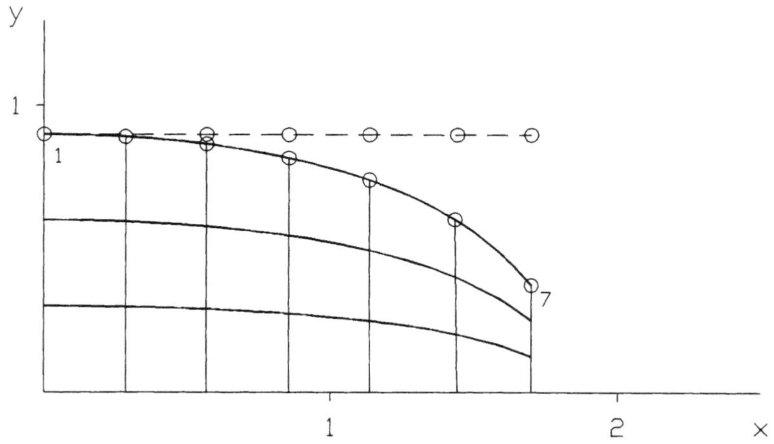

Figure 4. Result for the cylindrical drop. 7 geometry points. 18 elements.
Dashed line: starting surface; solid line: result.

obtained in four iterations when the calculations were started with a horizontal surface and with a uniform geometry point distribution. As boundary conditions to Equation (13) the position and tangent vector direction were fixed at the point x = 0 m (point 1 in Figure 4). Point 1 has also been the control point.

The analytical and finite element values of z and ϕ are shown in Table 1. The error in the finite element results for z and ϕ is at the most 0.3 %.

The second example is the dynamic case without gravity where the fluid flows along concentric circles in a horizontal plane. In polar coordinates

Table 1. Cylindrical drop. Analytical and finite element results.

	x/m	z/m		ϕ	
		anal.	FEM	anal.	FEM
1.	0.000 00	0.100 00	0.100 00	0.000 00	0.000 00
2.	0.284 00	0.108 02	0.108 05	0.057 23	0.057 42
3.	0.566 97	0.133 36	0.133 45	0.123 65	0.124 00
4.	0.853 78	0.181 31	0.181 54	0.212 23	0.212 80
5.	1.132 65	0.258 56	0.259 01	0.335 56	0.336 45
6.	1.431 57	0.396 53	0.397 40	0.544 16	0.545 66
7.	1.700 00	0.625 33	0.627 14	0.894 12	0.896 83

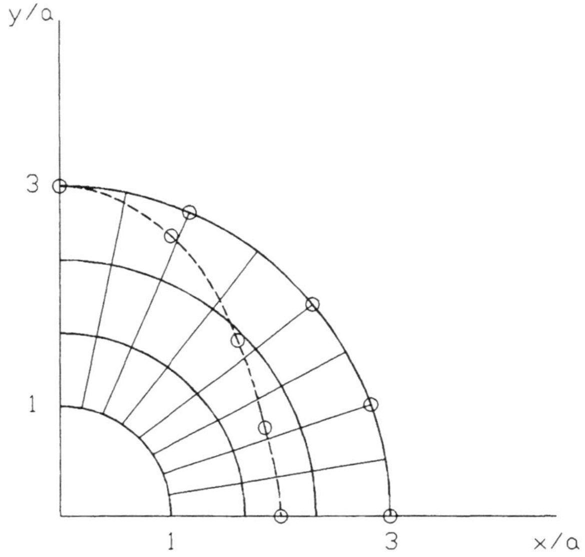

Figure 5. Flow with surface tension and without gravity. 5 geometry points.24 elements. Dashed line: starting free surface; solid line: result.

the velocity components are $u_r = 0$ and $u_\theta = A/r$. The Bernoulli equation is now written most conveniently in the form

$$\frac{p}{\rho} + \frac{1}{2}q^2 = C \tag{21}$$

where C is a known constant. The calculations have been performed in the domain $a \leq r \leq 3a$, $a = 2$ m, with given values of C (= 1 m^2/s^2), σ and ρ ($\sigma/\rho = 17/3\ m^3s^{-2}$) corresponding to the value 2 m^2s^{-1} of the constant A. The result shown in Figure 5 has been obtained in eight iterations from the guessed free surface shown in the figure. The following inflow conditions are assumed to be known at the section x/a = 0: the position and tangent vector direction of the surface together with the magnitude of the velocity at the point on the surface (the control point).

The largest error in the calculated free surface positions occurs at the outflow section where the error is -0.30 %. The error in the tangent directions is evenly distributed among the points on the surface. Maximum difference between the calculated and analytical values is 0.0049 rad. The errors in the calculated pressures are also very small, 0.26 % at the most. If the same flow is treated with an element mesh having only eight elements, but the same number of geometry points (five) the results are still very good. The largest error in the free surface position rises only to the value -0.69 %.

It can be noted that the element meshes do not become distorted although the geometry points are moved perpendicularly to a current free surface. This desirable behaviour is clearly seen from Figures 4 and 5 where the original positions of the geometry points have been marked.

CONCLUSIONS

The excellent results obtained confirm the applicability of the proposed model. The two special cases treated in the numerical examples demonstrate especially the properties of the free surface description. It is expected that the model gives good results also in more complicated flows where analytical results are not available.

ACKNOWLEDGEMENT

The writer wishes to thank Associate Professor E.-M. Salonen for his valuable comments on the manuscript.

REFERENCES

1. Räsänen, S.M. and Salonen E-M. A u,v-Formulation for Free Surface Flow, in Numerical Methods in Laminar and Turbulent Flow (Ed. Taylor, C., Olson, M.D., Gresho, P.M. and Habashi, W.G.), pp. 1393-1401, Proceedings of the 4th Int. Conf. on Numerical Methods in Laminar and Turbulent Flow, Part 2, Swansea, U.K., 1985. Pineridge Press, Swansea, 1985.
2. Räsänen, S.M. A Velocity Formulation for Flow past a Symmetric Profile with a Wake, International Journal for Numerical Methods in Fluids, Vol. 9, pp. 1113-1119, 1989.
3. Räsänen, S.M. and Salonen, E-M. Free Surface Flows with Floating Bodies, Journal of Engineering Mechanics, Vol. 116, pp. 1305-1316, 1990.
4. Räsänen, S.M. and Salonen E-M. A Model for Understanding Free Surface Flow, in Numerical Methods for Non-Linear Problems (Ed. Taylor, C., Hinton, D. and Owen, D.R.J.), pp. 849-860, Proceedings of the 2nd Int. Conf. on Numerical Methods for Non-Linear Problems, Barcelona, Spain, 1984. Pineridge Press, Swansea,1984.
5. Flügge,W. Stresses in Shells, Springer-Verlag, New-York,1966.
6. IMSL FORTRAN Library, International Mathematical & Statistical Libraries, Inc.

Dam-Break Wave Routing by Shock Fitting Method

F.C.B. Mascarenhas, R.C. Vieira da Silva

Center of Technology, Federal University of Rio de Janeiro, UFRJ, Brazil

ABSTRACT

The objective of this paper is to present a shock fitting model for routing long waves in rivers with a discontinuity on the free surface which behaves as a moving boundary. Outside the shock zone the flow is governed by the Saint-Venant equations, while in the shock section the Hugoniot-Rankine equations are applied. An algorithm based on the method of characteristics is used to locate the discontinuity section and the correspondent values of the model dependent variables are used as downstream moving boundary conditions. This model has been applied to the study of the wave caused by a hazardous breach failure of the Serra da Mesa dam, located in the Tocantins river in the Amazon basin, Brazil. An instantaneous vertical total breach is assumed, so a discontinuity must appear soon after the failure due to the high values of the dam height and storage capacity. The shock fitting model showed to be the most suitable for that situation when compared to conventional routing methods.

INTRODUCTION

The routing of a wave caused by a dam failure can be distinguished from the natural flood waves by the appearing of a shock in the wave front. This phenomenon can affect the choice of the routing process because the resulting moving boundary causes the development of a discontinuity on the free surface. The Saint-Venant equations which describe the routing of long waves in rivers and channels are no longer valid at the shock zone and are replaced by relations known as Hugoniot-Rankine equations.

The method of characteristics is the most suitable for the solution of problems related to the routing of waves with discontinuities because it allows what in mathematical language is called front tracking approach. In this work an algorithm based on the method of characteristics is used for the shock location where the Hugoniot-Rankine equations are applied, while at the remaining sections the Saint-Venant equations apply.

The developed model is applied to the study of a hazardous failure of Serra da Mesa dam, located in the Tocantins river, in the Center-West of Brazil. This dam, 130 m high and a projected storage of 33,000 x 10^6 m^3, has aspects that allow the prediction of a shock wave appearing soon in the initial sections of the routing reach. The adopted hypothesis of failure predicts the development of a total vertical instantaneous breach in the sense that its evolution time is negligible in comparison with the emptying time of the reservoir, and the breach erosion develops faster in the vertical direction than in the lateral one.

The Mathematical Model

The mathematical model for the routing process of a wave caused by the failure of a dam can be described by the following set of equations and boundary conditions :

- Saint-Venant equations

These equations, which are valid at all the flow sections where there is no occurrence of shocks, written in the so called non divergent form, are

$$\frac{\partial h}{\partial t} + h \frac{\partial v}{\partial x} + v \frac{\partial h}{\partial x} = 0 \tag{1}$$

$$\frac{\partial v}{\partial t} + v \frac{\partial v}{\partial x} + g \frac{\partial h}{\partial x} = g (S_0 - S_f) \tag{2}$$

where

x,t - independent variables of space and time
h,v - mean depth and velocity at the flow section
g - gravity acceleration
S_0 - average fluvial bottom slope
S_f - energy line slope of the flow

- Hugoniot-Rankine equations

These equations, valid only at the shock section, for which the generical configuration is showed in the figure 1, are :

$$v_1 = v_2 + (A_1 - A_2)\left[\frac{g}{A_1 A_2}\,\frac{(A_1\bar{h}_1 - A_2\bar{h}_2)}{(A_1 - A_2)}\right]^{1/2} \tag{3}$$

$$\xi = \frac{A_1\ v_1 - A_2\ v_2}{A_1 - A_2} \tag{4}$$

where

v_1,v_2 - conjugate velocities upstream and downstream the wave front
$\bar{h}_1,\bar{h}_2$ - centroids of the upstream and downstream wetted areas of the wave front
A_1,A_2 - upstream and downstream wetted areas of the wave front
ξ - propagation celerity of the wave front

The fundamental condition for the solution of the dam-break wave equations is the initial hydrograph Q(0,t) known as failure hydrograph. This hydrograph can be obtained through structural hypothesis made about the failure process and the emptying time of the reservoir. In the present study it was made the hypothesis of failure associated with a total instantaneous breach of width b smaller than the dam width B. The failure hydrograph is assumed to be triangular with maximum discharge given by the expression :

$$Q_{max} = \frac{8}{27} (B/b)^{1/2} \, b \sqrt{g} \quad H^{3/2} \tag{5}$$

where H is the mean reservoir depth

The Numerical Scheme of Solution

In the mathematical model it was adopted a numerical scheme that uses the Hugoniot-Rankine equations together with the method of characteristics (eg. Abbott [1]) for the front wave location at each time step. One of the dependent variables at the upstream section to the front is used as downstream boundary condition for the routing between the dam and the wave front, which is made through an implicit finite difference scheme for the discretization of the Saint-Venant equations. Because the wave front moves in downstream direction the problem is of the moving boundary type and the re-discretization of the routing reach is needed for each shock position. Mathematically, supposing that the initial conditions are known at time t=0 as well the failure hydrograph as upstream boundary condition one has to determinate five unknowns in the equations (3) and (4); v_1, h_1, ξ, v_2 and h_2. It can be showed (eg. Stoker [9]) that the discontinuity results from the crossing of two positive characteristics, one upstream and another downstream to the wave front, and there is only one negative characteristic downstream to that front. After the transformation of the equations (1) and (2) into their characteristics forms, the application of two of them (one positive and the other negative) at the downstream section to the front gives, at a considered time step, the values of the variables v_2 and h_2 (or A_2). The equations (3) and (4) together with a positive characteristic equation downstream to the front are used for the determination of v_1, h_1 (or A_1) and ξ. Using the Saint-Venant equations adjusted for the flood plain situation (eg. French [4]) the following characteristics equations are obtained (eg. Chen and Armbruster [2]) :

$$(dx/dt)_{\pm} = 0.5\ (1 + \alpha)\ v \pm 0.5\ R_1 \tag{6}$$

$$(R_2 \pm R_3)\,\frac{dh}{dt} + \frac{dv}{dt} = g\,(S_0 - S_f) - (R_2 \pm R_3)\,\frac{A_x^h}{B} \tag{7}$$

where

$$R_1 = 0.5\sqrt{(1 - \alpha)^2\ v^2 + 4\ g\ \alpha\ D} \tag{8}$$

$$R_2 = -\ 0.5\ (1 - \alpha)\ v\ /(\alpha D) \tag{9}$$

$$R_3 = \sqrt{\left[0.5\ (1 - \alpha)\ v\ /\ (\alpha D)\right]^2 + g\ /(\alpha D)} \tag{10}$$

α - relation between the main channel and surface widths
D - hydraulic depth (A/B)
A_x^h - rate of change of A with x for h constant

For the solution of the equations (6) and (7) is used a linear explicit scheme with specified time interval, described by Cunge, Holly and Verwey [3]. In this scheme, and according to Figure 2, it can be written :

$$x_L = x_P - (v_L + \alpha\ \lambda_L^+\ D_L)\Delta t \tag{11}$$

$$v_P + \lambda_L^+\ h_P = v_L + \lambda_L^+\ h_L + F_L^+\ \Delta t \tag{12}$$

$$x_R = x_P - (v_R + \alpha\ \lambda_R^-\ D_R)\Delta t \tag{13}$$

$$v_P + \lambda_R^- h_P = v_R + \lambda_R^- h_R + F_R^- \Delta t \tag{14}$$

where

$$\lambda^{\pm} = (R_2 \pm R_3) \tag{15}$$

$$F^{\pm} = g(S_0 - S_f) - \alpha \, \lambda^{\pm} \, v \, A_x^h / B \tag{16}$$

The position of the wave front for each time step can be approximated by :

$$x_S(t+1) = x_S(t) + \xi(t) \, \Delta t \tag{17}$$

The equation (17) defines the new routing reach that is then solved by the Preissmann [7] implicit finite difference scheme, where any dependent variable f and its derivatives are given by:

$$f(x,t) \cong (\theta/2) \, (f_{i+1}^{j+1} + f_{i+1}^{j}) + \left[(1-\theta)/2\right](f_i^{j+1} + f_i^{j}) \tag{18}$$

$$\partial f/\partial t = (f_{i+1}^{j+1} - f_i^{j+1} + f_{i+1}^{j} - f_i^{j})/(2\Delta t) \tag{19}$$

$$\partial f/\partial x = (1/\Delta x) \left[\theta(f_{i+1}^{j+1} - f_{i+1}^{j}) + (1-\theta)(f_i^{j+1} - f_i^{j})\right] \tag{20}$$

where θ is a weighing factor between 0.6 and 1.0 for numerical stability, according to Liggett and Cunge [5], and i and j are indicial notations for time and space. The Newton-Raphson iterative method is used to solve the discretized system where the unknowns are the variables with subscripts i+1.

THE CASE STUDY

The developed computational model related to the numerical scheme has been applied to the hypothetical failure of Serra da Mesa dam, in early stage of construction in the State of Goias, Brazil. The dam is about 130 m high and its maximum storage capacity is of 33,000 x 10^6 m^3. The routing reach is limited downstream by the future backwater of Canabrava reservoir, located at about 64 km from the Serra da Mesa dam. The downstream portion of the Tocantins river below the dam presents an approximated regular topography and its sections can be considered having asymmetrical trapezoidal forms at some flood plain zones. The average bottom slope is about 0.00054 and the section widths vary between 200 and 400 meters. The available topographical data are sufficient for the estimative of the global flood plain caused by the dam failure. At near sites to the dam the geometrical characteristics are known with more detail, allowing the analysis of the free surface profiles at those sections for small time intervals after the failure, by the shock fitting approach. The total simulation time corresponds to the reservoir emptying time, determined by the failure hydrograph. It is important to be pointed out that the total and instantaneous failure hypothesis favours the safety criterion and the formation of discontinuous wave fronts with shocks at the sections nearest to the dam. This fact justifies the adopted mathematical front tracking approach.

RESULTS ANALYSIS

The free surface profiles obtained by the shock fitting approach are shown in Figure 3, with time intervals of 12 seconds between each of them, for the nearest sections to the Serra da Mesa dam. An important result given by the numerical scheme is the location of the wave front dissipation section and its time of occurrence after the dam failure. In the present application it was verified that the wave front is dissipated about 9 km downstream to the dam after approximately 6.5 minutes from its failure. After this dissipation the flow over the entire reach becomes gradually varied and the dependent variables can be obtained only through the solution of the Saint-Venant equations discretized by the Preissmann scheme. The flow depth hydrographs for various sections downstream the dam , obtained by the adopted finite difference scheme, are showed in Figure 4. With the aid of some topographic charts of the region the global flood plain caused by the dam failure is shown in Figure 5, estimated with the maximum depths values in the hydrographs of Figure 4. The free surface profiles near the dam (Figure 1) are important for the small scale zoning of the flooded sections at those sites and for the prediction of future safety requests. An important result is that it was not possible to evaluate the routing only with the Saint-Venant equations discretized by the implicit scheme before the occurrence of the wave front dissipation given by the shock fitting method. This may indicate that due to the high kinetic energy of the flow before the wave front dissipation the implicit method does not seem to be unconditionally stable as stated by Mascarenhas [6].

CONCLUSIONS

The shock fitting method showed to be usable for the prediction of the discontinuous free surface profiles behaviour at times immediately after the dam failure at the nearest sections to the dam. Because of the large height of the dam and also of the failure hypothesis adopted the routing based only on the unsteady gradually varied flow equations is not able to give the wave front behaviour neither the time and section of its dissipation. On the other hand, after the occurrence of that dissipation the entire flow becomes gradually varied and the values obtained in the shock fitting method are used as initial conditions for the conventional routing. The instability of the implicit scheme before the shock dissipation is related to the large values of the Vedernikov numbers for flows subjected to supercritical conditions at some sections, as explained by Samuels and Skeels [8]. Such numbers are quite high before the wave front dissipation as a consequence of large flow velocities occurring at those times. While the numerical scheme is set to the shock fitting approach the time increment Δt must be maintained small enough for the satisfaction of the Courant numerical stability condition in the method of characteristics. After the computational front wave dissipation test results positive,the increments in the implicit finite difference scheme can be enlarged for computational time saving in the solution of the Saint-Venant equations, which are then valid in the whole flow domain.

REFERENCES

1. Abbott, M.B. An Introduction to the Method of Characteristics, American Elsevier Publishing Company, New York, 1966.

2. Chen, C.L. and Armbruster, J.T. Dam-Break Wave Model: Formulation and Verification, Journal of the Hydraulics Division ASCE, Vol. 106, No. HY5, pp. 747-767, 1980.

3. Cunge, J.A., Holly, F.M. Jr. and Verwey, A. Practical Aspects of Computational River Hydraulics, Pitman Advanced Publishing Program, London, 1980.

4. French, R.H. Open-Channel Hydraulics, McGraw-Hill Book Company, New York, 1985.

5. Liggett, J.A. and Cunge, J.A. Numerical Methods of Solution of the Unsteady Flow Equations. Chapter 4, Unsteady Flow in Open Channels, (Ed. Mahmood, K. and Yevjevich, V.), Vol. 1, pp. 89-179, Water Resources Publications, Colorado, 1975.

6. Mascarenhas, F.C.B. Mathematical Modelling of Waves Caused by Breaking of Dams, Ph.D. Thesis, COPPE-UFRJ, Rio de Janeiro, Brazil, 1990.

7. Preissmann, A. Propagation des Intumescences dans les Canaux et Riviéres, pp. 433-442, 1ére Congrés de l'Association Française de Calcul, Grenoble, France, 1960.

8. Samuels, P.G. and Skeels, C.P. Stability Limits for Preissmann's Scheme, Journal of Hydraulic Engineering ASCE, Vol. 116, No. 8, pp. 997-1012, 1990.

9. Stoker, J.J. Water Waves, Interscience Press, New York, 1957.

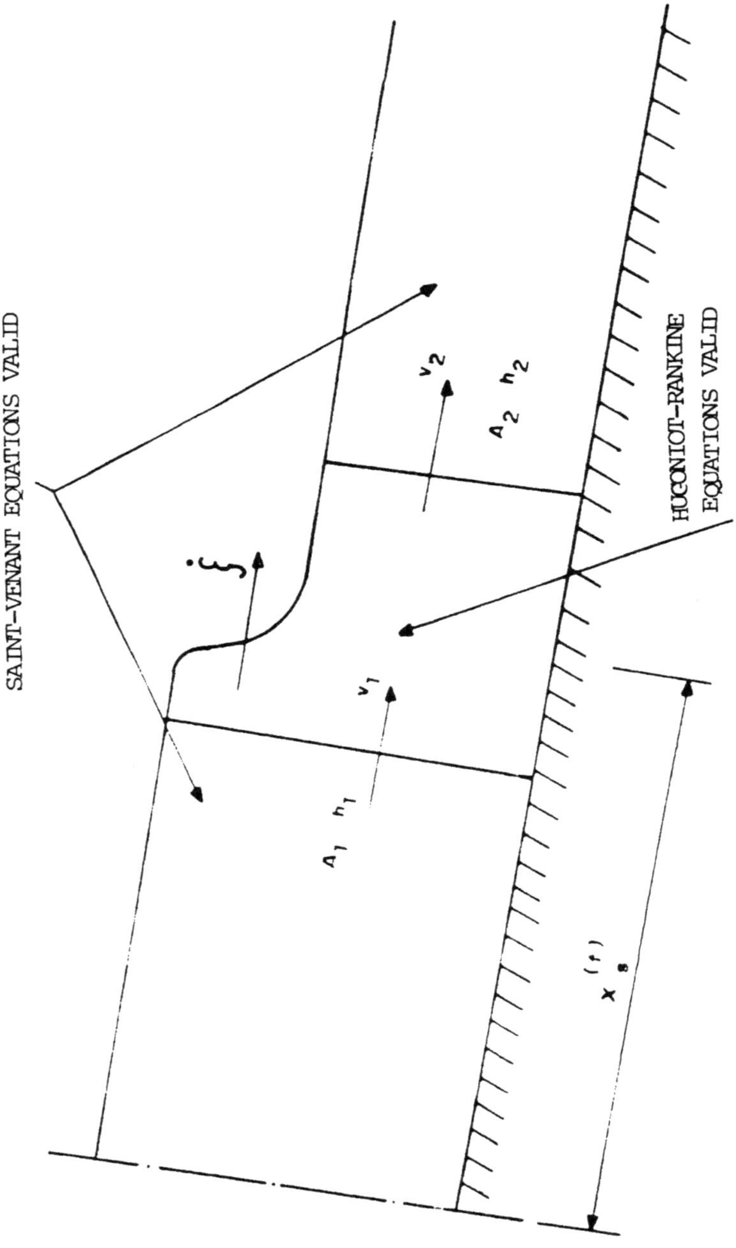

Figure 1 - Generical Configuration of the Shock Front

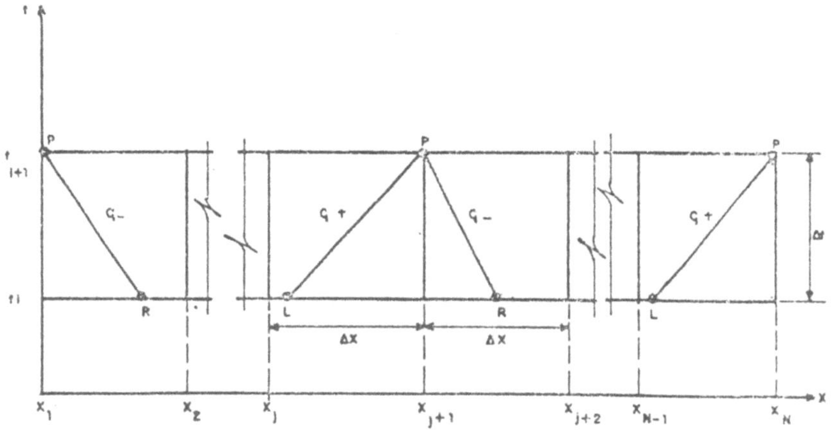

SUBCRITICAL FLOW

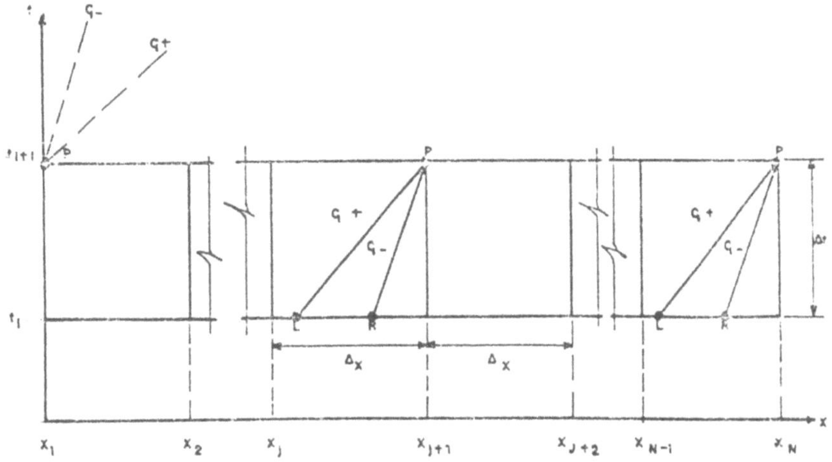

SUPERCRITICAL FLOW

Figure 2 - Explicit Finite Difference Mesh With Specified Time Interval and Characteristics Directions

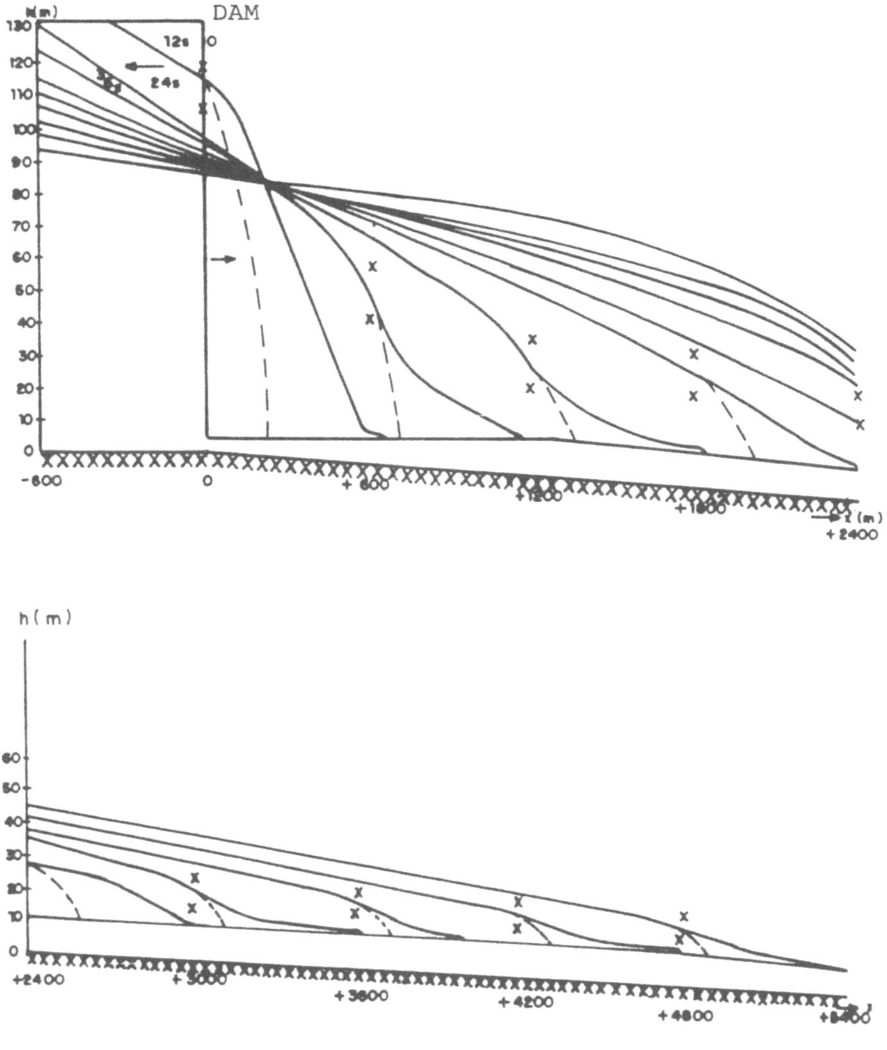

Figure 3 - Free Surface Profiles for Sections Near the Dam

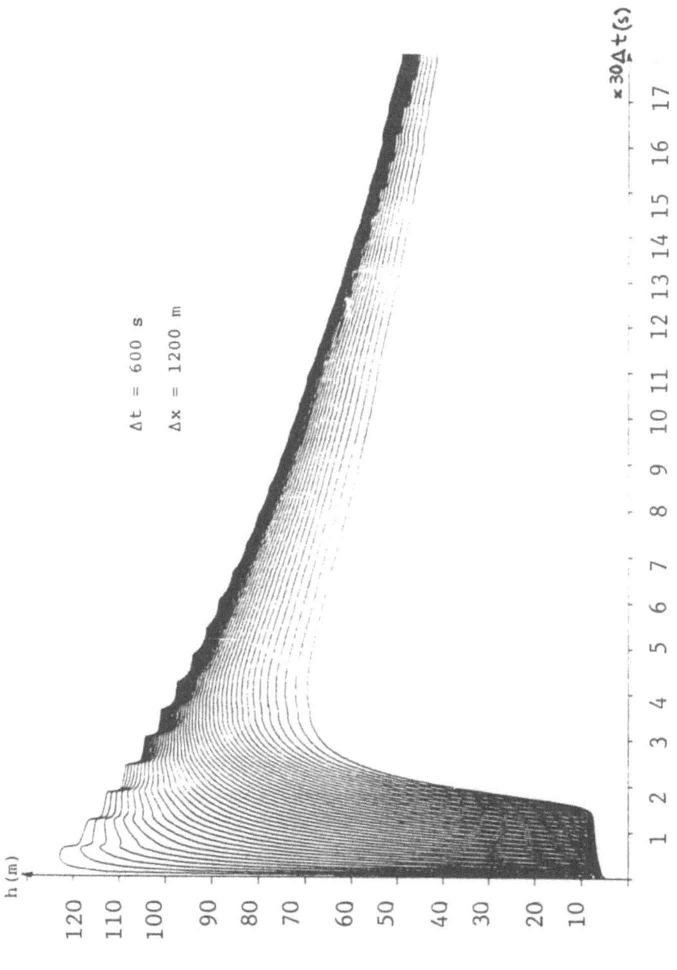

Figure 4 - Flow Depth Hydrographs for Sections Downstream the Dam

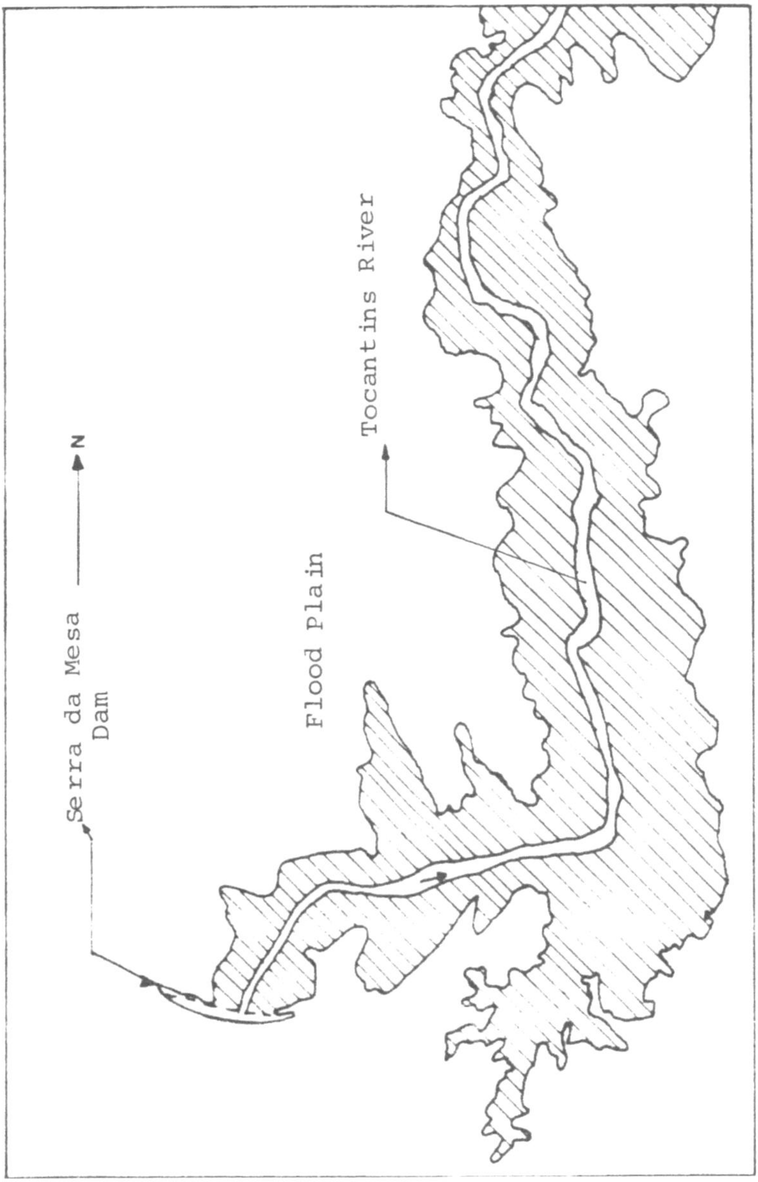

Figure 5 - Estimated Flood Plain After the Dam Failure

The Surface Marker Method

S. Chen, D.B. Johnson, P.E. Raad
Department of Civil and Mechanical Engineering, Southern Methodist University, Dallas, Texas 75275-0335, USA

ABSTRACT

Fundamental modifications of the marker and cell method are presented which result in significant reductions of the computer time and mass storage required for the solution of transient, free surface fluid flow problems. New marker movement and cell reflagging techniques are described that employ the use of only one row of markers along the fluid free surface and require the reflagging of only the cells along and adjacent to the free surface in each computational cycle. The advantages of the use of the new techniques are demonstrated by reference to the solutions of a broken dam problem and two cavity filling problems.

INTRODUCTION

All occurrences of waves, including ocean waves running up on a beach, fuel sloshing in a container, the filling of a closed cavity during die casting, and the filling of bottles on an assembly line are examples of free surface fluid flow. A distinguishing feature of these fluid flows is the existence of free surfaces that continuously change shapes and locations. In each new time step of a computer simulation of a free surface fluid flow, the location of the free surface must be determined before the values of any of the field variables, such as velocity and pressure, can be computed. A solution procedure for free surface fluid flow problems must include a method for the determination of the free surface locations and the treatment of the free surface boundary conditions in addition to the steps that must be included in a solution procedure for other incompressible fluid flow problems. The method used to overcome the difficulties introduced by the presence of free surfaces is independent of the numerical techniques employed to approximate and solve the differential equations that govern the fluid flow.

The marker and cell (MAC) method for the solution of transient, incompressible, free surface fluid flow problems was developed at the Los Alamos National Laboratory in the late 1960s. Since that time, many authors have presented extensions to and modifications of the MAC method. The simplified marker and cell (SMAC) method [1] represents the most significant revision of the MAC method and serves as the point of departure for the research reported in this paper.

In a recent study [2], the SMAC method was used to solve several different free surface fluid flow problems in order to demonstrate that parallel processing can be used to reduce wall clock computing time. It also was determined that approximately three-fourths of the total computing time was required to (a) solve the pressure potential Poisson equation and (b) move the markers and reflag the finite difference cells. A significant reduction in the time required to solve the pressure potential Poisson equation was achieved simply by the replacement of the successive overrelaxation method with the preconditioned conjugate gradient method. The movement of markers and the reflagging of cells, however, cannot be characterized by a specific type of mathematical equation, and it is not possible to achieve a similar reduction in the time required to move markers and reflag cells simply by replacing one mathematical method with a more efficient one. Since marker movement involves explicit calculations, a reduction in the time required for these calculations was achieved by use of parallel processing.

In the present paper, new marker movement and cell reflagging techniques are presented which are fundamentally different than those used in the SMAC method. The use of these new techniques enables significant reductions in the time required to locate and track the movement of fluid free surfaces. The complete solution procedure that incorporates the new marker movement and cell reflagging techniques is referred to in the remainder of this paper as the surface marker (SM) method.

In the SMAC method, the changes in the shape of the region occupied by fluid are tracked by markers which initially are evenly distributed throughout the fluid. The number of markers changes only if fluid either enters or exits the computational domain. Each finite difference cell is assigned a flag. The extent of the fluid filled region is defined by the finite difference cells that possess certain flags. During each computational cycle, all markers must be moved. All of the cells that contained markers before the current time step as well as any additional cells that contain markers after marker movement must be reflagged.

In the SM method, markers are not distributed throughout the fluid. Instead, only one string of markers that stretches along the free surface is required. Markers can be added at any point along the free surface at any time. In addition, it is only necessary to reflag the cells that lie along the free surface and some of their neighbor cells in each computational step. As a result of these differences, the time required to move markers and reflag cells is significantly less than in the SMAC method.

In what follows, the SMAC method is briefly reviewed, the SM method is introduced, three specific free surface fluid flow problems are solved by both methods, and comparisons are made between the solutions obtained by use of the two methods.

THE SIMPLIFIED MARKER AND CELL METHOD

In the SMAC method, the Navier-Stokes and continuity equations for two-dimensional, incompressible fluid flows are expressed as

$$\frac{\partial u}{\partial t} = -\frac{\partial u^2}{\partial x} - \frac{\partial uv}{\partial y} - \frac{\partial \phi}{\partial x} + \nu\left(\frac{\partial^2 u}{\partial y^2} - \frac{\partial^2 v}{\partial x \partial y}\right) + g_x \tag{1}$$

$$\frac{\partial v}{\partial t}=-\frac{\partial v^2}{\partial y}-\frac{\partial uv}{\partial x}-\frac{\partial \phi}{\partial y}+\nu\left(\frac{\partial^2 v}{\partial x^2}-\frac{\partial^2 u}{\partial x \partial y}\right)+g_y \tag{2}$$

$$\frac{\partial u}{\partial x}+\frac{\partial v}{\partial y}=0 \tag{3}$$

where u and v are the velocity components in the x and y directions, respectively; ϕ is the ratio of the pressure to the fluid density; g_x and g_y are the components of gravity in the x and y directions, respectively; and ν is the fluid kinematic viscosity. The tentative velocities $\tilde{u}$ and $\tilde{v}$ and the pressure potential ψ are defined such that

$$\frac{\partial \tilde{u}}{\partial t}=-\frac{\partial u^2}{\partial x}-\frac{\partial uv}{\partial y}-\frac{\partial \theta}{\partial x}+\nu\left(\frac{\partial^2 u}{\partial y^2}-\frac{\partial^2 v}{\partial x \partial y}\right)+g_x \tag{4}$$

$$\frac{\partial \tilde{v}}{\partial t}=-\frac{\partial v^2}{\partial y}-\frac{\partial uv}{\partial x}-\frac{\partial \theta}{\partial y}+\nu\left(\frac{\partial^2 v}{\partial x^2}-\frac{\partial^2 u}{\partial x \partial y}\right)+g_y \tag{5}$$

$$\frac{\partial \psi}{\partial t}=\phi-\theta \tag{6}$$

where θ is an arbitrary function referred to as the pseudopressure. It can be shown that the equation governing the pressure potential ψ is the following Poisson equation

$$\frac{\partial^2 \psi}{\partial x^2}+\frac{\partial^2 \psi}{\partial y^2}=\frac{\partial \tilde{u}}{\partial x}+\frac{\partial \tilde{v}}{\partial y} \tag{7}$$

and Equations (1) and (2) can be replaced, respectively, by

$$\frac{\partial u}{\partial t}=\frac{\partial \tilde{u}}{\partial t}-\frac{\partial}{\partial t}\left(\frac{\partial \psi}{\partial x}\right) \tag{8}$$

$$\frac{\partial v}{\partial t}=\frac{\partial \tilde{v}}{\partial t}-\frac{\partial}{\partial t}\left(\frac{\partial \psi}{\partial y}\right) \tag{9}$$

The first step in the SMAC solution strategy is to calculate the tentative velocities $\tilde{u}$ and $\tilde{v}$ explicitly by use of Equations (4) and (5), respectively. The next step is to compute the pressure potential ψ by use of Equation (7). Then, the final velocities u and v are obtained by use of Equations (8) and (9), respectively, and the markers imbedded in the fluid are moved by use of this final velocity field. Finally, each cell is reflagged based on the new locations of the markers, and the new cell flags are utilized to define the fluid region for the next computational cycle.

The SMAC computational region is composed of a rectangular mesh of cells. Five different types of cells are defined. Cells of different types are assigned different "flags" and are handled differently. The five cell types are BND, EMP, FUL, SUR, and OB. BND represents a cell adjacent to but outside the boundary of the computational region; EMP represents a cell inside the computational region that contains no markers; FUL represents a cell that contains at least one marker and is not adjacent to an EMP cell; SUR represents a cell that contains at least one marker and is adjacent to at least one EMP cell; and OB represents a cell that defines the location of a rigid, stationary obstacle. In each computational cycle, reflagging of FUL, SUR, and EMP cells is performed during and after the movement of markers. Markers are moved sequentially. If a marker moves into an EMP cell, the cell immediately becomes a SUR cell. After all markers have been moved, a sweep is made through all the markers to ascertain which cells contain markers and which cells do not. All the SUR cells that no longer contain markers become EMP cells. Then, another sweep is made through the cells that contain markers. All the FUL cells that now are adjacent to an EMP cell become SUR cells. All the SUR cells that now are not adjacent to an EMP cell become FUL cells. The reflagging of the cells is the last step in a computational cycle.

THE SURFACE MARKER METHOD

According to the cell flag definitions of the SMAC method, the flags of BND and OB cells, the cells that represent rigid boundaries and obstacles, are fixed. The flags of all of the other cells in the computational domain can be either EMP, SUR, or FUL and can change as the solution progresses. Two of the flagging rules are that a FUL cell cannot have a directly adjacent EMP neighbor, and a SUR cell must have at least one directly adjacent EMP neighbor. After the reflagging step in each computational cycle, the region that contains fluid is defined by the FUL and SUR cells.

The finite difference forms of the Navier-Stokes equations are based on the assumption that convective momentum flux is exchanged only between adjacent cells. Thus, for a two dimensional problem in Cartesian coordinates, no fluid is allowed to travel in one time step a distance in either the x or y direction which is larger than the cell size in that direction. Specifically, both of the conditions $U\delta t < \delta x$ and $V\delta t < \delta y$ must be satisfied, where U and V are the maximum fluid speeds in the x and y directions, respectively; δx and δy are the cell spacings in the x and y directions, respectively; and δt is the time increment. In other words, a marker cannot move in either the x or y direction in one time step a distance greater than the cell size in that direction. Therefore, the SUR cells and their neighbors are the only cells whose flags can change in one time step. The flag of a FUL cell that is surrounded by FUL neighbors or of an EMP cell that is surrounded by EMP neighbors cannot change in one time step.

Based on the preceding observations, it can be concluded that only the SUR cells and their neighboring cells need to be considered in the cell reflagging process. Furthermore, the markers in all of the FUL cells which have no SUR neighbors are superfluous to the determination of the fluid filled region at the end of a computational cycle. The surface marker method was developed based on these conclusions. In the SM method, only SUR cells and some of their neighbors are reflagged, and markers are distributed only in one row or string along the free surface. In the following, the portions of the SM method that are different from their counterparts in the SMAC method, those associated with marker management and cell reflagging, are described in detail.

In the SM method, surface markers initially are equally spaced along the free surface of the fluid. If the fluid is adjacent to an OB or a BND cell, a marker is placed in the OB or the BND cell. This added marker is the image of the marker which is on the free surface and nearest to the OB or the BND cell. The markers are consecutively numbered, and the distance between any two adjacent markers is equal to a pre-defined distance D. At the end of each computational cycle, after all markers have been moved, the distance between each pair of consecutively numbered markers is compared with D. If the distance is larger than D, a marker is added midway between the two markers in question.

The SM method requires the introduction of a new flag, the CHK flag. Before moving the markers, all cells containing markers are flagged as CHK cells. CHK is a temporary flag, and only CHK cells can change to EMP cells after marker movement and marker addition. After the markers have been moved and markers have been added as required, all of the cells that now contain markers are flagged SUR temporarily. These SUR cells must be either FUL or SUR cells in the next computational cycle. If a remaining CHK cell has a directly adjacent EMP neighbor, its flag is changed to EMP; otherwise, it is changed to FUL. Finally, if any SUR cell now is not directly adjacent to at least one EMP cell its flag is changed to FUL.

Since the new marker management and cell reflagging schemes of the SM method require the movement of fewer markers and the reflagging of fewer cells in each computational cycle than are required in the SMAC method, the SM method is significantly more efficient than the SMAC method, and there is no compromise of accuracy. In the next section, these claims are substantiated by direct comparisons of the solutions obtained by use of both the SM and SMAC methods of three specific free surface fluid flow problems.

COMPARISON OF THE SM AND SMAC METHODS

Three different example problems are considered. These three examples are referred to as Cases 1, 2, and 3. Case 1 is a broken dam problem, and Cases 2 and 3 are cavity filling problems. For the cavity filling problem of Case 2, it is assumed that the flow takes place in a horizontal plane, and gravity does not affect the flow. For the cavity filling problem of Case 3, the flow takes place in a vertical plane, and gravity does affect the flow. The solutions for each case are illustrated by figures that display the fluid filled regions at successive times. The SM and SMAC solutions are both displayed in each figure. In the figures, the thick solid line represents a rigid boundary, the thin dotted lines indicate the finite difference grid, and the larger dots represent markers. In the illustrations of SMAC solutions, markers are distributed throughout the fluid filled region. In the illustrations of SM solutions, a string of markers surrounds the fluid filled region.

For each case, sixteen markers are initially located in each cell in the SMAC solution; the maximum distance D between consecutively numbered surface markers is $\delta x/4$ in the SM solution; δx and δy are both equal to 0.5 in; the time increment is equal to 0.02 s; the convergence criterion for the successive overrelaxation solution of Equation 7 is 10^{-4}; and the fluid kinematic viscosity is 0.2 in^2/s. In Cases 1 and 3, the acceleration of gravity is 1 in/s^2. The free-slip boundary condition is applied along each rigid boundary. The distance between adjacent markers is initially the same (0.125 in) for both the SM and SMAC methods.

The broken dam problem is Case 1. Initially, as shown in Figure 1a, a vertical fluid column at rest is located in the lower left section of the computational region. The dimensions of the initial fluid column are 2.5 in by 5.0 in, and the dimensions of the rectangular container are 10.0 in by 7.5 in. Gravity acts downward, and nothing constrains the fluid on the right side of the fluid column. Consequently, the fluid begins immediately to flow down and to the right. Figures 1b, 1c, and 1d illustrate the numerical solutions at 1, 5, and 10 s, respectively. As far as the predicted motion of the fluid is concerned, there are no discernible differences between the SM and SMAC solutions illustrated in Figures 1b,1c, and 1d.

An interesting feature of the SM method is displayed by the SM solution illustrated in Figure 1d. After ten seconds have elapsed, the fluid that ran up the right boundary has fallen back down and rejoined the bulk of the fluid. Consequently, a section of the string of surface markers has become engulfed by the fluid. The markers in this section continue to move with the fluid in a manner similar to the movement of interior markers in the SMAC method. The ability to track markers that once were on the surface but now are in the interior could potentially be useful. If such markers are considered undesirable, however, they are eliminated.

Case 2 is a two-dimensional cavity filling problem in which the flow takes place in a horizontal plane. As shown in Figure 2a, fluid enters a rectangular enclosure, or cavity, with a uniform velocity through an opening, or gate, in the center of one side of the rectangle. The gate velocity is 1 in/s; the dimensions of the rectangle are 8 in and 5 in, and the width of the gate is 2 in. Figures 2a, 2b, and 2c illustrate the numerical solutions at 2.5, 5, and 10 s, respectively, while Figure 2d shows the SMAC solution at 18.86 s and the SM solution at 18.76 s. The time given for each of the solutions in Figure 2d is the fill time for the associated solution (the time when all cells become FUL), and the cause for the difference between fill times is discussed later. The filling of the cavity proceeds with the fluid first flowing from the gate toward the opposite boundary. Then, the fluid spreads to the sides, after which it proceeds back toward the gate. As in Case 1, there are no discernible differences between the fluid boundaries illustrated by the SM and SMAC solutions at any particular time. Note that all of the markers in the SM solution shown in Figure 2d are contained in cells that have been flagged as FUL. All of these markers, therefore, are similar to those that have been engulfed in the fluid in the SM solution shown in Figure 1d.

For Case 3, the filling of the cavity takes place in a vertical plane. This is the only difference between Cases 2 and 3. The cavity filling pattern in Case 3, however, is dramatically different than that in Case 2. Figures 3a, 3b, and 3c illustrate the SM and SMAC solutions at 2.5, 5, and 10 s, respectively, while Figure 3d shows the solutions at the SMAC fill time of 18.44 s and the SM fill time of 17.72 s. Figures 3a, 3b, and 3c can be compared directly with Figures 2a, 2b, and 2c, respectively, since the first three times are the same in both cases. In Case 3, gravity causes the flow to spread rapidly as it rises, resulting in the filling of the cavity from the bottom to the top. Once again, there are no discernible differences between the fluid boundaries illustrated by the SM and SMAC solutions at any particular time.

For any SM solution, the distance separating any two adjacent surface markers at any instant of time is less than or equal to the original separation distance D. Consequently, the free surface boundary of the fluid at each instant of time is

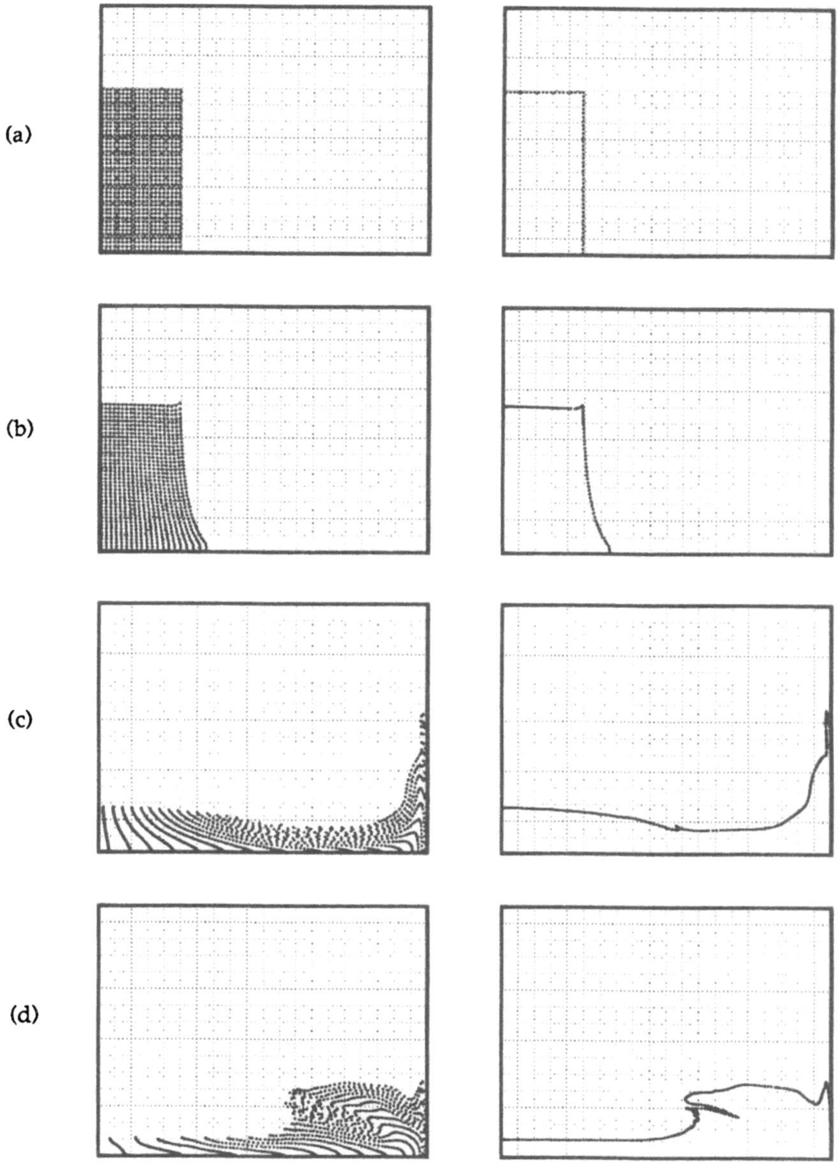

Figure 1. Case 1: Broken Dam Solutions at (a) 0 s, (b) 1 s, (c) 5 s, and (d) 10 s

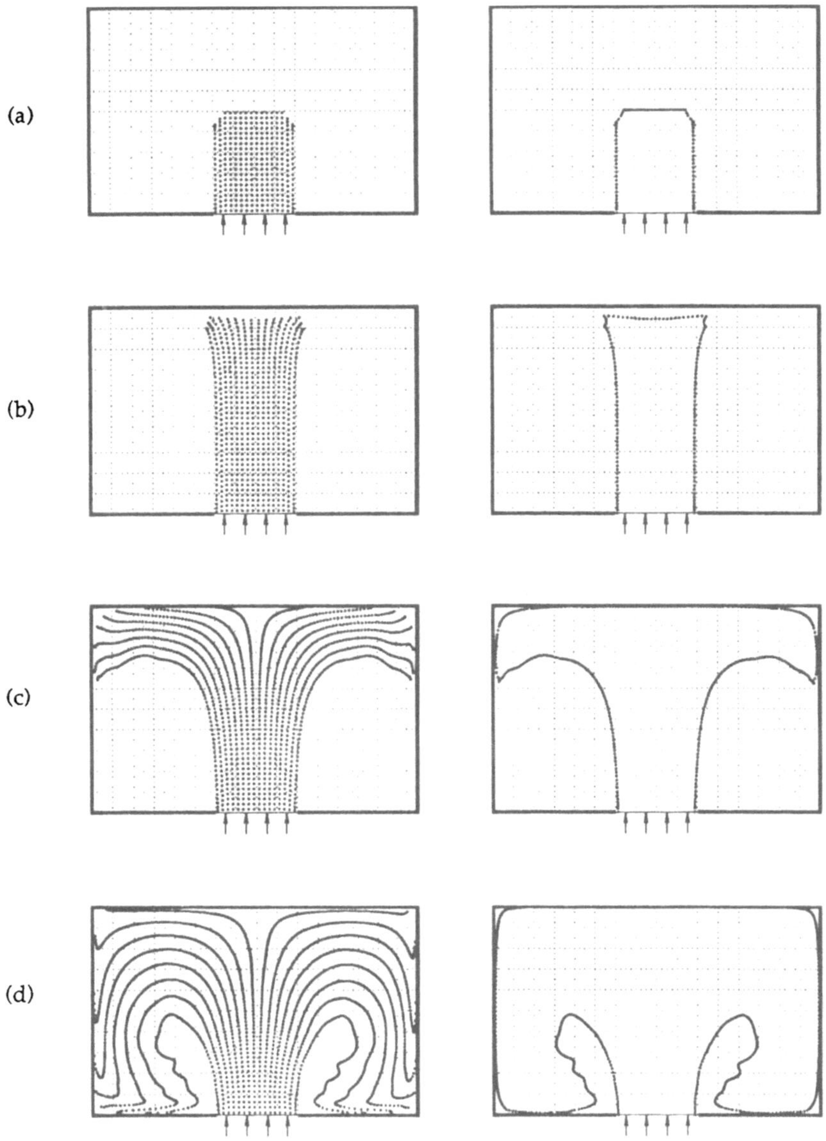

Figure 2. Case 2: Horizontal cavity filling solutions at (a) 2.5 s, (b) 5 s, (c) 10 s, and (d) 18.86 s (SMAC) and 18.76 s (SM)

(a)

(b)

(c)

(d)

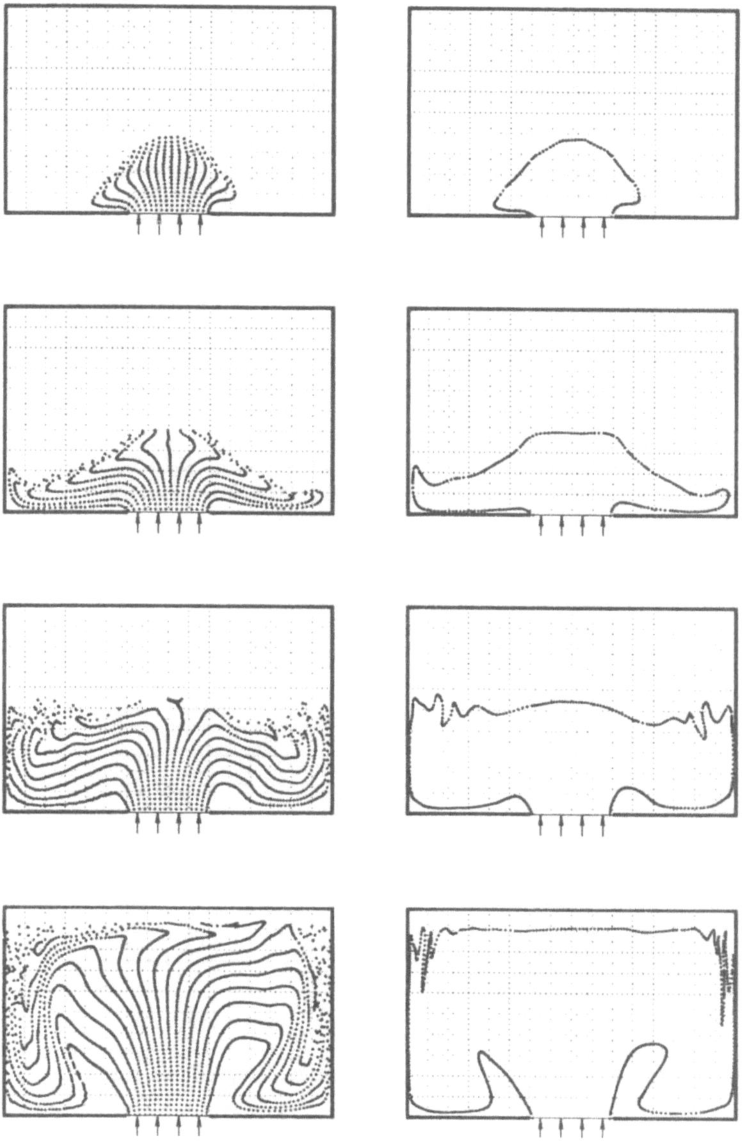

Figure 3. Case 3: Vertical cavity filling solutions at (a) 2.5 s, (b) 5 s, (c) 10 s, and (d) 18.44 s (SMAC) and 17.72 s (SM)

illustrated very clearly in an SM solution. The distance separating adjacent markers in a SMAC solution is dependent on the time histories of the velocity fields in the neighborhoods of the markers in question. For the SMAC solutions of Figures 1c, 1d, 2c, 2d, 3c, and 3d, the distances between adjacent markers are greater than the cell width in many instances.

The results of computing time analyses of Cases 1, 2, and 3 are presented in Table 1. For each case, the amount of time consumed in each subroutine and the total time are indicated for both the SMAC and SM solutions. The two subroutines of particular interest here are MVMARK and FLAG. MVMARK represents the marker movement subroutine, and FLAG represents the cell reflagging subroutine. The same subroutine names are used in both the SMAC and SM methods, although the MVMARK and FLAG subroutines themselves are substantially different in the two methods. The differences between the amounts of time spent in the two methods in the subroutines other than MVMARK and FLAG are not significant.

Inspection of Table 1 reveals that significantly less time is spent in the MVMARK subroutine in the SM solutions than in the SMAC solutions. For Case 1, marker movement requires only 185 s in the SM solution, a reduction of almost 40% from the 304 s required for marker movement in the SMAC solution. For Cases 2 and 3, the amounts of time, 193 s and 166 s, required for marker movement in the SM solutions are 59% and 63% less, respectively, than the associated amounts of time, 475 s and 454 s, required in the SMAC solutions.

Table 1. Computing time analysis

Subroutine	Time Consumed (seconds)					
	Case 1		Case 2		Case 3	
	SMAC	SM	SMAC	SM	SMAC	SM
MVMARK	303.68	184.64	475.47	192.88	454.02	166.20
PLOT	177.40	166.35	127.28	101.80	123.57	98.53
PRESS	94.00	95.98	165.47	185.17	187.07	165.77
FLAG	32.32	25.08	41.12	12.57	40.68	10.83
VTILDE	27.77	25.95	35.70	31.60	34.58	28.93
BNDVEL	13.45	13.30	9.30	10.37	7.67	7.18
SURVEL	10.58	10.28	6.45	7.05	5.52	5.22
THSBR	10.58	11.03	4.80	5.35	4.22	4.12
VFINAL	6.95	7.15	7.02	7.53	6.57	6.35
DTILDE	5.53	5.25	3.85	4.92	4.63	4.08
SAVE	2.42	2.95	1.35	1.23	1.48	1.27
INIT	0.57	0.13	0.15	0.12	0.15	0.15
OBVEL	0.02	0.02	1.88	2.17	1.88	1.68
System Calls	0.53	0.43	0.52	0.45	0.62	0.52
Total Time	685.80	548.55	880.35	563.20	872.65	500.83

The time spent in cell reflagging is much less in all instances than the time spent in marker movement, but comparison of the SM and SMAC solutions reveals, nevertheless, that the percentage reductions of the amounts of time spent in the FLAG subroutine are also very substantial. Cell reflagging requires 22%, 69%, and 73% less time in the SM solutions of Cases 1, 2, and 3, respectively, than it does in the SMAC solutions.

Several apparent shortcomings in the SMAC solutions of the three example problems are also faithfully reproduced by the corresponding SM solutions. In Case 1, it can be seen in Figure 1b that the markers in the extreme upper right corner of the fluid column have not fallen as far during the first second as the neighboring markers; and, in Figure 1c, a thin jet can be observed extending up from the main body of fluid along the right boundary. In Case 2, since it is assumed that the fluid enters a vacant cavity, the fluid should move as plug flow until the opposite boundary is reached. However, Figure 2a reveals that both corners of the leading edge of the flow appear to have been retarded. A potentially more serious shortcoming is apparent in Figure 2b where part of the fluid column has widened even though the fluid front has not yet reached the opposite boundary. Figure 2c shows that the continuation of the simulation allows the fluid to eventually reach the opposite boundary. The theoretical fill time for Case 2 is 20 s. In the SMAC and SM simulations of Case 2, however, all cells have been flagged as FUL after only 18.86 s and 18.76 s, respectively. The difference in fill times for the two methods is caused by the fact that markers on the surface in the SMAC solution become too sparse for all cells along the surface to be flagged correctly. Consequently, some cells in the SMAC solution are not flagged correctly until markers from the interior reach the cells in question. On the other hand, in the SM solution, adjacent surface markers at any time are separated by a distance less than or equal to the original separation distance D. Therefore, surface cells are never flagged incorrectly in the SM method due to sparsity of markers. Another shortcoming, not related to marker sparsity, is displayed in the fill time solutions of Figure 2d in which there are areas in all four corners of the cavity and on both sides of the gate that markers have not yet reached. In Case 3, the flow should be symmetrical about a vertical line through the center of the gate. However, asymmetry is already apparent in Figure 3a after only 2.5 s have elapsed. Asymmetry is also apparent in Figures 3b, 3c, and 3d. The theoretical fill time for Case 3 is also 20 s. In the SMAC and SM simulations of Case 3, however, all cells have been flagged as FUL after only 18.44 s and 17.72 s, respectively, despite the fact that markers have barely entered any of the finite difference cells along the top boundary of the cavity. Investigation of these apparent shortcomings is not an objective of the present paper. They are investigated in Reference [3] where methods for overcoming them are described which incorporate the new marker movement and cell reflagging techniques presented here.

CONCLUSIONS

Very significant improvements in the computational efficiency of the SMAC method in simulating free surface fluid flow problems were achieved by using new marker movement and cell reflagging techniques to track and define the fluid free surface. Since the free surface tracking and defining scheme employed in a finite difference numerical simulation method is independent of the mathematical methods employed to approximate and solve the associated differential equations, the new techniques can be used in any finite difference method for free surface fluid flow simulation.

ACKNOWLEDGMENT

This research was supported by the Defense Advanced Research Projects Agency of the US Department of Defense as Projects 5-25084 and 5-25085 of Contract MDA093-86-C-0182.

REFERENCES

1. Amsden, A. A. and Harlow, F. H., "The SMAC Method: A Numerical Technique for Calculating Incompressible Fluid Flows," Los Alamos Scientific Laboratory Report LA-4370, 1970.
2. Johnson, D. B., Raad, P. E. and Rausch, R. E., "The Application of Parallel Processing to the Simplified Marker and Cell Method," Proceedings of the Second IEEE Symposium on Parallel and Distributed Processing, pp. 357-362, Dallas, Texas, December 1990.
3. Chen, Shea,"The SMU Method: A Numerical Scheme for Calculating Incompressible Free Surface Fluid Flows by the Surface Marker Utility," Ph.D. Dissertation, Southern Methodist University, Dallas, Texas, May 1991.

Development of Conveyance Routing

M. Karmegam (*), V. Rangapathy (**),
S. Haribabu (*)
(*) Centre for Water Resources, Anna University, Madras - 600 025, India
(**) Department of Civil Engineering, Annamalai University, Annamalai Nagar - 608 002, India

ABSTRACT

A complete solution to flood routing problems is obtained by hydraulic routing methods of which the implicit routing method is popular because of its unconditional stability. This method requires a large amount of geometric and hydraulic data related to the channel reach through which the flood is routed. The geometric and hydraulic factors of the channel reach signify the conveyance of the flood through the reach. Hence, if the reach is described directly by its conveyance, it is possible to eliminate the requirement of a large volume of geometric data and the resulting model will be the conveyance routing model. Such a model has been developed and tested satisfactorily. The conveyance routing model maintains the accuracy of the implicit routing model. It saves large amount of space, computes faster and is less costly compared to the conventional implicit routing.

INTRODUCTION

Flood routing problems require the solution of the Saint Venant equations of unsteady gradually varied flow. These equations

$$\frac{\partial Q}{\partial x} + B\frac{\partial y}{\partial t} = 0 \qquad (1)$$

and $$\frac{\partial Q}{\partial t} + \frac{\partial}{\partial x}\left(\frac{Q^2}{A}\right) + gA\left(\frac{\partial y}{\partial x} - S_o + S_f\right) = 0 \qquad (2)$$

are respectively due to the principles of conservation of mass and conservation of momentum. Here Q is flow in the channel, y is depth of flow, B is top width of flow, A is cross-sectional area of flow, g is acceleration due to gravity, S_o is bed slope of the channel, S_f is friction slope, x is distance along the reach and t is time. Q and y are the dependent variables while x and t are the independent variables. Hydrological methods of routing solve only the equation of continuity but a complete solution to the flood routing problem is obtained in hydraulic routing methods which solve both Equations (1) and (2).

These equations are first order nonlinear hyperbolic partial differential equations having no closed form integration. The advent of digital computers enabled quick solutions with numerical integration. A large number of finite difference techniques were developed of which the characteristics methods were the earliest. Stoker [1,2], Ligget and Woolhiser [3] and Strelkoff [4] have done extensive work on this method. The development of explicit methods and their application to unsteady flow can be traced to Isaacson et al. [5]. The various explicit schemes differ from each other basically in the grid formation and the method of calculation of derivatives in the numerical form. However, both the characteristics and the explicit methods were plagued by stability considerations which were governed by the Courant condition. The characteristics method possessed the added disadvantage of the solution proceeding in an irregular grid form.

The very early works of Thomas [6] and later of Isaacson et al. [5] led to solutions from implicit techniques. Despite complex computations and not withstanding high computational cost and larger computer memory requirements, the implicit methods have become popular because of their inherent unconditional stability.

Amein's four point finite difference scheme is presently being used widely in implicit flood routing problems. The finite differnece scheme in the x-t plane can be explained with reference to Figure 1. At the point M, which is in the centre of the grid, the average values and the partial derivatives of any function α are

$$\alpha(M) = \frac{1}{4}\left(\alpha_i^j + \alpha_{i+1}^j + \alpha_{i+1}^{j+1} + \alpha_i^{j+1}\right) \tag{3}$$

$$\frac{\partial \alpha(M)}{\partial x} = \frac{1}{2\Delta x}\left[\left(\alpha_{i+1}^j - \alpha_i^j\right) + \left(\alpha_{i+1}^{j+1} - \alpha_i^{j+1}\right)\right] \tag{4}$$

$$\frac{\partial \alpha(M)}{\partial t} = \frac{1}{2\Delta t}\left[\left(\alpha_i^{j+1} - \alpha_i^j\right) + \left(\alpha_{i+1}^{j+1} - \alpha_{i+1}^j\right)\right] \tag{5}$$

where subscript i represents the space section and superscript j represents the time level.

Applying the above scheme to Equations (1) and (2), the following two finite difference equations are obtained

$$\frac{1}{2\Delta x}\left(Q_{i+1}^j - Q_i^j + Q_{i+1}^{j+1} - Q_i^{j+1}\right) +$$
$$\frac{1}{4}\left(B_i^j + B_{i+1}^j + B_{i+1}^{j+1} + B_i^{j+1}\right)\frac{1}{2\Delta t}\left(y_i^{j+1} - y_i^j + y_{i+1}^{j+1} - y_{i+1}^j\right) = 0 \tag{6}$$

$$\frac{1}{2\Delta t}\left(Q_i^{j+1} - Q_i^j + Q_{i+1}^{j+1} - Q_{i+1}^j\right) +$$
$$\frac{1}{2\Delta x}\left[\frac{(Q_{i+1}^j)^2}{A_{i+1}^j} - \frac{(Q_i^j)^2}{A_i^j} + \frac{(Q_{i+1}^{j+1})^2}{A_{i+1}^{j+1}} - \frac{(Q_i^{j+1})^2}{A_i^{j+1}}\right] +$$
$$g\frac{1}{4}\left(A_i^j + A_{i+1}^j + A_{i+1}^{j+1} + A_i^{j+1}\right) \tag{7}$$
$$\left[\frac{1}{2\Delta x}\left(y_{i+1}^j - y_i^j + y_{i+1}^{j+1} - y_i^{j+1}\right) - So + \frac{1}{4}\left(S_i^j + S_{i+1}^j + S_{i+1}^{j+1} + S_i^{j+1}\right)\right] = 0$$

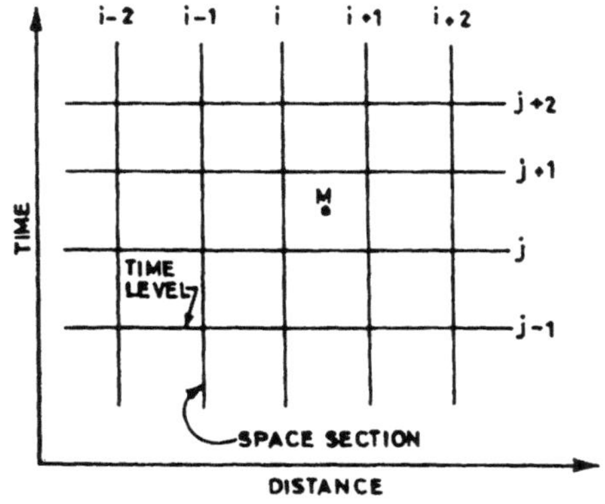

FIG.1 RECTANGULAR NET OF POINTS FOR THE IMPLICIT SCHEME.

Here the friction slope S_f can be evaluated by Manning's formula at the respective nodal points. If the total reach is defined by N number of sections, there will be (N-1) grids and successive application of Equations (6) and (7) at two adjoining grids will result in (2N-2) equations in 2N unknowns. The upstream boundary condition, which may be the discharge hydrograph at i=1 described by

$$Q_1^{j+1} = f_1 \ (t^{j+1}) \tag{8}$$

and the downstream boundary condition at i=N, which may be a rating curve described by

$$Q_N^{j+1} = f_N \ (y_N^{j+1}) \tag{9}$$

provide the remaining two equations for the system of nonlinear equations to be adequately specified.

With the initial conditions (at j=1) known, these values form the first guess for routing to the next time level (at j=2). Adopting Newton-Raphson method, the values of variables Q and y are found by iteration. In each iteration the system of nonlinear simultaneous equations is reduced to a system of linear simultaneous equations which is then solved by any method but most preferably by Gauss elimination procedure. For detailed information regarding the scheme and procedure, reference can be made to Amein [7], Fang [8] and Amein and Fang [9].

DATA FOR IMPLICIT ROUTING

To solve the Saint Venant's equations numerically, one has to define the values of Δx, the subreach length and Δt, the routing period. In order that the cross-section is defined as accurately as possible and in order that the inflow hydrograph is reproduced as correctly as possible, the values of Δx and Δt must be as small as feasible, in routing real floods. This requirement places some restraints on practical and computational aspects.

The length of the reach considered in a real time flood routing problem may be in hundreds of kilometres and so the reach may have to be divided into a very large number of subreaches so that the value of Δx can be kept very low. Consequently, cross-section survey at all these space sections will have to be carried out which is not only costly but also time consuming. In fact, in third world countries such cross-section details may not be available for most rivers. The flood period in many floods may extend over a couple or more weeks with the result that the routing will have to be carried out over a large number of time levels because Δt values are small.

The procedure requires that the cross-section data, which include the cross-sectional area of flow A, top width of flow B and wetted perimeter of flow P, have to be stored for each section of the river reach under study. These data have to be specified discretely for a large number of depths of flows, y. For a deep river with irregular cross-section, the geometric data A,B and P have to be specified for very small increments in y so that the cross-section is defined as accurately as possible. Thus an implicit routing problem requires a high computer storage.

The routing procedure being an iterative one, a large number of computations have to be made. Also, if the number of space sections of the river reach is N, there will be (2N-1) equations to be solved simultaneously. A matrix of size (2N-1) x (2N-1) will have to be stored, occupying further storage space. Even if the matrix is compressed by taking into account the zero elements as done by Karmegam [10], its size will not be less then (2N-1) x 5. Solving (2N-1) equations iteratively at every time level takes up a lot of computer time also. This, in turn, will raise the computing cost. Thus the computational time, cost and computer storage requirements are very high for an implicit routing procedure.

It should be noted that even the best efforts to obtain a true representation of the geometry of the reach may still have an inadequacy which may impart errors in the results of routing. There is also another factor which causes errors in the routed results. Any finite difference scheme has an inherent truncation error and obviously solution by implicit routing will reflect this error. If the routing should produce accurate results, this truncation error must be eliminated.

Karmegam [10] tried to solve the problem by optimizing the geometric and hydraulic parameters of the reach. He considered the base width of the channel, the side slope, the Manning's n and the power of hydraulic radius as the four model parameters to be optimized. Adopting a combination of Influence Coefficient Algorithm (Yeh and Becker [11]) with linear programming he obtained the required solution. It is noted that this procedure will entail high computational costs.

The four model parameters can be classified into two categories : (1) physical (bed width and side slope) and (2) hydraulic (Manning's n and power of R). The first group signifies the size of the waterway while the second group signifies the resistance it will offer to the flow. Manning's equation combines both these groups as

$$Q = (A/n)\, R^{2/3}\, (S_f)^{1/2} \tag{10}$$

or

$$Q = K\, (S_f)^{1/2} \tag{11}$$

where

$$K = (1/n)\, AR^{2/3} \tag{12}$$

The term K is the conveyance of the channel section. It is a measure of the carrying capacity of the channel section. Optimizing the four model parameters (after Karmegam) is in effect optimizing the conveyance of the channel.

It follows that if the conveyance of the channel section can be properly defined, the implicit routing model can be modified with the conveyance term incorporated in it. Such a model will dispense with the enormous cross-section data which would be otherwise necessary.

FORMULATION OF CONVEYANCE ROUTING MODEL

General observations show that if a plot is made of A Vs y for a natural channel, a second degree polynomial can be fitted. Also a plot of $AR^{2/3}$ Vs y is very nearly a straight line in a log-log paper. As a first step the cross-section data at different sections are taken up to plot the two relationships A Vs y and $AR^{2/3}$ Vs y. These discrete values are then regressed to fit in a best fit polynomial for A and a power equation for $AR^{2/3}$. Thus one obtains

$$A = a_1 y^2 + b_1 y + c_1 \tag{13}$$
$$AR^{2/3} = a_2 y^{b2} \tag{14}$$

for a space section. Similar equations can be developed for all space sections.

As no values of B will be read in the computer program in the conveyance routing, the continuity equation (Equation 1) is employed in the form

$$\frac{\partial Q}{\partial x} + \frac{\partial A}{\partial t} = 0 \tag{15}$$

and the momentum equation is retained as such (Equation 2). The friction slope is evaluated thus

$$Z = AR^{2/3} \tag{16}$$
$$K = Z/n \tag{17}$$
and $$S_f = Q^2/K^2 \tag{18}$$

where the value of Manning's n is suitably chosen. Thus the computer program for conveyance routing model is obtained from suitably modifying the implicit routing model.

When the area of flow equations (Equation 13) and 'section factor' equations (Equation 14) are used in routing, they should produce fairly accurate results subject to some variations imposed by the differences between the discrete values of A and $AR^{2/3}$ and the continuous values obtained from regressed equations.

With the view to simplify the data requirements this study used only one equation for the section factor $AR^{2/3}$ in all sections. It now becomes necessary to decide which of the section factor equations be used commonly for all sections.

A sensitivity analysis studied by Periakaruppan [12] shows that cross-section data and bed slope values at the downstream section and subreach are very sensitive to the implicit routing. Even errors upto 30% at upstream sections do not affect the routed results abnormally, whereas even a 10% error induced in the downstream boundary section introduces very high errors in the routed results. Following this finding one can use the section factor equation of the downstream boundary section as a single equation for all the sections. Any marginal errors in results can be minimized by suitably altering the coefficients in the equation by a few trials or by optimizing the coefficient.

APPLICATION OF CONVEYANCE ROUTING

The conveyance routing model was tested in two hypothetical floods in a prismatic channel.

Geometric data

The geometry of the reach was as follows :

Length of the reach	= 36000m
Number of sections	= 4
No. of subreaches	= 3
Subreach lengths	= 10000m, 12000m and 14000m
Bed widths	= 100m, 110m, 120m and 130m
Side slope of the channel	= 1 in 1 everywhere
Bed slopes of the subreaches	= 0.0003, 0.0004, 0.00035

Hydraulic and flow data

Manning's n was assumed as 0.030. The inflow at the upstream section was generated by the exponential equation as given in [13]

$$Q(t) = Q_b + Q_o\left\{\frac{t}{t_p}\ \exp(1 - \frac{t}{t_p})\right\}^{\beta} \qquad (19)$$

where Q(t) is inflow at time t, Q_b is base flow, Q_a is amplitude of flow, t_p is time to peak and ß is a factor deciding the curvature of the hydrograph around the peak. Two floods were generated; Flood No.1 had Q_b = 200 cumecs, Q_a = 4000 cumecs, t_p = 40 hours and ß = 8. The flood period was adopted as 160 hours. In Flood No.2, Q_a was taken as 10000 cumecs and ß was 12, keeping other values the same.

Downstream boundary condition

These two floods were then routed by the conveyance routing model. The solution requires that the downstream boundary condition be specified. In routing real floods, this is provided by a rating curve at the downstream gauging station. However, such directly applicable conditions cannot be evolved in routing hypothetical floods. Fread [14] has mentioned the use of a normal flow equation like Manning's as a downstream boundary condition. Karmegam [10] and Rangapathy [15] have also used Manning's equation as downstream boundary condition in both hypothetical and real floods.

However, applying a normal flow equation in flood situation cannot be supported by theory as the flood is unsteady varied flow. This difficulty can be overcome if the reach is extended beyond the downstream section. Application of normal flow equation as downstream boundary condition at the extended section should not unduly affect the discharges computed at the real downstream section. The concept of real and extended downstream section is shown in Figure 2. This figure shows extension of the reach by twice its length. In a similar manner the reach can be extended to thrice its length if needed.

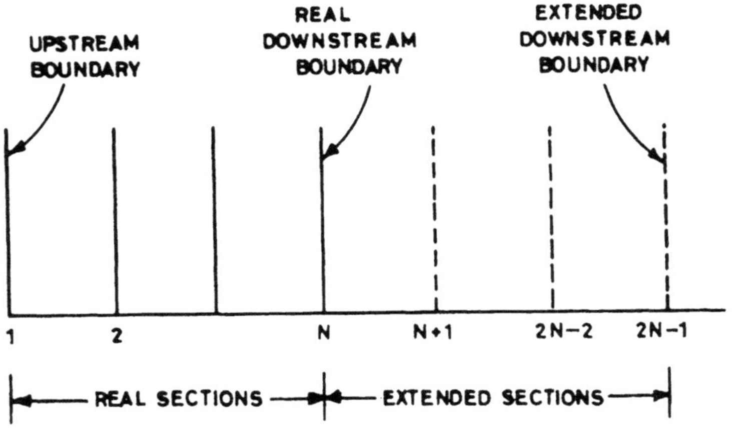

FIG.2 REAL AND EXTENDED DOWNSTREAM BOUNDARY

In this study the reach was extended as described here. There were three subreaches each of length 10000m, 12000m and 14000m and thus there were 4 sections. These four sections had four different trapezoidal cross-sections. The last subreach was extended six more times; thus the original reach length of 36000m has now increased to 120000m. Thus 6 additional sections were added making a total of ten sections. The bed slope of the last (third) real subreach was adopted for the six extended subreaches also. The cross-section of the last (fourth) real section was repeated for the six extended sections also. In effect a long length of uniform cross-section and bed slope conditions were introduced.

In this study the area equations used are

For section 1, $A = 100y + y^2$ (20)
For section 2, $A = 110y + y^2$ (21)
For section 3, $A = 120y + y^2$ (22)
For section 4, $A = 130y + y^2$ (23)

For sections 5 to 10, Equation (23) was applied.

The equation used for the section factor at all sections was prepared by using the geometric data of the downstream boundary, i.e. Section 4. It was

$$AR^{2/3} = 129\, y^{1.67} \qquad (24)$$

RESULTS AND CONCLUSION

The outflow hydrographs obtained from conveyance routing model are compared with those obtained from the conventional implicit routing model. The peak flow predicted by the conveyance routing model is compared with that predicted by the implicit routing model. Also the time of occurrence of peak flow as predicted by the two models can also be compared. These results are presented in Table 1.

Figures 3 and 4 present the outflow hydrographs predicted by the two models for Flood No.1 and Flood No.2 respectively.

In Flood No.1, the two hydrographs almost coincide at all time levels and where there is a deviation, it is very negligible. If the conveyance routing result is compared with that of the implicit routing (assuming the latter to be the true results), the conveyance routing shows a very negligible error (0.41%) in predicting the peak flow. Both models have predicted the time of occurrence of peak flow identically. In Flood No.2 the error in peak flow is 5.13% and the time of occurrence of peak flow is predicted to be in advance by two hours to that predicted by the

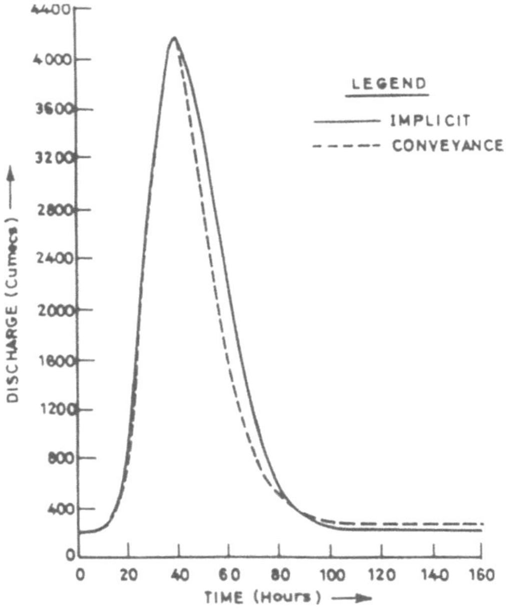

FIG.3 COMPUTED OUTFLOW HYDROGRAPHS OF FLOOD No.1.

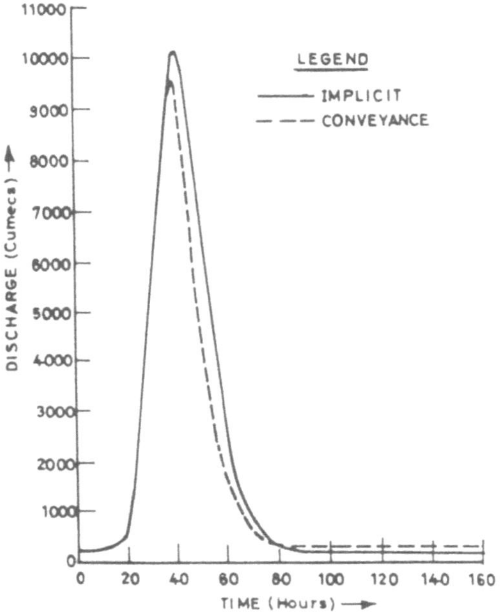

FIG.4 COMPUTED OUTFLOW HYDROGRAPHS OF FLOOD No.2.

conveyance routing model. These errors may be considered small and insignificant in a real time routing problem.

Flood No.1 is a low peaked flood with its curvature at the peak flow region some what flat. Flood 2 is a high peaked flood with its curvature at the peak flow region a little steeper. The latter flood has shown higher values of error. A fairly accurate downstream boundary condition cannot be obtained for hypothetical floods in hypothetical channels. Application of a normal flow equation at the downstream boundary is justified only if the flow has attained near uniform conditions at the boundary. In this study, the reach was extended to more than thrice the real length and it was assumed that the errors may be minimum. The results in Table 1 may be interpreted to justify this assumption in the case of Flood No.1. Perhaps the high peak flow in Flood No.2 has made it difficult for the errors to be minimum. It is possible that if the boundary is further extended, there would be an improvement in the results of Flood No.2 also.

Table 1 Comparison of the results between implicit and conveyance routing models

Flood	Computed peak outflow (cumecs)		Error in peak (%)	Time of occurence of peak (hours)		Error in time of occurence of peak
	Implicit	Conveyance		Implicit	Conveyance	
Flood No.1	4188.33	4171.14	0.41	41	41	0
Flood No.2	10128.28	9608.23	5.13	41	39	-2

NOTE : The error values are calculated assuming the Implicit routing results as standard.

Again, the section factor equation used in the reach is really valid only from the third subreach onwards, but was used throughout the length of the reach in the computations. The coefficient of this equation (Equation 24) may be optimized for minimum errors in routing. Such an optimized section factor equation can be believed to improve the accuracy of routing.

Table 2 shows the computer storage requirements for the two routing models. Implicit routing model requires more storage space than the conveyance routing model. But the difference is only marginal. However, in real flood routing, the difference may be much more. In this hypothetical flood routing the geometric data used are very minimum. For example, the cross-section of the channel at each space section is defined by just five discrete values of y and B. It means, the dimension of y,A,B

Table 2 Computer storage requirements

Method	Program and data (bytes)	Objective file (bytes)	Executive file (bytes)
Implicit	21680	35440	74868
Conveyance	17911	29787	69956

and P would be only 5 here. However, in natural channel routing the dimension of these data may be in the order of hundreds. In such a case, the storage required in conveyance routing model will be far less than that required by implicit routing model.

The time required for the computations with the two routing models on a PC-XT is presented in Table 3. Conveyance routing model requires less time both for compilation and linking and for execution. The saving in time is not appreciable as seen in this table. However, these are time requirements for routing a hypothetical flood in a length of small reach divided into just 10 space sections. Also the routing computations were carried to only 160 time levels. In real time routing problems, the number of space sections will be much more and the flood period may be a prolonged one necessitating routing through a large number of time levels. Then, the computer time saved by conveyance routing model will be substantial.

Table 3 Computer time requirements

Method	Compilation and linking	Running time	
		Flood No.1	Flood No.2
Implicit	$5^m\ 13^s$	$2^m\ 42.8^s$	$2^m\ 41.5^s$
Conveyance	$4^m\ 41^s$	$2^m\ 26.0^s$	$2^m\ 24.1^s$

The section factor equation is developed from the data of the downstream reach and then it is optimized for minimum errors in routing. This procedure opens up the possibility for reducing the requirement of accurate geometric survey, limiting it to the downstream section only. The geometric survey at other sections could be approximate. This facility results in enormous saving of survey time and survey costs.

This study enables the following conclusions to be arrived at.

A conveyance routing model has been developed and tested successfully. This model is a modification of implicit routing model in that the geometric data is mostly dispensed with and is replaced by area of flow and section factor equations. There is substantial saving in computer storage and computer time requirements. In real time routing computations, the conveyance routing model will be faster and less costly than implicit routing.

LIST OF REFERENCES

1. Stoker, J.J. The Formation of Breakers and Bores, Common. Appl. Maths., Vol.1, pp. 1-87, 1948.

2. Stoker, J.J. Water Waves, Interscience Publications, New York, 1957.

3. Ligget, J.A. and Woolhiser, D.A. Difference Solution of the Shallow Water Equation, Journal of the Engineering Mechanics Division, A.S.C.E, Vol.93, EM.2, pp. 39-71, 1967.

4. Strelkoff, T. Numerical Solution of Saint-Venant Equations, Journal of the Hydraulics Division, A.S.C.E., Vol.96, Hy.1, pp. 223-252, 1970.

5. Isaacson, E.J., Stoker, J.J. and Troesch, B.A. Numerical Solution of Flood Prediction and River Regulation Problems, Report No.IMM, NYU-205, Institute of Mathematical Sciences, New York University, 1956.

6. Thomas, H.A. Hydraulics of Flood Movements in Rivers, Engineering Bulletin, Carniege Institute of Technology, Pittsburgh, Pennsylvania, 1934.

7. Amein, M. An Implicit Method for Numerical Flood Routing, Water Resources Research, Vol.4, No.4, pp. 719-726, 1968.

8. Fang, C.S. Mathematical Solution of the Complete Equations of Unsteady Flow in Open Channels, Thesis presented to the North Carolina State University, Raleigh, U.S.A., in partial fulfillment of the requirements for the degree of Doctor of Philosophy, 1968.

9. Amein, M. and Fang, C.S. Implicit Flood Routing in Natural Channels, Journal of the Hydraulics Division, A.S.C.E., Vol.96, Hy.12, pp. 2481-2500, 1970.

10. Karmegam, M. Parameter Identification in Flood Routing, Thesis presented to the University of London, London, in partial fulfillement of the requirements for the degree of Doctor of Philosophy, 1981.

11. Yeh, W.W.G. and Becker, L. Linear Programming and Channel Flow Identification, Journal of the Hydraulics Division, A.S.C.E., Vol.99, Hy.11, pp. 2013-2021, 1973.

12. Periyakaruppan, PR. Data Accuracy Requirements in Implicit Flood Routing, Thesis presented to Anna University, Madras, India, in partial fulfillement of the requirements for the degree of Master of Engineering, 1988.

13. Flood Studies Report, Vol.III, National Environmental Research Council, 27, Charing Cross Road, London, 1975.

14. Fread, D.L. Discussion on Comparison of Four Numerical Methods for Flood Routing by Price, R.K., Journal of the Hydraulics Division, A.S.C.E., Vol.101, Hy.3, pp. 565-567, 1975.

15. Rangapathy, V. An Analysis of Flood Routing Methods and Identification of Routing Parameters, Thesis presented to the Anna University, Madras, India, in partial fulfillment of the requirements for the degree of Doctor of Philosophy, 1988.

Solution of Problem of Unknown Boundary in Free Surface Flow Calculations

K. Kosorin

Institute of Hydrology and Hydraulics, Slovak Academy of Sciences, Bratislava, CSFR, Czechoslovakia

1 INTRODUCTION

The hydrodynamic problem of unknown boundary frequently occurs at numerical simulation of steady, unsteady and waves phenomena of gravitational flows with free surface. There belongs wind and tsunami waves and their coast approaching, flood and dam break waves in extensive inundation areas, flow singularities on river and channel hydrotechnical systems as well as many types of filtration motion with significant time and space changes. The problem stems from the fact that the free surface as a part of material boundary of flow domain is unknown and has to be found out simultaneously and together with solution of hydrodynamic task as a whole. In that sense the problem is "mixed".

The classic hydrodynamics brought two approximate theories as very effective solutions of the problem, namely the small disturbances (linear) theory and shallow water theory, see Stoker [12]. The same could be said about Stokes and Boussinesq results regarding water waves and filtration motion; Lamb, 1943, [6].

This achievement, of course, does not solve the problem entirely being limited to phenomena, where the accepted simplifying assumptions hold and correspond to used theory or solution.

Unknown boundary shows to be unsuitable even for modern numerical methods such as the method of finite elements or finite differences.

The paper contains theoretical base and practical results of a new approach to solution of the problem, which stems from the so-called hydrodynamic theory of boundaries, see Kosorin, [4]. Main product of this theory is the method for transformation of N dimensional hydrodynamic task into N-1 dimensional one. Its application yields:

i.) Motion equations for material boundaries of nonviscous **flow** including equation of free surface of water waves. For the given vorticity it is shown that obtained system can be completed and in such a way the unknown boundary problem is solved entirely. The method allows next transition i.e. from two-dimensional material boundary to one-dimensional one. Result of this step gives the dynamic equation of a shore line which represents an one-dimensional boundary among three media-solid bottom, moving liquid and air.

ii.) Nonvorticity conditions and wave equation for boundary potential for flows with velocity potential. This nonlinear boundary wave equation takes the linear form of the classic wave equation of Helmholtz if the curvature of material boundary is negligible, [4]. Moreover, for such case the derived equation of fluid motion along material boundary are integrated giving the first so-called boundary Cauchy-Lagrange's integral. Using Hamilton's variational principle the same result has been obtained by Miles but only for free surface [7].

iii.) For filtration motion (in nonhomogeneous and unisotropic underground) the boundary velocity field in dependence on boundary values of the pressure function. Since the pressure function on free surface $y=h(x,z,t)$ coincides with the value $h(x,z,t)+$ $+\text{const}$, the solution of this unknown boundary problem becomes very effective because it leads to mere one equation for free surface (without any simplifying assumptions). On the other hand, there is a proved possibility to determine whole filtration velocity field if the free surface (and all hydrogeological characteristic of a flow domain, of course) are known. It is very interesting fact not only from theoretical but even more from practical point of view.

2 HYDRODYNAMIC THEORY OF BOUNDARIES

The main aim and product of this theory is method for transformation of N dimensional task into N-1 dimensional one. It should allow simultaneously to divide an originally mixed problem into two separate tasks. In order to get up formalism for the theory we have to define and explain some terms. Let the N-1 dimensional relation.

$$y= h(x,z,t) \tag{1}$$

defines the moving surface ∂G which is a part of G and belongs to N dimensional domain G of independent variables $x_i= x,y,z,t;\ i=1,2,3,4\ (N=4)$. One has to seek the function f(x,y,z,t) as the solution of mixed hydrodynamic task in domain G. Let us mark by trip those parts of the function f and their partial derivatives, which are evaluated in points of surface (1). For f it means the formal notation

$$\overline{f} = f(x,h,(x,z,t),z,t) \tag{2}$$

which transforms f into N-1 dimensional "boundary function" f(x,z,t) because on the surface ∂G originally independent variable y becomes dependent one due to relation (1). Differentiating the eq. (2) with respect to N-1 variables x_i one obtains

$$\overline{f}_{x_i} = \overline{f_{x_i}} + \overline{f_y} h_{x_i} \tag{3}$$

where two kinds of partial derivatives relating to surface ∂G have occurred. The "inner" $\overline{f}_{x_i} = \partial\overline{f}/\partial x_i$ as the partial derivatives of the boundary function $\overline{f}$ and the outer $\overline{f_{x_i}} = \overline{\partial f/\partial x_i}$, $\overline{f_y} = \overline{\partial f/\partial y}$ defined as the boundary values of the partial derivatives of f. Generally $\overline{f_{x_i}} \neq \overline{f}_{x_i}$ except for h=const. Formalism of the type (3) can be spread on the higher order derivatives, e.q.

$$(\overline{f_{x_i}})_{x_j} = \overline{f_{x_i x_j}} + \overline{f_{x_i y}} h_{x_j}$$

Terms inner and outer derivatives are not new, see Courant-Hilbert [1] or Smirnov [11]. Nevertheless, strict distinction between them is necessary for next treatment. Their geometrical interpretation is shown in fig. 1. As regards the hydrodynamic sense, the inner derivatives of $\overline{f}$ describe behavior of fluid on the surface ∂G while outer derivatives of f do it in near neighborhood on left or right of ∂G. Here "inner" means that it is or belongs to surface ∂G while outer means that it is beside but out of ∂G. The surface ∂G can be geometrical or material (boundary).

The basic differential equations govern the flow phenomena into domain G including points of its boundary ∂G. Marking all the members of these equations by trips one obtains the algebraic relations among outer derivatives of the dependent variables. In order to get the differential equations valid along surface ∂G it is necessary to exchange all the outer partial derivatives in the mentioned algebraic relations for the inner partial derivatives. For this purpose the treatment with total material derivatives of the velocity field is important. Written for surface ∂G they are also of two kinds; the outer total material derivative of f on ∂G is

$$\overline{\frac{df}{dt}} = \overline{f_t} + \overline{u}\,\overline{f_x} + \overline{v}\,\overline{f_y} + \overline{w}\,\overline{f_z} \tag{4}$$

while the inner total material derivative on ∂G ($y=h(x,z,t)$) contains only inner partial derivatives:

$$\frac{d\bar{f}}{dt} = \bar{f}_t + \bar{u}\bar{f}_x + \bar{w}\bar{f}_z \qquad (5)$$

The relation between $\overline{df/dt}$ and $d\bar{f}/dt$ is interesting for the wanted transition process. It shall occur applying the formalism (3) in (4). Then

$$\overline{\frac{df}{dt}} + \tau f_y = \frac{d\bar{f}}{dt} \qquad (6)$$

where

$$\tau = h_t + \bar{u}h_x + \bar{w}h_z - \bar{v} \qquad (7)$$

is factor the value of which determines the character of surface ∂G. If ∂G is geometrical boundary, then $\tau \neq 0$. If ∂G given by $y=h(x,z,t)$ denotes some material boundary then $\tau=0$ and (7) represents kinematic condition on such unpenetrating boundary. It can be a free surface or solid bottom (wall) which bound the moving fluids.

In the case that $\tau=0$ the eq. (6) takes the simple form

$$\overline{\frac{df}{dt}} = \frac{d\bar{f}}{dt} \qquad (8)$$

It allows to transform the Euler dynamic equations for barotropic gas or incompressible fluid into differential equations valid along material boundary of the flow.

The relation (8) is crucial for our transformation process because it works on material boundaries which behave hyperbolically as the so-called characteristic surfaces for the Euler eq. As it is known, on such surfaces it is impossible to determine the outer derivatives $\overline{f_{x_i}}$, $\bar{f}_y$ depending on inner values $\bar{f}$ and $\bar{f}_{x_i}$ given on ∂G in sense of the Cauchy formulation of boundary problem, see [1] or [11]. If behavior of the basic system on G is elliptical then the needed transformation is made by means of direct elimination of outer derivatives.

It is important, of course, that in both cases such transformation can be successful. Naturally, these considerations are limited to sufficiently smooth boundaries $y=h(x,z,t)$ having partial derivatives of the desired order.

3 MOTION EQUATIONS OF FLUID BOUNDARY

Let us use the above procedure in the treatment on some hydrodynamic tasks. If one writes the complete Euler equations for boundary $y=h(x,z,t)$ of incompressilble fluid and uses eq. (8) for inertial terms, then he removes all outer derivatives of velocity components. Further, if by means of eq. (3) he eliminates the outer derivatives of pressure p, then obtains dynamic equations of material boundaries of non-viscous flow, see Kosorin [4]

$$\frac{d\bar{u}}{dt} + \left(g + \frac{d\bar{v}}{dt}\right)h_x + \frac{1}{}\bar{p}_x = 0$$
$$\frac{d\bar{w}}{dt} + \left(g + \frac{d\bar{v}}{dt}\right)h_z + \frac{1}{}\bar{p}_z = 0 \qquad (9)$$

Modifying (9) the equations of free surface ($\bar{p}$=const) of water waves follow from (9) in the form

$$\bar{u}_t + \bar{u}\bar{u}_x + \bar{w}\bar{u}_z + \Omega h_x = 0$$
$$\bar{w}_t + \bar{u}\bar{w}_x + \bar{w}\bar{w}_z + \Omega h_z = 0 \qquad (10)$$

with the second order terms

$$\Omega = (g+h_{tt}+2uh_{xt}+2wh_{zt}+u^2h_{xx}+2uwh_{xz}+w^2h_{zz})/(1+h_x^2+h_z^2) \qquad (11)$$

which is reduced into the gravitational acceleration g (Ω=g) for long waves of small curvature. In such case eqs.(10) get the known form of the shallow water equations. Further, for simpler two-dimensional case we get the equation of free surface motion in the form

$$\bar{u}_t + \bar{u}\bar{u}_x + h_x(g+h_{tt}+2\bar{u}h_{xt}+\bar{u}^2h_{xx})/(1+h_x^2) = 0 \qquad (12)$$

interesting for comparison with the known approximate water waves equation of Boussinesq.

3.1 Dynamic equations of shore line

The above shown transformation process from the three--dimensional domain into its two-dimensional boundary can proceed up to one dimensional boundary. Such a curve with unknown coordinates is created, for example, by boundary among three medias e.i. water, air and solid bottom. Its governing equations stem from eq. (3), (8) and (9) through the same procedure, see Kosorin [4] The equation of motion of the so-called shore line $\partial\partial G$

$$\bar{u}_t + \bar{u}\bar{u}_x + ((g+P_1)\bar{y}_x + P_2\bar{z}_x)/\bar{k}^2 = 0$$
$$(P_1=\bar{y}_{tt}+2\bar{u}\bar{y}_{xt}+\bar{u}^2\bar{y}_{xx},\ P_2=\bar{z}_{tt}+2\bar{u}\bar{z}_{xx}+\bar{u}^2\bar{z}_{xx},\ \bar{k}^2=1+\bar{y}_x^2+\bar{z}_x^2) \qquad (13)$$

is obtained where u(x,t) is the x-th velocity component on line $\partial\partial G$ of the coordinates $y=\bar{y}(x,t)$ and $z=\bar{z}(x,t)$. Next two components of fluid velocity on this space curve are given by the kinematic conditions $\bar{v}=\bar{y}_t+\bar{u}\bar{y}_x$, $\bar{w}=\bar{z}_t+\bar{u}\bar{z}_x$.

As regards a number of the needed equations, the

above derived boundary systems (9) or (10) are not complete . The same is to be said about the 1-D system (13) in spite the fact that one further eq. can be written for shore line. It follows from eq. of solid bottom $y=h_o(x,z)$, which contains the shore line $\partial\partial G$ common with the free surface $y=h(x,z,t)$. Due to this fact the mentioned relation has the form $y=h_o(x,z(x,t))$. The above results are valid, if the space curve $\partial\partial G$ is continuous and smooth, of course.

3.2 The Cauchy-Lagrange Integral for boundary potential as the first integral of the boundary motion equations

Potential flow in G satisfies the Laplace eq. for potential and the Cauchy-Lagranges integral for energy, the kinetic part of which is expressed through the space derivatives of φ. Writing the Cauchy-Lagrange's integral for material boundary ∂G it takes the form of the algebraic condition containig the outer derivatives of the potential φ on ∂G. There is a need to exchange them for inner derivatives of φ. It is possible due to ellipticity of the inner task in G. Writing (3) for potential φ and (7) with $\tau=0$ one can determine the outer derivatives on $y=h(x,z,t)$

$$\overline{\varphi_y} = (h_t+h_x\,\overline{\varphi}_x+h_z\,\overline{\varphi}_z)/k^2 \qquad \overline{\varphi_x} = \overline{\varphi}_x-\overline{\varphi_y}h_x$$
$$\overline{\varphi_t} = \overline{\varphi}_t-\overline{\varphi_y}h_t \qquad \overline{\varphi_z} = \overline{\varphi}_z-\overline{\varphi_y}h_z \tag{14}$$

where $k^2=1+h_x^2+h_z^2$.

The desired first integral of dynamic system (9) can be constructed by gradual integration process by means of (8) and (14). However, the same result can be reached by direct use of (14) in the Cauchy-Lagrange's integral, i.e. by elimination of outer derivatives of potential. We obtain the following integral:

$$\overline{\varphi}_t + \frac{1}{2}(\overline{\varphi}_x^2+\overline{\varphi}_z^2) + gh + \frac{\overline{p}}{\rho} = \frac{1}{2k^2}(h_t+h_x\overline{\varphi}_x+h_z\overline{\varphi}_z)^2 \tag{15}$$

which represents the nonlinear differential equation of the first order for boundary potential $\overline{\varphi}(x,z,t)$. From inner domain to its material boundary we proceeded by transformation of dynamic condition to dynamic differential equation for this boundary.

Using Hamilton's principle, Miles derived equation of type (15) for free surface with constant outer pressure (p=const). However, the validity of (15) is more general and it regards an arbitrary, sufficiently "smooth" material boundary of potentional flow of incompressible fluid. By introduction of pressure function $P(x,y,z,t)$ this validity is extended also to potential flow of compressible fluid. It follows from the fact that relations (15) and transformation of Euler dynamic equations to fluid material boundary are valid concurrently, regardless of the fact whether this fluid is compressible or not.

4 COMPLETING OF BOUNDARY MOTION EQUATIONS SYSTEM

The obtained boundary motion eqs. (9) or (10) are not complete. Practical use desires their completion. It can be performed through the solution of inner task in various degree of approximation. For example, the first step of such approximate completing of eq. (10) should be the continuity equation $h_t+(uH)_x+(wH)_z=0$ from the shallow water approach. One possibility to complete the system of eq. (9) or eq. (10) follows from the knowledge of vorticity (α,β,γ) of flow in original domain of solution. In this case the inner task for domain G between upper boundary $y=h(x,z,t)$ and lower one $y=h_o(x,z)$ is formulated for the system

$$u_x + v_y + w_z = 0 \qquad (16)$$

$$u_y - v_x = 2\alpha \;; \qquad w_y - v_z = 2\beta; \quad u_z - w_x = 2\gamma \qquad (17)$$

Elipticity of this system allows to determine the outer derivatives of the velocity field on h or h_o as functions of their inner derivatives. In order to do so it is sufficient to write continuity eq.(16) and the first two eq. of (17) for boundary of $\underline{G}$ and six relations of eq. (3) written for $\bar{u}$, $\bar{v}$ and $\bar{w}$. Solving in this manner obtained linear algebraic system (for zero vorticity) one can get the outer y-th derivatives

$$\begin{aligned} \overline{v_y} &= -(\bar{u}_x+\bar{w}_z-h_x\bar{v}_x-h_z\bar{v}_z)/k^2 \\ \overline{u_y} &= \bar{v}_x-h_x\overline{v_y}; \quad \overline{w_y} = \bar{v}_z-h_z\overline{v_y} \end{aligned} \qquad (18)$$

It is possible, in the same way, to find out the outer y-th derivatives of the n-order

$$u_n = \partial^n u/\partial y^n; \quad v_n = \partial^n v/\partial y^n; \quad w_n = \partial^n w/\partial y^n \qquad (19)$$

as the functions of the inner n-order derivatives built upon outer derivatives of the n-1 order because of: i.) valitidy of the form (3) for arbitrary order of the velocity derivatives (10); ii.) there is possibility to write eqs. of the type (16) and (17) for y-derivatives of the n-order (19) differatiating eqs. (16), (17) step by step with respect to y. Then using the symbolism $\partial u_n/\partial x=u_{nx}$, $\partial w_n/\partial z=w_{nz}$ one obtains y-derivatives (19) in the form (18) i.e. recurrently

$$v_{n+1} = -(u_{nx}+w_{nz}-h_x v_{nx}-h_z v_{nz})/k^2$$
$$u_{n+1} = v_{nx}-h_x v_{n+1}; \quad w_{n+1} = v_{nz}-h_z v_{n+1} \tag{20}$$

Consequently, one can construct solution of velocity field in G in the form of the Taylor s series ($u=1,2,\ldots,N$)

$$u=\bar{u}+\sum_{n=1}^{N} u_n(y-h)^n/n! \quad v=\bar{v}+\sum_{n=1}^{N} v_n(y-h)^n/n!$$
$$w=\bar{w} \quad \sum_{n=1}^{N} w_n(y-h)^n/n! \tag{21}$$

Finally, the desired completing of system for free surface (10) is reached after integration of continuity eq. (16) along co-ordinate y from h_o to h using velocity field (21) and known kinematic condition on solid bottom h_o and surface h. It gives the wanted equation

$$h_t + \phi_x + \psi_z = 0. \tag{22}$$

Horizontal streams ϕ and ψ are functions of h_o, h, u, w and of their space and time derivatives. The system of eq. (10) and eq. (22) of N+2 order contains 3 unknowns: horizontal velocities u and w and free surface h; $\bar{v}$ can be eliminated through the kinematic condition $\tau = 0$ in eq. (7) . So it represents solution of the mentioned unknown boundary problem.

Streams ϕ and ψ used in (21) follow from (20) and (21) in various degree of approximation. In the zero degree the streams $\phi=\bar{u}H$, $\psi=\bar{w}H$ ($H=h-h_o$ is water depth) respond to shallow water theory where horizontal field does not depend on y.

The first degree approximation yields the linear velocity distribution along y giving the horizontal streams

$$\phi = H(\bar{u} - \frac{H}{2}u_1); \quad u_1 = (k^2\,\bar{v}_x+h_x(\bar{u}_x+\bar{w}_z-h_x\bar{v}_x-h_z\bar{v}_z))/k^2$$
$$\psi = H(\bar{w} - \frac{H}{2}\,w_1); \quad w_1 = (k^2\,\bar{v}_z+h_z(\bar{u}_x+\bar{w}_z-h_x\bar{v}_x-h_z\bar{v}_z))/k^2 \tag{23}$$

As it can be seen, even the result of the first degree of approximation does not seem to be simple. The completing eq. (22) becomes simpler only for such waves surface at which the second order terms in (23) can be neglected. Then eq. (22) will have the form

$$h_t+(H(\bar{u}-\frac{H}{2}\bar{v}_x))_x + (H(\bar{w} - \frac{H}{2}\bar{v}_z))_z=0 \tag{24}$$

The second order approximation in (21) gives more accurate expressions for streams, i.e.

$$\begin{aligned} \phi &= H(\bar{u} - \frac{H}{2} u_1 + \frac{H^2}{6} u_2) \\ \psi &= H(\bar{w} - \frac{H}{2} w_1 + \frac{H^2}{6} w_2) \end{aligned} \tag{25}$$

where, of course, boundary derivatives of the second order being expressed through u_1 and w_1 from (20) or (23) are

$$\begin{aligned} u_2 &= (k^2 v_{1x} + h_x(u_{1x} + w_{1z} - h_x v_{1x} - h_z v_{1z}))/k^2 \\ w_2 &= (k^2 v_{1z} + h_z(u_{1x} + w_{1z} - h_z v_{1x} - h_z v_{1z}))/k^2 \\ v_1 &= -(u_x + w_z - h_x v_x - h_z v_z)/k^2 \\ \bar{v} &= h_t + \bar{u} h_x + \bar{w} h_z \end{aligned} \tag{26}$$

i.e. quite complicated.

Even if neglecting the terms of higher order for free surface of fairly small curvature, the result completing eq.

$$\begin{aligned} & h_t + (H(\bar{u} - \frac{H}{2} \bar{v}_x - \frac{H^2}{6} (\bar{u}_x + \bar{w}_x)_x))_x + \\ & + (H(\bar{w} - \frac{H}{2} \bar{v}_z - \frac{H^2}{6} (\bar{u}_x + \bar{w}_z)_z))_z = 0 \end{aligned} \tag{27}$$

is still intricate. Nevertheless, obtained results can be more suitable even for numerical simulation of some hydraulics and hydrodynamics phenomena than other approach.

5 MODEL OF FILTRATION MOTION. CALCULATION OF FILTRATION VELOCITY FIELD IF FREE SURFACE IS KNOWN

Filtration motion is described by eq. (16) and dynamic equations

$$P_x + u/k_1 = 0; \; P_y + v/k_2 = 0; \; P_z + w/k_3 = 0 \tag{28}$$

for the pressure function $P = p/g + y + \text{const}$ and velocity vector (u, v, w); p is pressure, y vertical independent variable and (k_1, k_2, k_3) is the given filtration coefficient vector. Solution of the unknown boundary problem for eq. (28) seems to be even more effective because it leads to one equation for free surface only. Introducing dimensionless velocities $U = u/k_1$, $V = v/k_2$ and $W = w/k_3$ the continuity eq. (16) takes the form (assuming $k_1 \doteq k_3 \doteq k_o$)

$$U_x + aV_y + W_z = - F_o \qquad (29)$$

where $a = k_2/k_o$ and right hand side in (28)
$F_o = (Uk_{ox} + Vk_{2y} + Wk_{oz})/k_o$ is the function of velocity vector.

Eliminating P from eqs. (28) one gets relations which correspond to eqs. (17). Therefore, using the previous treatment the velocity field (21) is obtained. Moreover, it is simple to derive the boundary velocity field depending on boundary h. Use of eq. (3) for pressure function P on y=h and elimination of the outer derivatives P_x, P_y and P_z by means of eqs. (28) lead to two linear equations which together with kinematic condition on free surface give boundary velocity field

$$\bar{u} = -\bar{k}_o h_x(\bar{k}_2 + \bar{m}h_t)/D; \quad \bar{v} = \bar{k}_2(\bar{m}h_t - \bar{k}_o(h_x^2 + h^2{}_z))/D$$
$$\bar{w} = -\bar{k}_o h_z(\bar{k}_2 + \bar{m}h_t)/D; \quad D = \bar{k}_2 + \bar{k}_o(h_x^2 + h_z^2) \qquad (30)$$

as the functions of free surface derivatives.

Therefore, resulting equation (22) after elimination of the boundary velocities in (21) by means of eqs. (30) obtains the form

$$\bar{m}h_t = F(h_o, h, h_x, h_z, h_t, h_{xx}, h_{zz}, h_{xt}, h_{zt}, \ldots) \qquad (31)$$

Existence of eqs. (30) together with eqs. (21) allows to determine the filtration velocity field if the free surface h and m, k_1, k_2, k_3 are known what is surprising but important result.

6 CONCLUSIONS

Given solution of the unknown boundary problem at the same time evokes further questions. They are connected with the formulation of boundary conditions for obtained systems of boundary motion equations. For example eq. (31) represents an algebraic combination of quantities H and h and their space and time derivatives up to N+1 order. N is not, from theoretical aspect, limited. There appear, consequently, the problems of correctness and permissiveness of various forms of boundary conditions for the obtained systems. These have not yet been solved as a matter of principle. Above demonstrated possibility to state the filtration velocity vector in G as the function of time and space changes of free surface only is very interesting from practical point of wiew. Surprising matter is that such calculation does not require any other boundary conditions. It should be said that the mentioned result has been received through the treatment with basic equations of filtration motion without any simplifying assumptions. Its practical utilization has been prepared.

7 REFERENCES

[1] COURANT, R.--HILBERT,D.: Methoden der mathematischen Physik, 2 vol., New York, 1944.

[2] KORTEWEG, D.J.-de VRIES, G.: On the change of form of long waves advancing in a rectangular canal and a new type of long stationary waves. London, Dublin, Phil. Mag., ser. 5, 1985.

[3] KOSORIN, K.: Dam break wave singularities, XX. Congress IAHR, Moskva, 1983.

[4] KOSORIN, K.: Hydrodynamic theory of boundaries, DrSc thesis ÚHH SAV, Bratislava, 1983.

[5] KOSORIN, K.: Nonlinear waves approaching a shore and its numerical simulation. Proc. XXIII IAHR Congress, Ottawa, 1989.

[6] LAMB, H.: Hydrodynamics, Dover Publ., New York, 1945.

[7] MILES, J. W.: On Hamilton principle for surface waves. J.Fluid Mech., 83, 1977.

[8] OVSIANIKOV, L.V.: Ploskaja zadača o neustanoivšemisia dviženii židkosti so svodnymi granicami. Dinamika splošnoj sredy. Novosibirsk, 1972.

[9] PEREGRINE, D.H.: Long waves on a beach. J. Fluid Mech., 27, 1967.

[10] PRÜSSER, H.H.-ZIELKE, W.: An Extended Boussinesq--wave model for waves on current. XXIII. IAHR Congress, Ottawa, Proc. of Maritime hydraulics, 1989.

[11] SMIRNOV,

[12] STOKER, J.J.: Water Waves; Intesc. Publ., Inc. New York, 1957.

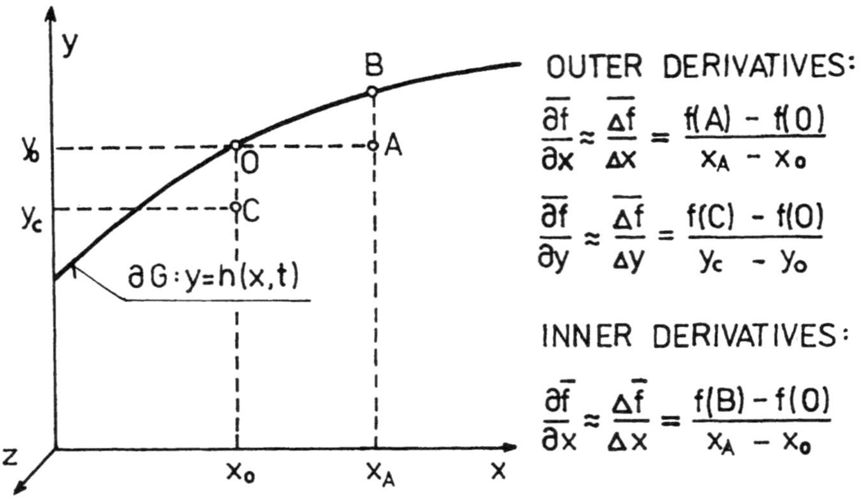

FIG.1 GEOMETRICAL INTERPRETATION OF INNER AND OUTER DERIVATIVES

The Free Surface Flow Induced by a Submerged Source or Sink from a Three-Layer Fluid

X. Wen, D.B. Ingham

Department of Applied Mathematical Studies, University of Leeds, Leeds LS2 9JT, England

ABSTRACT

In this paper the steady, irrotational, ideal fluid flow which is produced by a submerged source, or sink, from a region containing three homogeneous layers of fluid at different densities with gravity as the restoring force is considered. The mathematical theories that have been developed for two-layer fluids predict that there may be a stagnation or cusp point on the free surface immediately above the source, or sink. In the three-layer flow the situation is much more complicated and flows of the following type are possible: (i) stagnation point solutions on both of the free surfaces, (ii) cusp point solutions on both of the free surfaces, and (iii) a cusp point solution on one of the free surfaces and a stagnation point solution on the other free surface. In the present paper we consider the cases (i) and (ii). Based on the assumptions of steady, ideal, irrotational flow the problem is formulated as a system of nonlinear integral equations. In case (i) we obtain a solution which is valid for small values of the Froude number. In case (ii) it is found that there exist solutions for all values of the Froude number greater than unity but the position of the source, or sink, and the cusp attachment points are limited to a small region in space.

INTRODUCTION

Steady two-dimensional free surface flow produced by a submerged source, or sink, with gravity as the restoring force has been the subject of many papers. The situation in which a stagnation point exists directly above the source, or sink, in a two-layer fluid flow has been considered by Craya [1], Yih [2], Peregrine[3], Imberger [4], Vanden-Broeck, Schwartz and Tuck [5], Hocking and Forbes [6], and Mekias and Vanden-Broeck [8]. When a cusp point exists in the free surface directly above the source, or sink, in two-layer fluid flow then this has been considered by Tuck and Vanden-Broeck [7], Hocking [6], Collings [9], Vanden-Broeck and Keller [10], Hocking [11], King and Bloor [12] and Hocking [13].

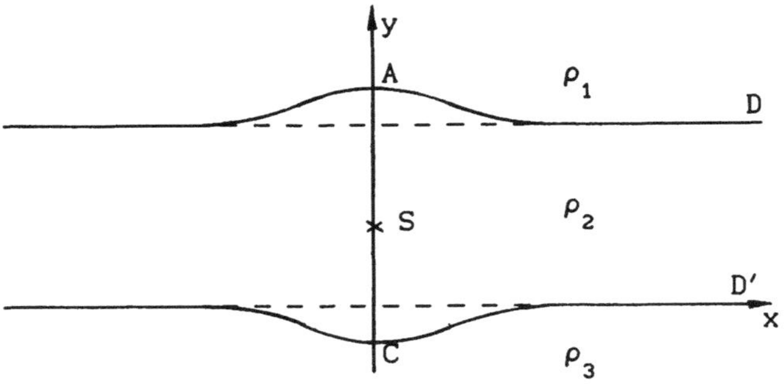

Fig. 1a Flows when there are stagnation points on both free surfaces (case(i)).

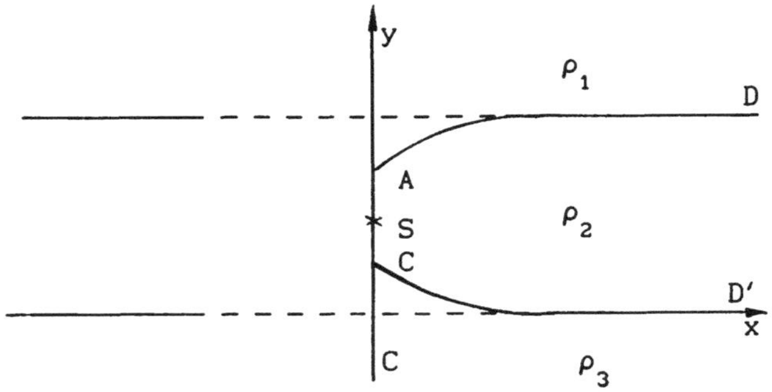

Fig. 1b Flows when there are cusp points on both free surfaces (case(ii)).

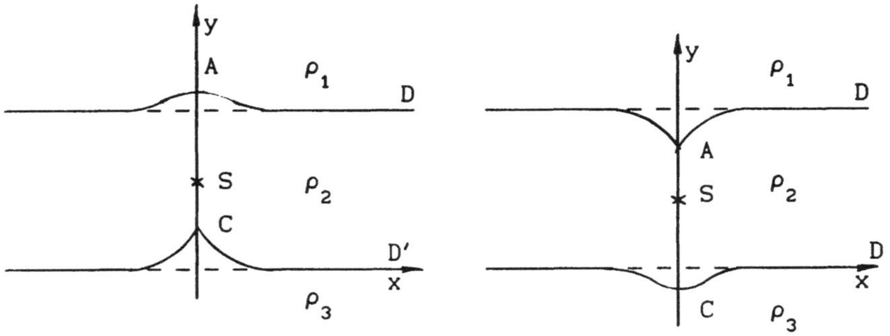

Fig. 1c Flows when there is cusp point on one free surface and a stagnation point on the other surface (case(iii)).

In this paper we consider the problem when the fluid is withdrawn from a fluid which consists of three homogeneous layers of fluid of different densities and which are separated by two interfaces. For the purposes of this paper we will assume that the fluid is withdrawn from the middle layer (see Fig. 1) but the mathematical technique can easily be extended to other situations. Although this problem has not previously been discussed there are many examples of where it occurs in practical situations, for example, the withdrawal of a warm liquid (water) which is above a cooler and more dense liquid (water) and for which there is air above the warm liquid (water). The warm liquid is bounded by two interfaces and it is withdrawn from a sink, S. We are interested in determining the behaviour of the flow so that we may assign the position of the sink so that only warm liquid is withdrawn. Because there are two interfaces then there exist three types of possible solutions, namely (i) stagnation point solutions on both of the free surfaces (see Fig. 1a), (ii) cusp point solutions on both free surfaces (see Fig. 1b), and (iii) a cusp point solution on one free surface and a stagnation point solution on the other free surface (see Fig. 1c). In the present paper we consider the cases (i) and (ii). In case (ii) it is found that there exist solutions for all values of the Froude number greater than unity, but the position of the source, or sink, and the cusp attachment points are limited to a small region in space. The results show that the three-layer problem is quite different from the two-layer problem.

PROBLEM FORMULATION AND SOLUTION

Because Bernoulli's theorem holds on both free surfaces then one has to enforce a boundary condition on the free surface which depends on the square of the speed of the fluid. Hence, once a solution has been obtained then the potential flow solution may be reversed and the results for a submerged source may be obtained from the solution for a submerged sink by merely a change of sign in the solution. For convenience we therefore consider here the flow due to a source.

We consider three layers of fluid each of which have different constant densities ρ_1, ρ_2 and ρ_3 ($\rho_1 < \rho_2 < \rho_3$). The light and heavy fluids occupy a semi-infinite region of space and are separated by an infinite strip of fluid of height H and density ρ_2 in the middle layer, see Fig. 2a. A source of strength 2q is now situated at the point S. We assume that on both of the free surfaces the attachment points occur vertically above and below the point S, at the points A and C, respectively. The flow is assumed to be steady, inviscid, incompressible and irrotational and only moves under the action of gravity.

A coordinate system $Z=X+iY$ is introduced with the X axis along the undisturbed level of the lower surface of the middle layer fluid and the Y axis through the source S. In Fig. 2a, H_s, H_A and H_C are the heights of the source, and the attachment points A and C, respectively, above the undisturbed level of the lower surface of the middle layer fluid. At large distances from the source the flow in the middle layer of fluid is uniform and has speed U away from the source. Since the flow is symmetrical we need only consider the right hand half-plane $X \geq 0$. The complex velocity potential is defined by $W = \Phi + i\Psi$, where Φ is the velocity potential and Ψ is the streamfunction. Without any loss of generality we choose $\Psi = 0$ to be the surface SCD′ and $\Psi = UH$ on the surface SAD, see Fig 2b.

We now choose a new non-dimensional coordinate $z = (x+iy)/H$, and Fig. 2a represents the flow in the physical plane where $h_A = H_A/H$, $h_s = H_s/H$, $h_C = H_C/H$. The complex velocity potential may be non-dimensionalised, $w = \phi + i\psi = (\Phi + i\Psi)/UH$, and the strip in Fig. 2b represents the flow in the W-plane and all the speeds have now been non-dimensionalised with respect to U.

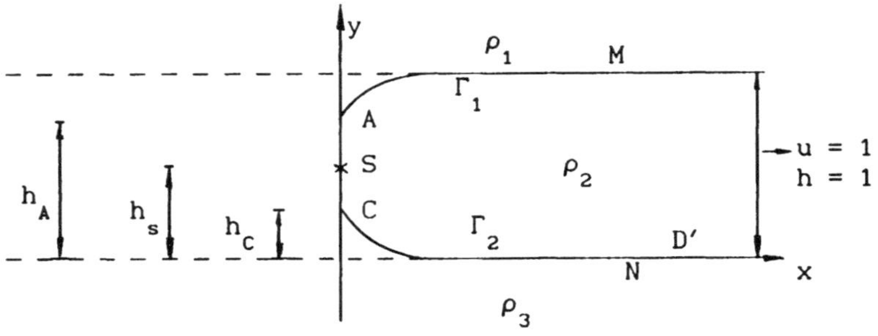

Fig. 2a The physical z plane

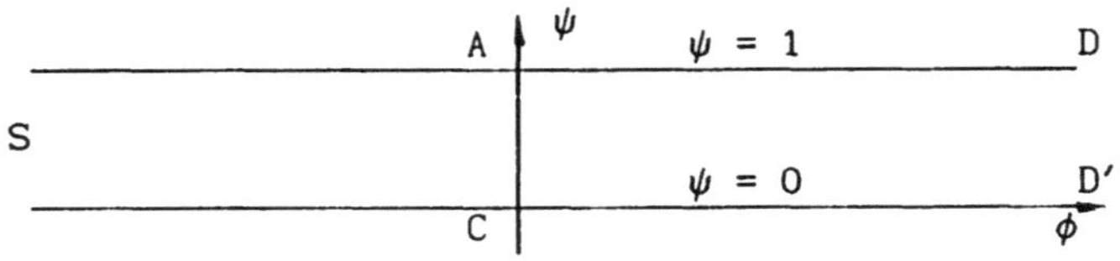

Fig. 2b The W-plane

Along the free surfaces AD and CD′ we apply Bernoulli's equation in the form

$$\frac{u^2}{2} F_{r1}^2 + y = \frac{F_{r1}^2}{2} + 1 \qquad \text{on AD} \qquad (1)$$

$$\frac{u^2}{2} F_{r2}^2 - y = \frac{F_{r2}^2}{2} \qquad \text{on CD}' \qquad (2)$$

where the Froude numbers are defined by $F_{r1} = U/\sqrt{g_1 H}$ $F_{r2} =$

$U/\sqrt{g_2 H}$, where $g_1 = g(\rho_2 - \rho_1)/\rho_2$ and $g_2 = g(\rho_3 - \rho_2)/\rho_2$. We next consider the complex velocity

$$\frac{dw}{dz} = ue^{-i\theta} \tag{3}$$

in which u is the non-dimensionalised fluid speed and θ is the angle that the velocity vector on the free surface makes with the positive x axis. Then the logarithm of the complex velocity may be written in the following form:

$$\Omega = i \, \mathrm{Ln}\left(\frac{dW}{dZ} \right) = \theta + i\tau \tag{4}$$

where $\tau = \mathrm{Ln}(u)$. We will now consider the cases (i) and (ii), namely when there are stagnation point solutions on both of the free surfaces (i), and when there are cusp point solutions on both free surfaces (ii).

(i) STAGNATION POINTS ON BOTH FREE SURFACES

We consider the case in which the attachment points A and C are stagnation points on both of the free surfaces. The physical plane and the complex potential velocity plane are shown in Figs. 3a and 3b, respectively. We now map the flow domain in the complex potential plane onto the upper half-plane of the auxiliary t-plane by

$$t = e^{\pi w} \qquad \text{or} \qquad w = \ln\left(\frac{t}{\pi}\right) \tag{5}$$

see Fig. 3c.

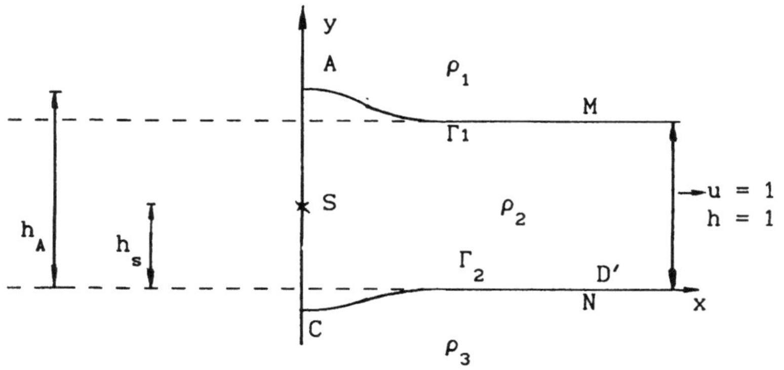

Fig. 3a The physical z plane

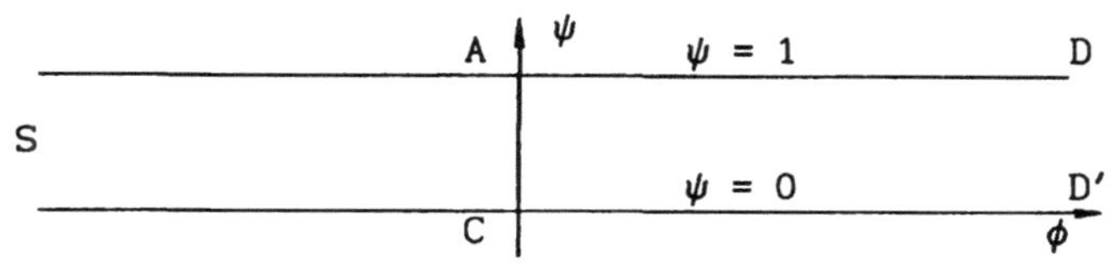

Fig. 3b The W-plane

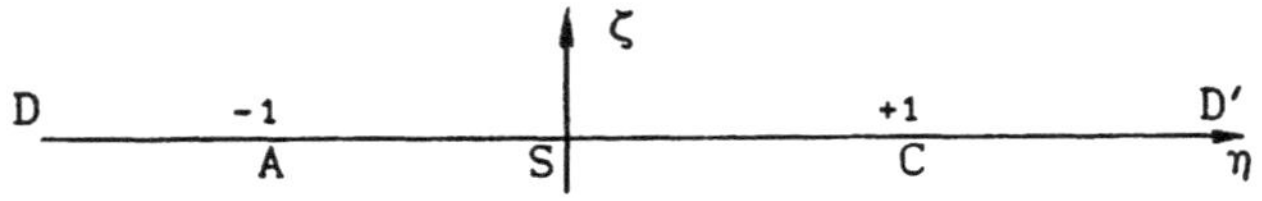

Fig. 3c The t-plane

The boundary conditions on the real η axis of the t-plane are

$$
\begin{aligned}
\mathrm{Re}\Omega(\eta) &= \tfrac{1}{2}\pi && \eta < -1 \\
\mathrm{Re}\Omega(\eta) &= \theta_1(\eta) && -1 < \eta < 0 \\
\mathrm{Re}\Omega(\eta) &= -\tfrac{1}{2}\pi && 0 < \eta < t_c \\
\mathrm{Re}\Omega(\eta) &= \theta_2(\eta) && t_c < \eta
\end{aligned}
\qquad (6)
$$

where θ_1 and θ_2 are the angles that the velocity vector on the free surfaces AD and CD′ make with the positive x axis. In the upper half t-plane the problem now reduces to a Dirichlet boundary-value problem in which the value of the real part of the analytic function Ω is given on the real axis η. Therefore in the upper half t-plane the value of Ω is given by the Schwartz formula [14], namely,

$$\Omega(t) = \frac{1}{\pi i}\int_{-\infty}^{+\infty} \frac{\mathrm{Re}\Omega(\eta)}{\eta - t}\, d\eta + ib \qquad (7)$$

where b is a real constant to be determined.

At large distances from the source the non-dimensional speed is unity and the free surface is horizontal, i.e.

$$\lim_{t\to\infty} \Omega(t) = 0, \quad \text{or} \quad \lim_{t\to\infty} \tau(t) = 0 \quad \text{and} \quad \lim_{t\to\infty} \theta(t) = 0 \qquad (8)$$

and this results in b=0.

On substitution of condition (6) into equation (7) yields

$$
\begin{aligned}
\Omega(t) = {} & \frac{1}{2i}\int_{-1}^{0} \frac{1}{\eta - t}d\eta + \frac{1}{\pi i}\int_{-\infty}^{-1} \frac{\theta_1(\eta)}{\eta - t}d\eta \\
& - \frac{1}{2i}\int_{0}^{1} \frac{1}{\eta - t}d\eta + \frac{1}{\pi i}\int_{1}^{+\infty} \frac{\theta_2(\eta)}{\eta - t}d\eta
\end{aligned}
\qquad (9)
$$

where we have taken t to be in the upper half plane and the Cauchy Principal Value as t approaches the real axis, η. On separating the real and imaginary parts of Ω then the speed u(t) on the free surface is given by is given by:

$$
\begin{aligned}
\mathrm{Ln}(u(t)) = {} & -\frac{1}{\pi} ⨍_{-\infty}^{-1} \frac{\theta_1(\eta)}{\eta - t}d\eta - \frac{1}{\pi} ⨍_{+1}^{+\infty} \frac{\theta_2(\eta)}{\eta - t}d\eta \\
& + \frac{1}{2}\mathrm{Ln}[(1+t)(1-t)/t^2]
\end{aligned}
\qquad (10)
$$

where $⨍$ denotes the Cauchy Principal Value.

Now consider the problem in the physical plane where the unknown functions are the profiles of the free surfaces AD and

CD'. These include the positions of the stagnation points, the speed of the fluid on SA and SC and the potential function on the free surfaces. On each of the free surfaces AD and CD' we choose the points A and C as the origin points and the arc length coordinate to be s. When we take the arc length s to be the independent variable then the angle, the speed and the potential function on the free surfaces are functions of s and from equation (5) we have

$$d\eta = \pi\eta(s)u(s)ds \tag{11}$$

and

$$\frac{d\phi(s)}{ds} = u(s) \tag{12}$$

$$\left.\begin{aligned}\frac{dx(s)}{ds} &= \cos\theta(s)\\ \frac{dy(s)}{ds} &= \sin\theta(s)\end{aligned}\right\}\quad \text{on the free surfaces} \tag{13}$$

$$\mathrm{Ln}(u(s)) = -\int_{\Gamma_1}\frac{\theta_1(l)\eta(l)u(l)}{\eta(l)-t(s)}dl - \int_{\Gamma_2}\frac{\theta_2(l)\eta(l)u(l)}{\eta(l)-t(s)}dl$$
$$+\ \frac{1}{2}\,\mathrm{Ln}[1+t(s)][1-t(s)]/t^2(s) \quad s \in \Gamma_1,\ \Gamma_2 \tag{14}$$

NUMERICAL METHOD AND ITERATIVE PROCEDURE

Along Γ_1 and Γ_2 the infinite integrals are truncated at some points M and N, say, (see Fig. 3a), which are so far from the stagnation points that the angles, $\theta_1(s)$ and $\theta_2(s)$, at these points on the free surfaces are less than some preassigned small value ε, say. In all the examples taken in this paper it is found that $\varepsilon = 10^{-6}$ is sufficiently small. The mesh nodes on Γ_1 and Γ_2 are distributed according to the magnitude in the change of the angle in a similar manner to that in the finite element and boundary element methods and the integration in each interval was evaluated by a Gaussian procedure.

In this paper we only study the case in which $Fr_1 = Fr_2$ and the values of the potential function at the stagnation points A and C are equal and, therefore, the flow pattern is symmetrical about the horizontal plane passing through the source point S. In each computation of the numerical scheme the values of Fr_1 and Fr_2 were fixed, the potential function and the profiles of the free surfaces were obtained through the following iterative procedure:

a. Assume the initial profiles of the surfaces Γ_1 and Γ_2,

b. Use expressions (1) , (2) and (12) to evaluate the speed $u^0(s)$ and the potential functions $\phi^0(s)$ on the free surfaces.

d. Insert $u^0(s)$ and $\phi^0(s)$ into the right hand sides of equation (14) to obtain the speed $u^1(s)$ on the free surfaces AM and CN.

e. From expressions (1) and (2) the new vertical coordinate y is determined and a spline technique is used to obtain the angles $\theta_1(s)$ and $\theta_2(s)$.

These values are then used as new approximations and the iterative procedure from step b to step e is repeated until the numerical procedure has converged.

RESULTS

We used the numerical procedure described in the above section to compute solutions for various values of the Froude number. The initial profiles of the free surfaces were taken to be horizontal planes. A solution was first obtained for a Froude number of 0.05 and solutions were then obtained for slightly larger values of the Froude number, using the previous solution for the angles of the free surfaces as an initial guess for the next solution. When the solutions were accurate to 3 significant figures the calculations are terminated. Solutions were obtained for values of the Froude number up to 0.3 and beyond this value it is difficult to obtain convergent solutions. In order to compare the behaviour of three-layer and two-layer flows we also obtained the solutions for the two-layer flow when the source is located at the undisturbed level of the bottom of the fluid in the middle layer. The free surface profiles for Froude numbers 0.1, 0.2, 0.3 are shown in Fig. 4. It is observed that the free surface is much steeper in the case of three-layer flows than in the corresponding case of two-layer flows. Therefore, the critical Froude number below which the solution exists should be smaller in three-layer flows than in two-layer flows.

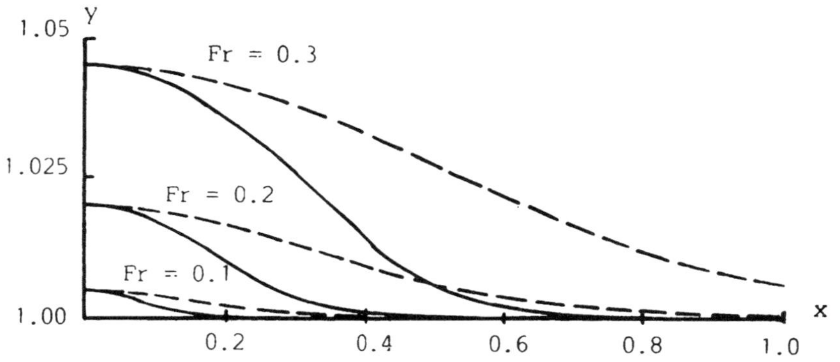

Fig. 4 A comparison of the free surfaces profiles for three layers flows ——, for the corresponding two-layer flows - - - -

(ii) CUSP POINTS ON BOTH FREE SURFACES

We now consider the case in which the attachment points A and C are both cusp points. The physical plane and the complex potential velocity plane are shown in Figs. 5a and 5b, respectively. The flow domain in the complex potential plane is now mapped onto the upper half-plane of the auxiliary plane by the transformation (5). In this case we use the technique appropriate to a boundary problem and to do this we use the following boundary conditions on the real η axis of the t-plane:

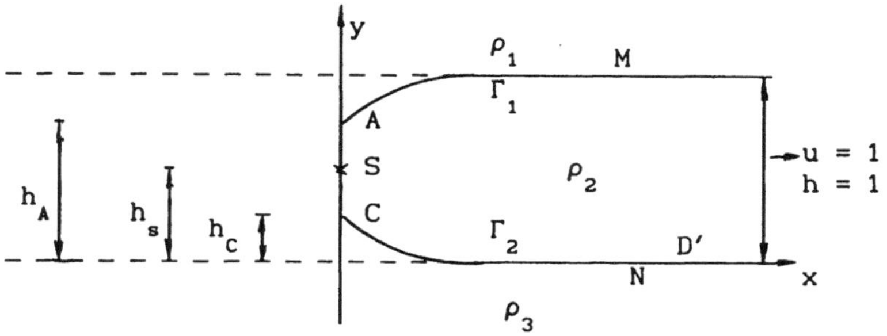

Fig. 5a The physical z plane

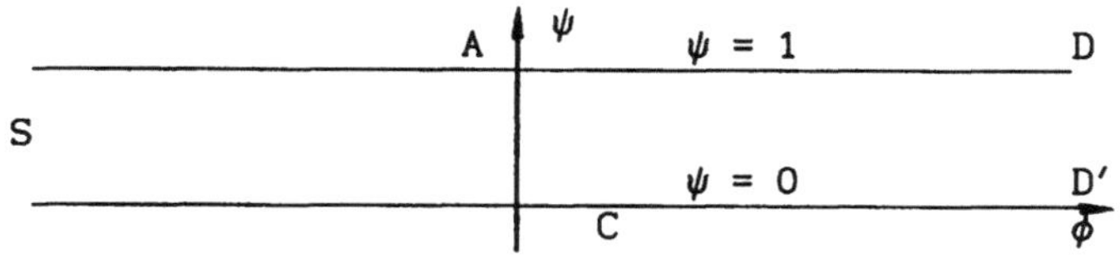

Fig. 5b The W-plane

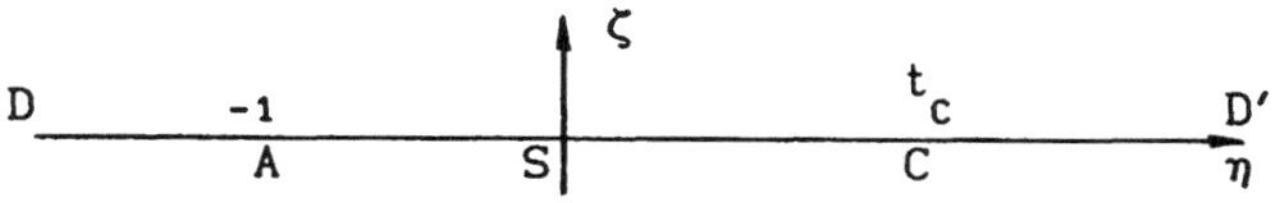

Fig. 5c The t-plane

$$\begin{aligned}
\mathrm{Im}\Omega(\eta) &= \tau(\eta) = \frac{1}{2}\,\mathrm{Ln}\left(1 + \frac{2[1 - y(\eta)]}{F_{r1}^2}\right) && \eta < -1 \\
\mathrm{Re}\Omega(\eta) &= \frac{\pi}{2} && -1 < \eta < 0 \qquad (15)\\
\mathrm{Re}\Omega(\eta) &= -\frac{\pi}{2} && 0 < \eta < t_c \\
\mathrm{Im}\Omega(\eta) &= \tau(\eta) = \frac{1}{2}\,\mathrm{Ln}\left(1 + \frac{2y(\eta)}{F_{r2}^2}\right) && t_c < \eta
\end{aligned}$$

The problem now reduces to a mixed boundary-value problem in the upper half t-plane. By referring to the general solution of the Riemann-Hilbert problem [14] we obtain the solution of Ω in the form

$$\Omega(t) = \frac{\sqrt{(t+1)(t-t_c)}}{\pi}\left(\int_{-\infty}^{-1} \frac{\tau(\eta)}{\sqrt{(\eta+1)(\eta-t_c)}\ (\eta-t)}\,d\eta\right.$$

$$+ \frac{\pi}{2}\int_{-1}^{0} \frac{1}{\sqrt{(\eta+1)(t_c-\eta)}(\eta-t)}\,d\eta - \frac{\pi}{2}\int_{0}^{t_c} \frac{1}{\sqrt{(\eta+1)(t_c-\eta)}(\eta-t)}\,d\eta$$

$$- \left.\int_{t_c}^{\infty} \frac{\tau(\eta)}{\sqrt{(\eta+1)(\eta-t_c)}(\eta-t)}\,d\eta\right\} \tag{16}$$

where $X(t)=\sqrt{(t+1)(t-t_c)}$ is a particular solution for the homogeneous equation in which the right hand side of the expression (15) is zero. We now choose the branch cut for X(t) such that

$$\begin{aligned} X^+(\eta) &= \sqrt{(\eta+1)(\eta-t_c)} && \eta < -1 \\ X^+(\eta) &= -i\sqrt{(\eta+1)(t_c-\eta)} && -1 < \eta < t_c \\ X^+(\eta) &= -\sqrt{(\eta+1)(\eta-t_c)} && t_c < \eta \end{aligned} \tag{17}$$

Now we consider solutions for $1 < F_{r1}, F_{r2} < \infty$ and in this case the nonlinear integral equation (16) has no closed form solution but we may obtain numerical solutions by an iterative method. First, we develop an iterative method to obtain the potential function at the point C and the profiles of the free surfaces and then we may determine the positions of the sink S and the cusp points A and C.

In equation (16) we take t to be in the upper half plane and then take the Cauchy Principal Value as it approaches the real axis, η. Separating the real and imaginary parts of Ω results in the angle $\theta(t)$ on the free surface AD to be given by:

$$\theta_1(t) = \frac{\sqrt{(t+1)(t-t_c)}}{\pi}\left\{\int_{-\infty}^{-1} \frac{\tau(\eta)}{\sqrt{(\eta+1)(\eta-t_c)}\,(\eta-t)}\,d\eta\right.$$

$$\left. - \int_{t_c}^{\infty} \frac{\tau(\eta)}{\sqrt{(\eta+1)(\eta-t_c)}(\eta-t)}\,d\eta\right\} + \arcsin \frac{t(1-t_c)-2t_c}{t(t_c+1)}$$

$$-\infty < t < -1 \tag{18}$$

and the angle on the free surface CD′ to be given by

$$\theta_2(t) = \frac{\sqrt{(t+1)(t-t_c)}}{\pi}\left\{-\int_{-\infty}^{-1} \frac{\tau(\eta)}{\sqrt{(\eta+1)(\eta-t_c)}\,(\eta-t)}\,d\eta\right.$$

$$\left. + \int_{t_c}^{\infty} \frac{\tau(\eta)}{\sqrt{(\eta+1)(\eta-t_c)}(\eta-t)}\,d\eta\right\} + \arcsin \frac{t(1-t_c)-2t_c}{t(t_c+1)}$$

$$t_c < t < \infty \tag{19}$$

The speed of the fluid on AS and CS are then given by

$$u(t) = \frac{\sqrt{(t+1)(t-t_c)}}{\pi}\left\{-\int_{-\infty}^{-1} \frac{\tau(\eta)}{\sqrt{(\eta+1)(\eta-t_c)}\,(\eta-t)}\,d\eta\right.$$

$$+ \int_{t_c}^{\infty} \frac{\tau(\eta)}{\sqrt{(\eta+1)(\eta-tc)}(\eta-t)} d\eta \Bigg\} + \mathrm{Ln} \frac{-t(1+t_c)}{[\sqrt{t_c - t} + \sqrt{t_c(t+1)}]^2}$$

$$-1 < t < t_c \qquad (20)$$

and the boundary conditions at infinity (8) give

$$\frac{1}{\pi} \left\{ \int_{-\infty}^{-1} \frac{\tau(\eta)}{\sqrt{(\eta+1)(\eta-tc)}} d\eta - \int_{t_c}^{\infty} \frac{\tau(\eta)}{\sqrt{(\eta+1)(\eta-tc)}} d\eta \right\} + \arcsin \frac{1-t_c}{t_c+1} = 0$$

$$(21)$$

In the physical plane we may rewrite expressions (18), (19) and (21) as follows:

$$\theta_1(s) = \sqrt{[t(s)+1][t(s)-tc]} \Bigg\{ -\int_{\Gamma_1} \frac{\tau(l)\eta(l)u(l)}{\sqrt{[\eta(l)+1][\eta(l)-tc]}\,[\eta(l)-t(s)]} dl$$

$$- \int_{\Gamma_2} \frac{\tau(l)\eta(l)u(l)}{\sqrt{[\eta(l)+1][\eta(l)-tc]}[\eta(l)-t(s)]} dl \Bigg\}$$

$$+ \arcsin \frac{t(s)(1-t_c) - 2t_c}{t(s)(t_c+1)} \qquad s \in \Gamma_1 \qquad (22)$$

$$\theta_2(s) = \sqrt{[t(s)+1][t(s)-tc]} \Bigg\{ \int_{\Gamma_1} \frac{\tau(l)\eta(l)u(l)}{\sqrt{[\eta(l)+1][\eta(l)-tc]}\,[\eta(l)-t(s)]} dl$$

$$+ \int_{\Gamma_2} \frac{\tau(l)\eta(l)u(l)}{\sqrt{[\eta(l)+1][\eta(l)-tc]}[\eta(l)-t(s)]} dl \Bigg\} + \arcsin \frac{t(s)(1-t_c) - 2t_c}{t(s)(t_c+1)}$$

$$s \in \Gamma_2 \qquad (23)$$

in which $t_C = e^{\pi\phi(C)}$, $\phi(C)$ is the potential function at point C, and on Γ_1 $\eta(l) = -e^{\pi\phi(l)}$ and $t(s) = -e^{\pi\phi(s)}$ whilst on Γ_2, $\eta(l) = e^{\pi\phi(l)}$ and $t(s) = e^{\pi\phi(s)}$.

The infinity boundary condition (21) may then be expressed in the form

$$\frac{1}{\pi} \left\{ -\int_{\Gamma_1} \frac{\tau(l)\eta(l)u(l)}{\sqrt{[\eta(l)+1][\eta(l)-tc]}} dl + \int_{\Gamma_2} \frac{\tau(l)\eta(l)u(l)}{\sqrt{[\eta(l)+1][\eta(l)-tc]}} dl \right\}$$

$$+ \arcsin \frac{1-t_c}{t_c+1} = 0 \qquad (24)$$

The potential functions on the free surfaces are

$$\phi(s) = \int_0^s u(s)ds \qquad \text{on } \Gamma_1 \qquad (25)$$

$$\phi(s) = \phi_C + \int_0^s u(s)ds \qquad \text{on } \Gamma_2 \tag{26}$$

and the profiles of the free surfaces are given by

$$\left.\begin{aligned} y(s) &= h_A + \int_0^s \sin\theta_1(s)ds \\ x(s) &= \int_0^s \cos\theta_1(s)ds \end{aligned}\right\} \qquad \text{on } \Gamma_1 \tag{27}$$

$$\left.\begin{aligned} y(s) &= h_C + \int_0^s \sin_2\theta(s)ds \\ x(s) &= \int_0^s \cos\theta_2(s)ds \end{aligned}\right\} \qquad \text{on } \Gamma_2 \tag{28}$$

in which h_A and h_c are given, $y_1(s) \to 1$ as $s \to \infty$ and $y_2(s) \to 0$ as $s \to \infty$. The directions of all the linear integrations along Γ_1 and Γ_2 are in the direction of the velocity. If the position of A or C is known then the position of the source may be given by the expressions

$$h_s = h_A - \int_{-1}^{0} \frac{1 + t_c}{[\sqrt{t_c - t} + \sqrt{t_c(t+1)}]^2} e^{-G(t)}dt \tag{29}$$

$$h_s = h_c + \int_{0}^{t_c} \frac{1 + t_c}{[\sqrt{t_c - t} + \sqrt{t_c(t+1)}]^2} e^{-G(t)}dt \tag{30}$$

in which

$$G(t) = \sqrt{[t+1][t_c - t]} \left\{ \int_{\Gamma_1} \frac{\tau(l)\eta(l)u(l)}{\sqrt{[\eta(l)+1][\eta(l)-tc]}\;[\eta(l)-t]} dl \right.$$

$$\left. - \int_{\Gamma_2} \frac{\tau(l)\eta(l)u(l)}{\sqrt{[\eta(l)+1][\eta(l)-tc]}[\eta(l)-t]} dl \right\} \qquad -1 < t < t_c \tag{31}$$

NUMERICAL METHOD AND ITERATIVE PROCEDURE

Along Γ_1 and Γ_2 the infinite integrals are truncated at some points M and N, say. The mesh nodes on Γ_1 and Γ_2 are distributed according to the change in the magnitude of the angle that the free surface makes with the horizontal. In the neighborhood of the cusp points A and C this change is rapid and therefore the intervals are smaller here than elsewhere. The integrations in each interval was evaluated by using a Gaussian procedure.

In each computation the values of F_{r1} and F_{r2} were fixed and then the potential function at the point C and the profiles of the free surfaces were obtained through the following iterative procedure:

a. Assume the initial profiles of the surfaces Γ_1 and Γ_2,

b. Use expressions (1) and (2) to evaluate the speed $u^0(s)$,

c. Assume the potential function at the point C, ϕ_C, and using expressions (25) and (26) calculate the potential functions on the free surfaces. Equation (26) is used to check if ϕ_C is the root. In our calculations the bisection method was used to find the root of equation (26) in addition to obtaining $\phi^0(s)$.

d. Insert $u^0(s)$ and $\phi^0(s)$ into the right hand sides of equation (22), in order to obtain the angle $\theta^0(s)$ on the free surface AM, and equation (23), to obtain the angle $\theta^0(s)$ on the free surface CN.

e. From expressions (30) and (31) the new approximate profiles for the free surfaces are determined. These values are then used as new approximations and the iterative procedure from step b to step e is repeated until the numerical procedure has converged.

After obtaining the free surface profiles the position of the source can be obtained by using either of the expressions (25) or (26).

RESULTS

From equations (1) and (2) it may be seen that the results are symmetrical in F_{r1} and F_{r2}, i.e. if the vertical coordinate is $y_1(s)$ for $F_{r1} = a$ and $F_{r2} = b$, where a and b are two real numbers which are greater or equal to unity, then for $F_{r1} = b$, $F_{r2} = a$, the vertical coordinate $y_2(s) = 1-y_1(s)$. Thus we need only consider one fixed value of F_{r2} and then investigate the effects of varying F_{r1}.

Figure 6 shows the profiles of the free surfaces in which $F_{r2} = 1.2$ and $F_{r1} = 1$, 1.2, 1.5, 2 and ∞, and it is observed that the shape of the free surface is not very dependent on the value of F_{r1}. Computations have been performed for other values of F_{r2} and a similar observation may be made. It is of interest to note that no surfaces which contain a hump have been found and this contrasts with what is found in the case of a two-layer configuration, see [10] and [13]. The results also show that the position of the source is limited to the region $0.4718 < h_s < 0.5282$ for $1 < F_{r1} < +\infty$ and $1 < F_{r2} < +\infty$. This behaviour is quite different from the solutions obtained for the two-layer flow in which the source may be situated at any level above the bottom if the Froude number is sufficiently large, see [13]. The results also show that the cusp point C is limited to the region $0.3184 < h_C < 0.3515$ whilst the point A lies in the range $0.6485 < h_A < 0.6817$.

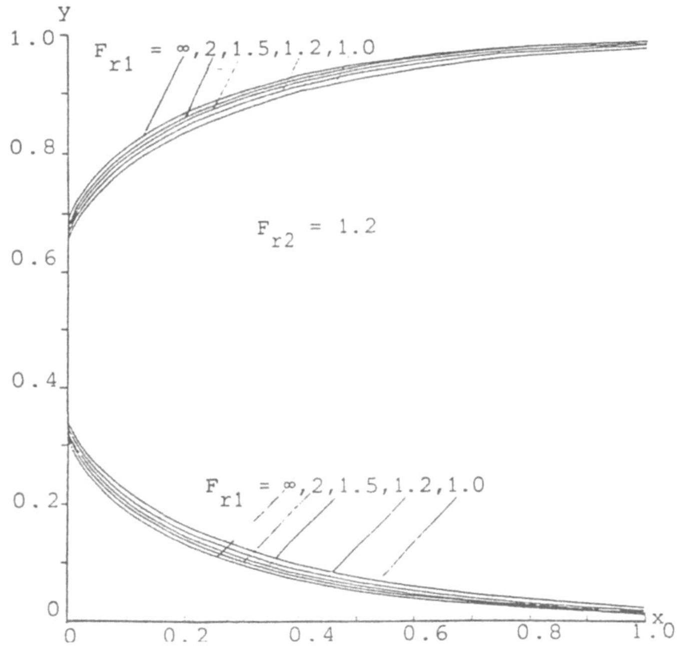

Fig. 6 The profiles of the free surfaces

CONCLUSION

In this paper we have presented some stagnation point and cusp-like solutions for the withdrawal of fluid by a submerged sink or source from a three-layer fluid. The results obtained show that cusp-like solutions on both the free surfaces exist for all values of the Froude numbers F_{r1} and F_{r2} which are greater than unity, but the position of the source is limited to a small region and the variation of the position of the cusp points are also small. These solutions indicate that the flow in a three-layer model is quite different from the flow in the two-layer model.

REFERENCES

1. A. Craya, Theoretical research on the flow of nonhomogeneous fluids, La Houille Blanche **4**, 44-55 (1949).
2. C. S. Yih, Dynamics of Nonhomogeneous Fluids, Macmillan, New York, (1965).
3. D. H. Peregrine, A line source beneath a free surface, University of Wisconsin Report 1248, (1972).
4. J. Imberger, Selective Withdrawal: a Review, 2nd Internat. Symp. on Stratified Flows, Trondheim (1980).
5. J.-M. Vanden-Broeck, L. W. Schwartz, and E. O. Tuck, Divergent low-Froude-number series expansion of non-linear free-surface flow problems, Proc. Roy. Soc. London Ser. **A**

136 207-224 (1978).
6. G. C. Hocking and L. K. Forbes, A note on the flow induced by a line sink beneath a free surface, J. Austral. Math. Soc. Ser. B **31**, (1989).
7. Tuck, E. O. and Vanden-Broeck J.-M., A cusp-like free-surface flow due to a submerged source or sink, J. Austral. Math. Soc. Ser. B **25**, 443-450 (1984).
8. H. Mekias and J.-M. Vanden-Broeck, Supercritical free-surface flow with a stagnation point due to a submerged source, Phys. Fluid A **1** (10), Oct. 1694-1697 (1989).
9. I. L. Collings, Two infinite Froude number cusped free surface flows due to a submerged line source or sink, J. Austral. Math. Soc. Ser. B **28**, 260-270 (1986).
10. J.-M. Vanden-Broeck and J. B. Keller, Free surface flow due to a sink, J. Fluid Mech. **175**, 109-117 (1987).
11. G. C. Hocking, Infinite Froude number solutions to the problem of a submerged source or sink, J. Austral Math. Soc. Ser. B **29**, 401-409 (1988).
12. A. C. King and M. I. G. Bloor, A note on the free surface induced by a surbmerged source at infinite Froude number, J. Austral Math. Soc. Ser. B **30**, 147-156 (1988).
13. G. C. Hocking, Critical withdrawal from a two-layer fluid through a line sink, 1989, J. Eng. Maths (In press).
14. N. I. Muskhelishvilli, Singular integral equations, (1966).

Free Surface Flow Formed by a Slot Suction

X. Wen, D.B. Ingham

Department of Applied Mathematical Studies,
The University of Leeds, Leeds LS2 9JT, England

ABSTRACT

The free surface flow formed by a slot suction in a channel is reexamined in this paper by solving a mixed boundary value problem. The fluid flow is assumed to be two-dimensional, steady, ideal and irrotational. The flow is sucked from the channel and a free surface is formed and the separation from the leading edge of the slot is smooth. The accurate results which are obtained in this paper are compared with those obtained in a previous study by Dewynne et al. [1] and show that there is a large discrepancy between the predictions of the location of the stagnation point.

INTRODUCTION

In the film cooling of gas turbine blades cool air is sucked from a channel inside the blade through a slot and this forms a film along the blade surface which insulates the surface from the ambient hot gases. The understanding of the behaviour of the fluid flow in the channel and slot is relevant to boundary-layer heat transfer calculations in order to determine the thermal behaviour of the turbine blade. A mathematical model for this flow may be described by a one-sided jet detaching from the edge of the slot in the channel and attaching to one side of the slot channel. This model was investigated by Dewynne et al. [1] where they studied the two-dimensional, irrotational flow of an inviscid and incompressible fluid from a slot in a channel.

Free streamline theory was first developed by Helmholtz [2] for treating two-dimensional jets. Compendia of this material are those of Birkhoff and Zarantonello [3], Gilbarg [4], Woods [5], Gurevich [6] and Milne-Thomson [7]. The complex variable technique has been employed and the problems were solved by using efficient numerical procedures. Classical hodograph techniques have been used and these require a series of mappings. This combination of maps requires knowledge of the shape of the hodograph domain and therefore this technique is of very restrictive use. For free surface problems the boundary of the fluid domain consists of a solid boundary on which the boundary conditions are given and a free surface on which the kinetic condition is also specified. In such circumstances we may transform this kind of problem into a Riemann-Hilbert mixed value boundary problem. Through a proper choice of a basic solution for the homogeneous problem we may express the problem in a form of integral equations and analytical solutions may be possible for some simple problems by using a singular integration technique [8]. The power of this method is not only that the technique is suitable for complex geometries but also that previous techniques, such as that described by Woods [5] or the Schwartz-Christoffel transformation technique, are special cases of this procedure.

In this paper we reexamine the free surface flow formed by a slot suction in a channel. The problem is solved by using an efficient and accurate numerical procedure and this has enabled us to obtain more accurate results than those obtained by Dewynne et al. [1].

PROBLEM FORMULATION AND SOLUTION

We consider the steady flow in a channel with a slot on one side. A co-ordinate system $z = x + iy$ is introduced with the origin at the separation point B and the geometry of the channel in the physical z-plane is shown in Fig. 1a. An ideal irrotational fluid flows along an infinitely long channel of width H, to which there is attached an infinitely deep slot of width L. The flow is uniform at large distances from the slot, i.e. $x \longrightarrow \pm\infty$, and the speed of the fluid at the points A and E are U_A and U_E, respectively. As the flow approaches the slot it divides and part of it goes through the slot. The dividing streamline meets the surface of the slot at the stagnation point, S, and the coordinates of this point is denoted by Z_s. The flow in the slot has a free surface, which originates at the detachment point B and extends to infinity, C. The x component of the coordinate of the point C is denoted by x_c and the fluid depth at C is denoted by H_c. The streamline BC is bounded on one side by a region of constant pressure, $p=p_s$, and the upstream static pressure is denoted by p_∞, with $p_\infty > p_s$. In this paper we neglect gravity and surface tension effects and therefore the speed of the fluid on the free surface BC is a constant, U_s.

Now define the complex velocity potential $W = \phi + i\,\psi$, where ϕ is the velocity potential and ψ is the streamfunction. The volumetric flux of fluid at large distances upstream is given by $q_1=HU_A$, and $q_2= U_sH_C$ denotes the mass flow rate into the slot. The conservation of mass then requires that

$$U_E / U_A = 1 - k \tag{1}$$

in which $k = q_2/q_1$.

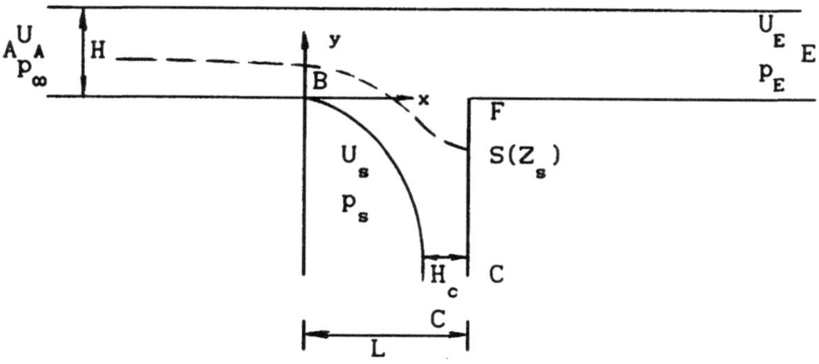

Fig. 1a The Physical Plane

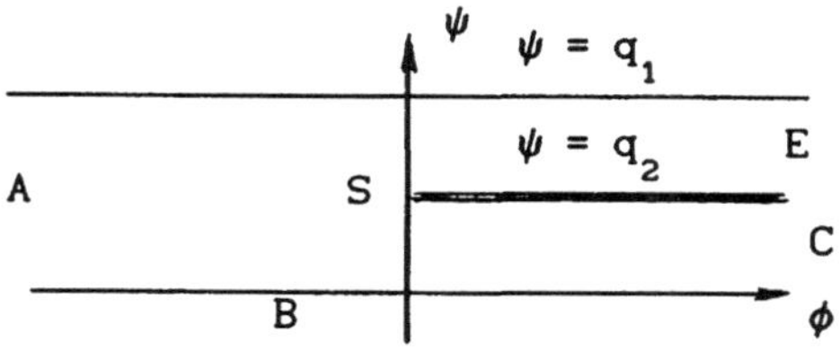

Fig. 1b The W-Plane

We may specify the value of the potential function at any point and therefore at the stangnation point S we enforce $\phi_s = 0$. The flow domain in the complex velocity potential plane is that as shown in Fig. 1b and it consists of an infinite strip of height q_1 with a slit starting at the point S, which is the stagnation point, and extending to $x = +\infty$. The amount of flow q_2 going into the slot is measured by the height between the bottom of the strip and the slit. We next consider the complex velocity

$$\frac{dW}{dz} = Ue^{-i\theta} \tag{2}$$

in which U is the fluid speed and θ is the angle that the velocity vector makes with the positive x axis. Then the logarithm of the complex velocity is

$$\Omega = \mathrm{Ln}\left(\frac{1}{U_s}\frac{dW}{dz} \right) = \tau - i\theta \tag{3}$$

where $\tau = \mathrm{Ln}(\frac{U}{U_s})$.

We will consider the three cases as discussed by Dewynne et al. [1], namely when the stagnation point is (i) at the trailing corner F, (ii) lies along the slot wall FC, and (iii) lies on the downstream channel wall, FE.

(i) THE STAGNATION POINT IS AT THE TRAILING CORNER

We begin by considering the case in which the stagnation point is at the trailing corner, F. The physical plane and the complex potential velocity plane are shown in Figs. 2a and 2b, respectively. We now map the flow domain in the complex potential plane onto the upper half-plane of the auxiliary t-plane by the transformation

$$W = \frac{q_1}{\pi}\Big[\, k \,\mathrm{Ln}\frac{k}{1-k} + \mathrm{Ln}(2-2k) + i\pi$$
$$- k\,\mathrm{Ln}(t+1) - (1-k)\mathrm{Ln}(t-1)\Big] \qquad (4)$$

in which

$$t_s = 2k - 1, \qquad -1 \le t_s \le 1, \qquad (5)$$

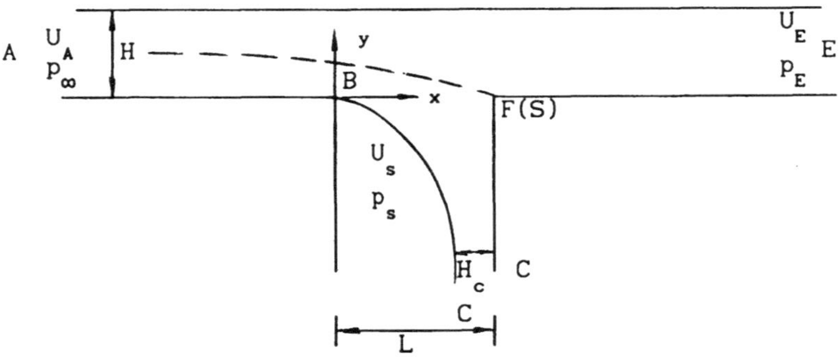

Fig. 2a The Physical Plane

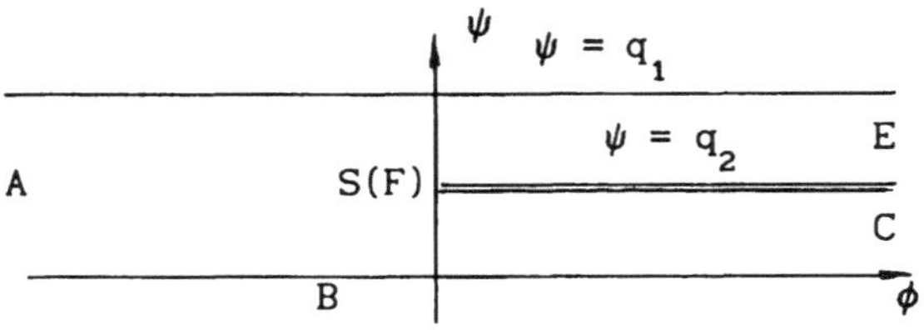

Fig. 2b The W-Plane

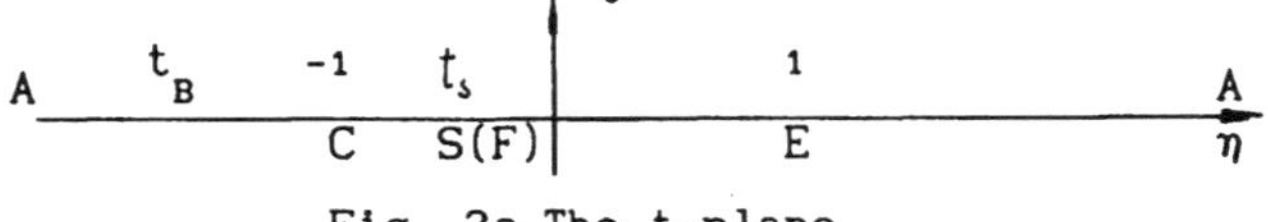

Fig. 2c The t-plane

so that the points A, C and E are mapped into the points ∞, -1 and 1 (see. Fig. 2c).

The transformation from the W-plane to the t-plane can be written in differential form as

$$\frac{dW}{dt} = \frac{t_s - t}{t^2 - 1} \tag{6}$$

On the free surface BC the velocity $U = U_s$, i.e. $\tau(\eta) = 0$. On the walls AE, AB and FE the direction of the velocity vector is such that $\theta = 0$, and on the wall FC, $\theta = -\frac{\pi}{2}$, so the boundary conditions for the function Ω on the real η axis of the t-plane may now be expressed as follows:

$$\begin{aligned}
&\mathrm{Im}\Omega(\eta) = 0 && -\infty < \eta < t_B \\
&\mathrm{Re}\Omega(\eta) = \tau = 0 && t_B < \eta < -1 \\
&\mathrm{Im}\Omega(\eta) = \frac{\pi}{2} && -1 < \eta < t_s \\
&\mathrm{Im}\Omega(\eta) = 0 && t_s < \eta < +\infty
\end{aligned} \tag{7}$$

These boundary conditions show that in the intervals $-\infty < \eta < t_B$ and $-1 < \eta < \infty$ the imaginary part of the complex function Ω is known, and in the interval $t_B < \eta < -1$ its real part is known. Therefore the problem has been reduced to a Riemann-Hilbert mixed value boundary problem. By referring to the general solution of the Riemann-Hilbert problem [9] we obtain the solution of Ω in the form

$$\Omega(t) = \frac{X(t)}{\pi}\int_{-1}^{t_s} \frac{\frac{\pi}{2}}{X^+(\eta)(\eta - t)}\, d\eta \tag{8}$$

where $X(t)=\sqrt{(t - t_B)(t + 1)}$ is a particular (or basic) solution for the homogeneous solution in which the right hand of the expression (7) is zero. Further, as t approaches the real η axis from above we choose a branch cut for X(t) such that

$$\begin{aligned}
&X^+(\eta) = \sqrt{(\eta - t_B)(\eta + 1)} && \eta < t_B \\
&X^+(\eta) = -i\sqrt{(t_B - \eta)(\eta + 1)} && t_B < \eta < -1 \\
&X^+(\eta) = -\sqrt{(t_B - \eta)(- \eta - 1)} && t_B < \eta < -1
\end{aligned} \tag{9}$$

By taking the boundary value on the real axis η and separating the real and imaginary parts of Ω, one obtains the speed of the fluid on SE and AE as follows:

$$\tau(t) = \mathrm{Ln}\left(\frac{U(t)}{U_s}\right) = \mathrm{Ln}\frac{\sqrt{(t - t_s)(-t_B - 1)}}{\sqrt{(t_s - t_B)(t + 1)} + \sqrt{(t_s + 1)(t - t_B)}} \tag{10}$$

or

$$\frac{U(t)}{U_s} = \frac{\sqrt{(t - t_s)(-t_B - 1)}}{\sqrt{(t_s - t_B)(t + 1)} + \sqrt{(t_s + 1)(t - t_B)}} \tag{11}$$

where $-1 < t < +\infty$.

Further, the direction on the free surface BC is given by

$$\theta(t) = -\frac{1}{2}\arcsin\frac{(t_s - t_B)(t + 1) + (t - t_B)(t_s + 1)}{(t_s - t)(-1 - t_B)} - \frac{\pi}{4} \tag{12}$$

where $t_B < t < -1$.

In the physical plane the profiles of the solid wall and the free surface are given by

$$dz = \frac{1}{U_s} e^{-\Omega(t)}\frac{dW}{dt}dt = \frac{1}{U_s}e^{-\tau(t)}[\cos\theta(t) + i\sin\theta(t)]\frac{dW}{dt}dt \tag{13}$$

On substituting expressions (6) and (12) into equation (13) and then integrating with respect to t from t_B to -1 yields the x coordinate of the point C, namely,

$$x_c = \frac{q_1}{\pi U_s}\sqrt{\frac{t_s - t_B}{-1 - t_B}}\,I + \frac{q_2}{U_s} = \frac{q_1}{\pi U_s}\left(\sqrt{\frac{t_s - t_B}{-1 - t_B}}\,I + k\right) \tag{14}$$

where

$$I = \int_{t_B}^{-1}\sqrt{\frac{t_s - t_B}{-1 - t}}\,\frac{1}{1 - t}\,dt = \mathrm{Ln}\frac{\sqrt{\frac{t_s - t_B}{-1 - t_B}} + 1}{\sqrt{\frac{t_s - t_B}{-1 - t_B}} - 1}$$

$$+ \sqrt{\frac{1 - t_s}{2}}\,\mathrm{Ln}\frac{\sqrt{\frac{t_s - t_B}{-1 - t_B}} + \sqrt{\frac{1 - t_s}{2}}}{\sqrt{\frac{t_s - t_B}{-1 - t_B}} - \sqrt{\frac{1 - t_s}{2}}} \tag{15}$$

Thus the width of slot is given by

$$L = x_c + \frac{q_2}{U_s} = \frac{q_1}{\pi U_s}\sqrt{\frac{t_s - t_B}{-1 - t_B}}\,I + \frac{q_2}{U_s} = \frac{q_1}{\pi U_s}\left(\sqrt{\frac{t_s - t_B}{-1 - t_B}}\,I + k\right) \tag{16}$$

When $t \to 1$ the expression (11) gives the velocity at the point E, namely,

$$\frac{U_E}{U_s} = \frac{\sqrt{(1 - t_s)(-t_B - 1)}}{\sqrt{2(t_s - t_B)} + \sqrt{(t_s + 1)(1 - t_B)}} \tag{17}$$

and as $t \to \infty$ the velocity at the point A is given by

$$\frac{U_A}{U_s} = \frac{\sqrt{-t_B - 1}}{\sqrt{(t_s - t_B)} + \sqrt{(t_s + 1)}} \tag{18}$$

The expressions (17), (18) and (1) yield the relationship between parameters t_B and t_s and

$$\frac{U_E}{U_A} = \frac{(\sqrt{t_s - t_B} + \sqrt{t_s + 1}\,)\sqrt{1 - t_s}}{\sqrt{2(t_s - t_B)} + \sqrt{(t_s + 1)(1 - t_B)}} = 1 - k = \frac{1}{2}\,(1 - t_s) \tag{19}$$

Using expressions (16) and (18) one obtains the result that

$$\frac{H}{L} = \frac{q_1/U_A}{L} = \frac{\pi\sqrt{-t_B - 1}}{\left(\sqrt{\dfrac{t_s - t_B}{-1 - t_B}}\;I + k\right)(\sqrt{t_s - t_B} + \sqrt{t_s + 1})} \tag{20}$$

It is observe that when the parameter k is given, i. e. t_s is given, then the parameter t_B is determined by expression (19) and then one can determine the speed U_s on the free surface through expression (18) and the width of the channel using expression (20).

Table 1 shows the results obtained for U_s/U_A and H/L as a function of q_2/q_1 as obtained by the present work by using double-precision arithmetic and the results as obtained by Dewynne et al. [1]. It is observed that the agreement between the two methods is extremely good.

q_2/q_1	U_s/U_A		H/L	
	Wen & Ingham	Dewynne et al.	Wen & Ingham	Dewynne et al.
0.00010	1.7320		6281.39	
0.00100	1.7312		627.857	
0.00626	1.7266	1.7266	99.9995	100.00
0.06084	1.6788	1.6788	9.99943	10.000
0.11770	1.6280	1.6280	5.00009	5.0000
0.44120	1.3130	1.3130	0.99984	1.0000
0.61880	1.1108	1.1109	0.49984	0.5000
0.70300	1.0019	1.0017	0.20378	0.2000
0.70400	1.0006	1.0000	0.17515	0.1000

Table 1 The variation of U_s/U_A and H/L as a function of q_2/q_1

As $t_B \to -\infty$ then expression (15) gives $I \to 0$, (18) gives $U_A/U_s \to 1$, (20) gives $H/L \to 0$, and (19) gives

$$\frac{1}{\sqrt{2} + \sqrt{t_s + 1}} = \frac{1}{2}\sqrt{1 - t_s} \qquad (21)$$

from which we obtain that

$$k \to [(\frac{19}{27} + \sqrt{\frac{11}{27}})^{1/3} + (\frac{19}{27} - \sqrt{\frac{11}{27}})^{1/3}]^2 \cong 0.7044025$$

This shows that the maximum value of q_2/q_1 is 0.7044025 and occurs when $H/L \to 0$. For a fixed value of H this requires that $L \to \infty$ and this results in the solution corresponding to the unique free surface which extends all the way upstream and on which the detachment point coincides with the point at infinity upstream.

(ii) THE STAGNATION POINT LIES ALONG THE SLOT WALL FC

Let us now consider the case in which the stagnation point lies along the slot wall, FC. The physical plane and complex potential velocity plane are shown in Figs 3a and 3b and we now map the flow domain in the complex potential plane onto the upper half-plane of the auxiliary t-plane by the transformation (4). Here the parameters t_s and t_f satisfy the condition that

$$-1 < t_s < t_f \qquad (22)$$

The boundary conditions for the function Ω on the real η axis of the t-plane are now expressed as follows

$$\begin{aligned}
&\mathrm{Im}\Omega(\eta) = 0 && -\infty < \eta < t_B \\
&\mathrm{Re}\Omega(\eta) = \tau = 0 && t_B < \eta < -1 \\
&\mathrm{Im}\Omega(\eta) = \frac{\pi}{2} && -1 < \eta < t_s \\
&\mathrm{Im}\Omega(\eta) = -\frac{\pi}{2} && t_s < \eta < t_f \\
&\mathrm{Im}\Omega(\eta) = 0 && t_f < \eta < +\infty
\end{aligned} \qquad (23)$$

The solution of Ω is now in the form

$$\Omega(t) = \frac{\sqrt{(t - t_B)(t + 1)}}{\pi}\left(\int_{-1}^{t_s} \frac{-\frac{\pi}{2}}{\sqrt{(t_B - \eta)(-1 - \eta)}\,(\eta - t)}\,d\eta + \int_{t_s}^{t_f} \frac{\frac{\pi}{2}}{\sqrt{(t_B - \eta)(-1 - \eta)}\,(\eta - t)}\,d\eta\right) \qquad (24)$$

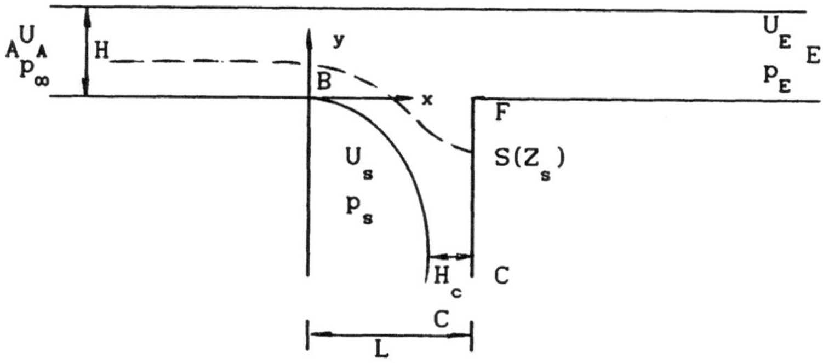

Fig. 3a The Physical Plane

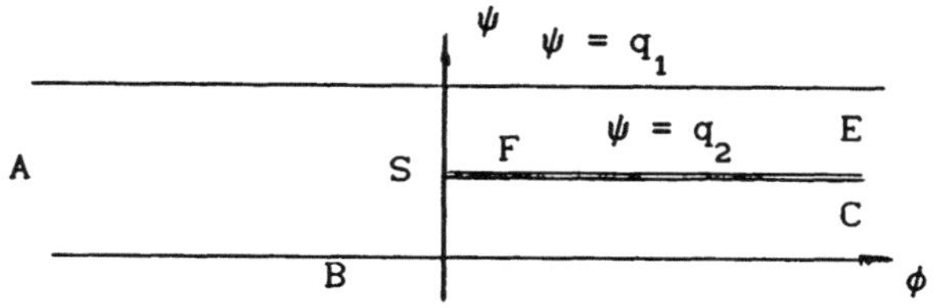

Fig. 3b The W-Plane

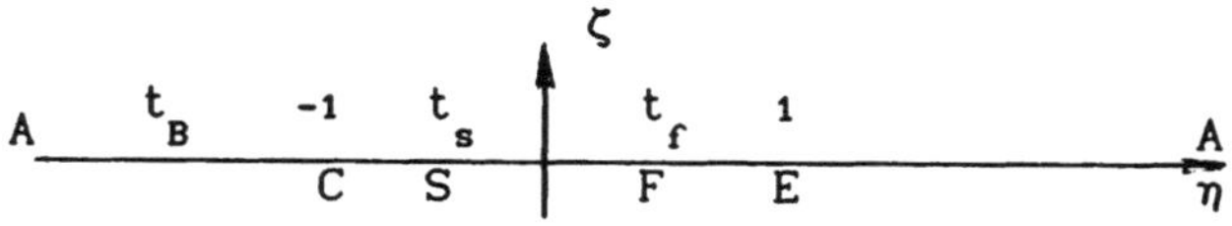

Fig. 3c The t-plane

The speed of the fluid on the boundaries SF, FE and AE is given by

$$\frac{U(t)}{U_s} = \frac{\sqrt{(-t_B - 1)(t - t_s)}}{\sqrt{(t - t_f)}} \times \frac{\sqrt{(t_f - t_B)(t + 1)} + \sqrt{(t - t_B)(t_f + 1)}}{[\sqrt{(t_s - t_B)(t + 1)} + \sqrt{(t - t_B)(t_s + 1)}]^2} \tag{25}$$

and the direction on the free surface BC is given by

$$\theta(t) = -\arcsin \frac{(t_s - t_B)(t + 1) + (t - t_B)(t_s + 1)}{(t_s - t)(-1 - t_B)} + \frac{1}{2}\arcsin \frac{(t_f - t_B)(t + 1) + (t - t_B)(t_f + 1)}{(t_f - t)(-1 - t_B)} - \frac{\pi}{4} \tag{26}$$

As $t \to 1$ the the expression (25) gives the speed of the fluid at the point E, i. e.

$$\frac{U_E}{U_s} = \frac{\sqrt{(-t_B - 1)(1 - t_s)}\,[\sqrt{2(t_f - t_B)} + \sqrt{(1 - t_B)(t_f + 1)}]}{\sqrt{(1 - t_f)}\,[\sqrt{2(t_s - t_B)} + \sqrt{(1 - t_B)(t_s + 1)}]^2} \tag{27}$$

and as $t \to \infty$, the speed of the fluid at the point A is given by

$$\frac{U_A}{U_s} = \frac{\sqrt{-t_B - 1}\,\,[\sqrt{t_f - t_B} + \sqrt{t_f + 1}\,]}{[\sqrt{t_s - t_B} + \sqrt{t_s + 1}\,]^2} \tag{28}$$

The expressions (27), (28) and (1) yield the relationship between the parameters t_B, t_s and t_f

$$\frac{U_E}{U_A} = \frac{\sqrt{2(t_f - t_B)} + \sqrt{(1 - t_B)(t_f + 1)}}{\sqrt{(1 - t_f)}\,[\sqrt{2(t_s - t_B)} + \sqrt{(1 - t_B)(t_s + 1)}]^2}$$

$$\times \frac{[\sqrt{t_s - t_B} + \sqrt{t_s + 1}\,]^2}{\sqrt{t_f - t_B} + \sqrt{t_f + 1}} = \frac{1}{2} \tag{29}$$

Substituting expressions (6) and (26) into (13) and integrating with respect to t from t_B to -1 yields the x coordinate of the point C, namely,

$$x_c = \frac{q_1}{\pi U_s}\int_{t_B}^{-1} \frac{t_s - t}{t^2 - 1}\cos\theta(t)\,dt = \frac{q_1}{\pi U}\,I_1 \tag{30}$$

where

$$I_1 = \int_{t_B}^{-1} \frac{t_s - t}{t^2 - 1}\cos\theta(t)\,dt \tag{31}$$

Substituting expressions (25) and (6) and using the fact that on the surface SF $\theta = -\frac{\pi}{2}$ into the expression (13) and integrating the resulting equation with respect to t from t_s to t_f yields the coordinate of stagnation point S, namely,

$$Z_s = L - \frac{iq_1}{\pi U_s}\,I_2 \tag{32}$$

where

$$I_2 = \int_{t_s}^{t_f} \frac{\sqrt{t_f - t}\,\,[\sqrt{(t_s - t_B)(t + 1)} + \sqrt{(t - t_B)(t_s + 1)}]}{\sqrt{-1 - t_B}\,\,[\sqrt{(t_f - t_B)(t + 1)} + \sqrt{(t - t_B)(t_f + 1)}](1 - t^2)}\,dt$$

Hence, the width of the slot is given by

$$L = x_c + \frac{q_2}{U_s} = \frac{q_1}{\pi U_s} I_1 + \frac{q_2}{U_s} = \frac{q_1}{U_s}\left(\frac{I_1}{\pi} + k\right) \tag{33}$$

On using expressions (28) and (33) one obtains the result that

$$\frac{H}{L} = \frac{q_1/U_A}{L} = \frac{[\sqrt{t_s - t_B} + \sqrt{t_s + 1}\,]^2}{\sqrt{-1 - t_B}\;[\sqrt{t_f - t_B} + \sqrt{t_f + 1}\,]\left(\frac{I_1}{\pi} + k\right)} \tag{34}$$

and from expressions (32) and (33) we obtain

$$\frac{Z_s}{L} = 1 - \frac{i\, I_2}{I_1 + k\pi} \tag{35}$$

When k and $\frac{H}{L}$ are specified we can then obtain the values of t_B and t_f which satisfy the expressions (29) and (34) and the condition (21). The values of U_s/U_A and Z_s/L may then be obtained using expressions (28) and (35).

The integrals I_1 and I_2 require numerical integrations and I_2 is evaluated by using Simpson's formula. The singularity at $t = -1$ in I_1 may be accommodated by means of a Gauss-type formula. We rewrite I_1 as

$$I_1 = \int_{t_B}^{-1} \sqrt{\frac{t - t_B}{-1 - t}}\, f(t)\, dt \tag{36}$$

where

$$f(t) = \frac{t_s - t}{\sqrt{(t - t_B)(- t - 1)}\;(1 - t)} \cos\theta(t) \tag{37}$$

We use the Tschebyscheff polynomials as the weighting functions and the numerical integration is given by

$$I_1 = \frac{2\pi(-1 - t_B)}{2n + 1} \sum_{i=1}^{n} \cos^2\left(\frac{2i - 1}{2n + 1}\frac{\pi}{2}\right)$$
$$\times f\left[t_B + (-1 - t_B)\cos^2\left(\frac{2i - 1}{2n + 1}\frac{\pi}{2}\right)\right] \tag{38}$$

Both integrations were evaluated by adaptive integration techniques with a preassigned tolerance of $\varepsilon = 10^{-6}$

In Table 2 a comparison of the results obtained using the method described in this paper and those obtained by Dewynne et al. [1] is given for H/L =1 and 10. It is observed that the two methods are in good agreement in predicting the mass of fluid leaving throught the slot as a function of q_2/q_1. However, there are discrepancies between the two methods in determining the location of the stagnation point, i.e. $(1 - Z_s/L)/i$, the distance between the points F and S.

q_2/q_1	U_s/U_A		$(1 - Z_s/L)/i \times 100$	
	Wen & Ingham	Dewynne et al.	Wen & Ingham	Dewynne et al.
H/L= 1				
0.44000	1.3120		0.00214	
0.40000	1.2764		0.42697	
0.30000	1.1912		2.22605	
0.25000	1.1512		3.17191	
0.20000	1.1135		3.88168	
0.18250	1.1009		4.05274	
0.15270	1.0803	1.0803	4.21671	4.48
0.10000	1.0471		3.97391	
0.05780	1.0241	1.0241	3.06801	4.45
0.01000	1.0032		0.78473	
0.00780	1.0025	1.0025	0.62812	1.14
0.00280	1.0008	1.0008	0.24263	0.46
0.0010	1.0003	1.0003	0.09228	0.18
H/L= 10				
0.04000	1.4196		2.10764	
0.03794	1.3947	1.3947	2.39614	1.96
0.03000	1.3005		3.48888	
0.02500	1.2430		4.08008	
0.02000	1.1874		4.49124	
0.01825	1.1685	1.1685	4.56940	5.32
0.01527	1.1369		4.59343	
0.01000	1.0840		4.16361	
0.00717	1.0575	1.0575	3.57391	5.36
0.00128	1.0087	1.0087	0.98844	1.77
0.00004	1.0024	1.0003	0.03696	0.08

Table 2 The variation of U_s/U and $(1 - Z_s/L)/i\times100$ as a function of q_2/q_1 for H/L = 1 and 10.

(iii) THE STAGNATION POINT LIES ALONG THE DOWNSTREAM CHANNEL WALL

In this case the stagnation point lies along the downstream channel wall and the physical plane, complex potential velocity plane and auxiliary t-plane are shown in Figs 4a, 4b and 4c. Here the parameters t_s, t_f satisfy the inequality

$$-1 < t_f < t_s \tag{39}$$

The boundary conditions for the function Ω on the real η axis of the t-plane are now expressed as

$$
\begin{array}{lll}
\mathrm{Im}\Omega(\eta) = 0 & -\infty < \eta < t_B & \\
\mathrm{Re}\Omega(\eta) = \tau = 0 & t_B < \eta < -1 & \\
\mathrm{Im}\Omega(\eta) = \frac{\pi}{2} & -1 < \eta < t_f & (40) \\
\mathrm{Im}\Omega(\eta) = \pi & t_f < \eta < t_s & \\
\mathrm{Im}\Omega(\eta) = 0 & t_f < \eta < +\infty &
\end{array}
$$

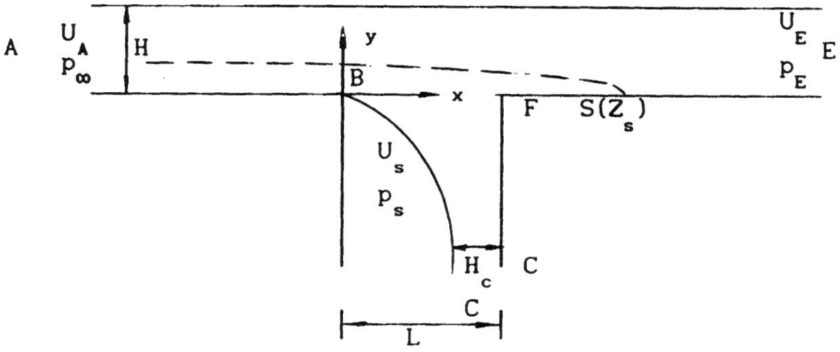

Fig. 4a The Physical Plane

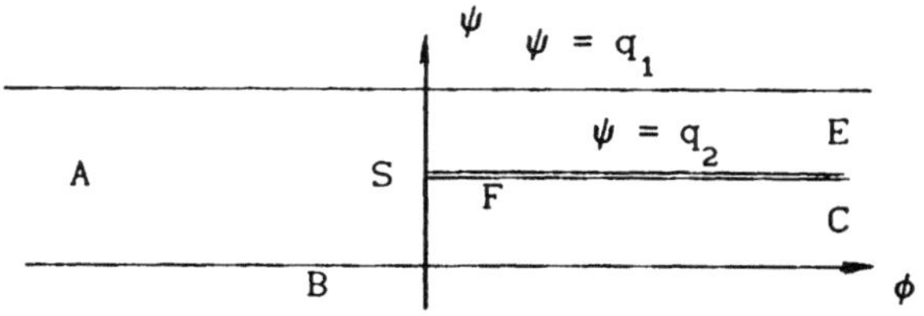

Fig. The 4b W-Plane

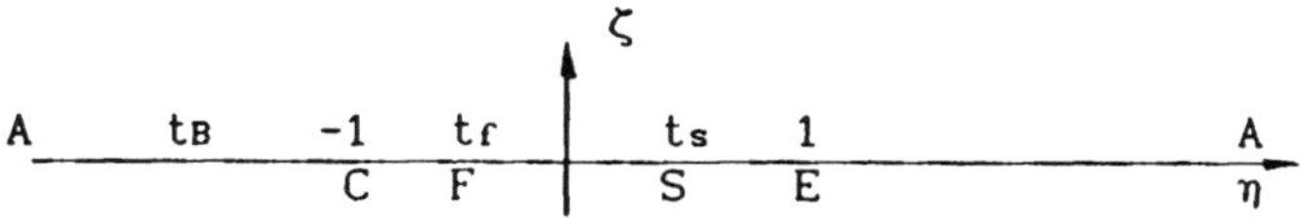

Fig. The 4c t-plane

and the solution for Ω is of the form

$$
\Omega(t) = \frac{\sqrt{(t - t_B)(t + 1)}}{\pi} \left(\int_{-1}^{t_f} \frac{-\frac{\pi}{2}}{\sqrt{(t_B - \eta)(-1 - \eta)}\,(\eta - t)} d\eta \right.
$$

$$
\left. + \int_{t_f}^{t_s} \frac{-\pi}{\sqrt{(t_B - \eta)(-1 - \eta)}\,(\eta - t)} d\eta \right)
$$

$$= \frac{\sqrt{(t - t_B)(t + 1)}}{\pi} \left(\int_{-1}^{t_s} \frac{-\frac{\pi}{2}}{\sqrt{(t_B - \eta)(-1 - \eta)}\ (\eta - t)} d\eta \right.$$

$$\left. + \int_{t_s}^{t_f} \frac{\frac{\pi}{2}}{\sqrt{(t_B - \eta)(-1 - \eta)}\ (\eta - t)} d\eta \right) \quad (41)$$

This solution has the same form as expression (24) and therefore the all formulae for this case are same as those when S is located along the slot wall. The only difference is that the parameters t_B and t_f must now satisfy the condition (39) rather than condition (21) and the numerical procedure is the same as that described in the last section.

The variation of U_s/U_A and Z_s/L as a function of q_2/q_1 for H/L=1 and 10 is given in Table 3 along with the results of Dewynne at al. [1]. These results again show that although the values of U_s/U_A are in good agreement the location of the stagnation point is quite different.

q_2/q_1	U_s/U_A		Z_s/L	
	Wen & Ingham	Dewynne et al.	Wen & Ingham	Dewynne et al.
H/L= 1				
0.73030	1.5874	1.5874	1.13233	1.4304
0.86680	1.7232	1.7232	1.32667	1.8615
0.99250	1.8503	1.8503	2.21423	3.6318
0.99620	1.8541	1.8541	2.42408	4.1162
0.99820	1.8562	1.8561	2.64860	4.6118
H/L= 10				
0.11706	2.4105	2.4105	1.09899	1.3733
0.20995	3.6503	3.6503	1.39305	2.0522
0.70702	10.366	10.366	3.48542	8.3508
0.90140	12.998	12.998	7.93949	15.281
0.96709	13.888	13.888	11.0412	22.265

Table 3 The variation of U_s/U_A and Z_s/L as a function of q_2/q_1 for H/L=1 and 10.

REFERENCES

1. Dewynne, J. N., Howison, S. D., Ockendon, J. R. Morland, L, C., and Watson, E. J. Slot Suction from Inviscid Channel Flow, J. Fluid. Vol. 200, pp. 265-282, 1989.
2 Helmholtz, H. On Discontinous Movements of Fluids, Phil. Mag., Vol. 36, pp. 337-346, 1868

3. Birkhoff G. and Zarantonello, E. H., Jets, Wakes and Cavities, Academic Press, New York, 1957.
4. Gilbarg, D., Jets and Cavities, Handbook, Phys., Vol. 9 pp. 311-445, 1960.
5 Woods, L. C., The Theory of Subsonic Plane Flow, Cambridge University Press, Cambridge, 1961.
6. Gurevich, The Theory of Jets in Ideal Fluid, Academic Press, New York, 1965.
7. Milne-Thomson, Theoretical Hydrodynamics, Macmillan, London, 1971.
8. Muskhelishivili, N. I., Singular Integral Equations, P. Noordhoff N. V. Groningen-Holland, 1953.

A Numerical Method for the Gravity Flow with Multiple Free Surfaces

Z.F. Yang (*), G.Q. Chen (**)
(*) Department of Mechanics, Peking University, Beijing 100871, P.R. China
(**) Institute of Mechanics, Academic Sinica, Beijing 100080, P.R. China

ABSTRACT

A numerical method, with conformal mapping techniques, is developed for gravity flow involving multiple free surfaces. Based on the extended Schwartz–Christoffel and the infinite–strip transformations, a nonlinear differential boundary–integral equation is derived to map the physical discharge plane directly onto the infinite strip plane, and consequently a mathematical model is proposed for the flow over a sharp–crested weir. The different element interpolating approximations and numerical procedures are also designed by the authors. The theoretical analysis and practical calculation indicate that the present method is superior to the existing ones in many aspects. The results for an example are given and consistences between them and measured data or semi–analytical valuc arc observed.

1. INTRODUCTION

The gravity flow is quite common in hydraulic engineering projects. Examples are the flow over weirs, gate flow and free gravity jet, which are one or more free surface flows strongly affected by gravitation. This has been a difficult problem of nonlinear boundary condition. The main difficulties arise not only from the nonlinear character of the boundary condition, but also from the fact that the discharge and the location of the free surfaces, and therefore the domain of computation, are not known in advance and they have to be found as parts of the solution.

To solve this problem, the transformation method, the finite difference method (FDM), the finite element method (FEM), the boundary element method (BEM) and the finite analytic method (FAM) have been applied successively, see, e.g., Benjamin 〔1〕, Southwell and Vaisey 〔2〕, Ikegawa and Washizu 〔3〕, Cheng et al 〔4〕 and Li et al 〔5〕. Almost all the work has been done on solution of one free surface problem of flow over a spillway crest or gated flow problem. In their studies, the trial free surface method was usually used, in which an assumed free surface was given initially and then adjusted through Bernoulli's principle untill the required adjustment in the free surface was considered negligible.

The sharp–crested weir is one of the three weir patters in hydraulic engineering. It is commonly employed as a device for flow measurement, and is also of fundamental importance as its free surface profile can provide the basis for the optimal design of the spillway. The flow over this weir involves upper and lower different free surfaces joined by no solid boundary, and therefore it is a more complicated problem so that the existing numerical methods as well as direct trial free surface method described as above are quite difficult to work well. An early numerical calculation of this flow was carried out by Han and Chow [6] through hodograph transformations. The method of hodograph transformation, like the FDM, FEM. and FAM, has to perform on finite whole domain during numerical computations and invariably involves with the determination of the hodograph boundary function and its derivatives. In fact, this is the only work in this area, to the authors' knowledge.

The conformal mapping methods are attractive since they involve only the solution of boundary integration requiring no out boundary conditions. If the solution in the interior of the flow field is desired, it can be analytically obtained by integration in arbitrary directions. Among those, the Schwartz–Christoffel and infinite–strip transformations have been of prime interest in the study of potential ideal flow. Trefethen [7] has presented a method for handling the numerical integration of the S–C transformation for pologons, i.e. the case where the sides are made up of straight lines, along with a review of other work in the area. Anderson [8] has used straight elements to do curved channel mappings also through the S–C transformation. An early work of the extension of the S–C transformation for pologons to curved boundaries can be found in Woods [9]. The most closely related work to the present is that of Davis [10] who has developed the extension contained in Woods to handle the curved elements and presented some example solution for flow past complex body shapes including free streamlines, but no method in this area can accout for the gravitative effect.

In this paper, we will first develop an one step transformation for curved boundaries in a more convenient manner for physically obvious reasons and numerical computation. Next we will show how we can use this approach to obtain solutions for the gravity–influenced free surface flow over a sharp–crested weir. Comparisons between the present results of calculation and available experimental data would substantiate the fact that the present method is effective in dealing with the multiple free surface flow problem strongly influenced by gravitation. Finally we will give some useful discussion and conclusion.

2. BASIC FORMULAE AND MATHEMATICAL MODEL

2.1. Schwartz–Christoffel Transformation and Its Extension

The Schwartz–Christoffel transformation is normally written in a form which is applicable only to pologons or boundaries which consists of straight line segments and is often given by

$$\frac{dz}{d\zeta} = M \prod_{i=1}^{n} (\zeta - a_i)^{\alpha_i} \tag{1}$$

which is used to map the upper half of the ζ plane into the interior of the pologon in the z plane. The quantity n relates to the number of the corners of the pologon, M is a complex constant to be determined, the a_is are real constants also to be determined, α_i's are relative supplementary angles in radians over π of the pologon circled in a counterclockwise sense and they are taken to be positive here for a clockwise rotation.

Equation (1) can obviously be rewritten as

$$\frac{dz}{d\zeta} = M exp \left[\sum_{i=1}^{n} \alpha_i ln(\zeta - a_i) \right] \tag{2}$$

Now consider a continuous curve rather than a pologon. Visualize the curve as being made of a large number of, i.e. n, straight line segments with the angles α_i replaced by $\triangle\beta_i$ as shown in Figure 1. In the limit as n $\to\infty$ the elements shrink to zero and the

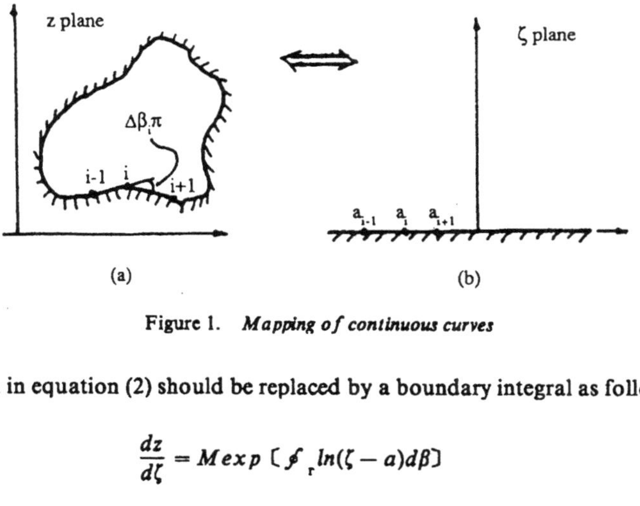

Figure 1. *Mapping of continuous curves*

summation in equation (2) should be replaced by a boundary integral as follows:

$$\frac{dz}{d\zeta} = M exp \left[\oint_r ln(\zeta - a) d\beta \right] \tag{3}$$

where r represents the boundary of mapped domain in the z plane, β is a relative angle at any point on r. The new definition of β will be given later.

Equation (3) is more general Schwartz–Christoffel transformation and includes (1). For example, if the surface of the body totally consists of straight line segments and corners, say a pologon, $d\beta = 0$ except at the corners where β is a step function with step α and equation (3) simplifies to equation (1) only for pologons.

2.2. Physical Flow Problem and Differential Boundary–integral Equation

The configuration of a sharp–crested weir is depicted in Figure 2 (a), where an initially uniform flow approaches the weir from the upstream infinity. The lower streamline stagnates at the point B and leaves the weir tangentially at the point C afterwards, forming the lower free surface. The top streamline drops continuously toward the downstream infinity as result of gravity. Both the free surfaces are not joined by any solid wall.

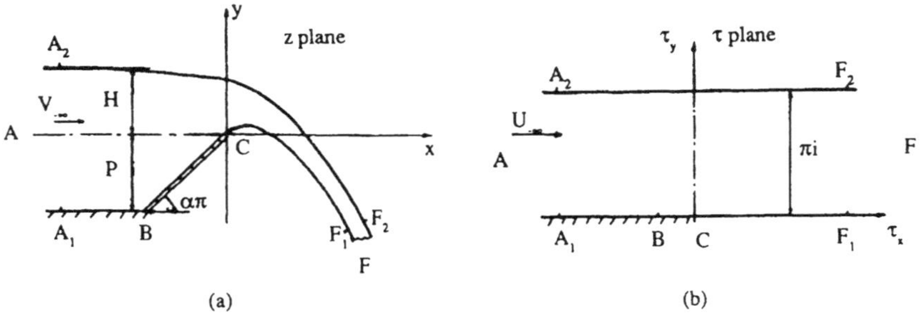

Figure 2. *A sharp–crested weir and its mapping*

Before we apply equation (3) to map the physical discharge plane as shown in Figure 2(a), here we choose to change, in a more convenient manner, the sign of the angles normally used in the Schwartz–Christoffel Theorem and now give the new definition that an angle at any point on the boundaries means the acute one between the tangential line at the same point and the horizontal line, and it is considered to be positive for a clockwise rotation. Hence the sloping angle of the weir shown in Figure 2(a) would be negative for the present sign convention.

For the flow problem with both straight and curved segments as shown in Figure 2(a), it is assumed that the upper streamline is parallel to the lower at the upstream and downstream infinities, and equation (3) becomes after lengthy algebra under the above definition of angles

$$\frac{dz}{d\zeta} = Mexp(-\alpha(\zeta - a_B) + \int_{-\alpha}^{\frac{1}{2}} ln(\zeta - a)d\beta^L - \int_0^{\frac{1}{2}} ln(\zeta - a)d\beta^U] \tag{4}$$

where the α is the relative sloping angle of the weir, a_B represents the location of the point B in the upper half of the ζ plane, and the β^L and β^U refer to the lower and upper free surfaces, respectively.

For the physically obvious reason, we take the mapping one step further and map an infinite strip of the τ plane as shown in Figure 2 (b) , which consists of a channel with parallel walls with distance π between them, also to the same ζ plane by equation (4). This step of the mapping can be done by the infinite–strip transformation given by

$$\zeta = -e^{-\tau} \tag{5}$$

The ζ plane is now eliminated to allow for a direct mapping from the z to the τ plane and the result is

$$\frac{dz}{d\tau} = f(\tau) = \frac{dz}{d\zeta} \cdot \frac{d\zeta}{d\tau} = \text{M.E.}e^{-i\beta\pi} \tag{6a}$$

where i in this case is the unit imaginary number and

$$E = exp\,[-\alpha ln|1-e^{-(\tau_{z\delta}-\tau)}| + \int_{-a}^{\frac{1}{2}} ln|1-e^{-(b-\tau)}|d\beta^{L} - \int_{0}^{\frac{1}{2}} ln|1-e^{-(b-\tau)}|d\beta^{U}] \tag{6b}$$

Equation (6) is a nonlinear differential boundary–integral equation through which the complicated case of free surface flow over the sharp–crested weir maps to a simple case of channel flow also with a uniform flow approaching and the boundaries in the transformed plane are both fixed as shown in Figure 2. In this equation, three constants can be chosen arbitrarily as those done in the Schwartz–Christoftel transformation. In the present case, we choose to map the infinities A and F in the z plane also to the infinities A and F in the τ plane and the top point C of the weir (it is also the origin of the z plane) to the origin of the τ plane. All those done above are of advantage in the next numerical analysis which will be apparent later.

2.3. Mathematical Model.

Based on the potential flow theory, the mathematical representation for the problem of free surface flow over a sharp–crested weir as shown in Figure 2(a) may be described as follows:

$$W = F(z) \qquad in\ \Omega \tag{7a}$$

$$Im(W) = Const_1 \qquad on\ CF_1F \cup AA_2F_2F \tag{7b}$$

$$\frac{dW}{dz} \cdot \frac{d\overline{W}}{dz} / 2g + Im(z) = Const_2 \qquad on\ CF_1F \cup AA_2F_2F \tag{7c}$$

$$\frac{dW}{dz} = V_{-\infty} \qquad at\ \Gamma_{-\infty} \tag{7d}$$

where W and $\overline{W}$ represent the complex potential and its conjugate function respectively, F(z) is an analytic complex function of z which needs to be determined, Im (z) is the imaginary part of z and Re(z) the real part, $Const_1$ and $Const_2$ are real constants, and $V_{-\infty}$ is the approaching flow speed.

Since the free boundaries are irregular and uncertain *a priori*, it is impossible for one to find out an exact expression of the complex potential F(z) in the z plane directly. It seems interesting that this complicated case in the z plane maps onto the simpler case of regular channel flow in the τ plane by equation (6), of which the complex potential is analytically known and normally given by

$$W(\tau) = U_{-\infty}\tau \tag{8}$$

where $W(\tau)$ is the complex potential in the τ plane, $U_{-\infty}$ is the upstream approaching flow speed also in the transformed plane and can be evaluated from the discharge conservation of the flows between both planes. Therefore

$$\frac{U_{-\infty}}{V_{-\infty}} = \frac{P+H}{\pi} \tag{9}$$

where P is the weir height and H is the upstream water depth over the weir as shown in Figure 2(a). It is obvious that if the tranformation (6) had been solved, the complex potential W(z) could be obtained by this transformation and equation (8). Unfortunately, in this case expressions of the type (6) can not been integrated exactly and this must be done numerically with determining the location of free surfaces. The problem of defining the complex potential function F(z) as described in equation (7) is so far converted into solving the equations (6) under the boundary conditions on the free surfaces and the weir, which will be shown later. With use of equations (5) and (6), in the z plane one has

$$V_{-\infty} = \lim_{z \to -\infty} \left(\frac{dW}{dz}\right) = \lim_{z \to -\infty} \left(\frac{dW}{d\tau} / \frac{dz}{d\tau}\right) = \frac{U_{-\infty}}{M}$$

and therefore

$$M = \frac{U_{-\infty}}{V_{-\infty}} \tag{10}$$

Substituting equations (9) and (10) into equation (6a), we obtain

$$\frac{dz}{d\tau} = \frac{P+H}{\pi} E e^{-i\beta\pi} \tag{11}$$

By virtue of equations (8)–(10), on the free surfaces one has

$$V_x - iV_y = Ve^{-i\beta\pi} = \frac{dW}{dz} = \frac{dW/d\tau}{dz/d\tau} = V_{-\infty} e^{-i\beta\pi} / E$$

and therefore

$$V = \frac{V_{-\infty}}{E} \tag{12}$$

where V is the speed on the free surfaces, and V_x and V_y are its components. Note that any point along the free surfaces in the physical plane is subject to a constant atmospheric pressure, so that according to the Bernoulli's equation from the energy conservation between the upstream infinity A and the top point C of the weir, the expression for $V_{-\infty}$ can be obtained as follows:

$$V_{-\infty} = \frac{E_1}{\sqrt{1-E_1^2}} \sqrt{2gH} \tag{13}$$

where E_1 denotes the E at the origin in the τ plane and is evaluated by equation (6b).

If the discharge over the weir is desired, it can be exactly evaluated from (13) or (9) (see Figure 2), in the physical plane as

$$Q = V_{-\infty}(P + H) \tag{14a}$$

or in the transformed plane as

$$Q = U_{-\infty}\pi \tag{14b}$$

So far we can choose to summarize and present a mathematical model, for the free surface problem of gravity flow over a sharp–crested weir in a more immediate and also simpler manner than that adopted in equation (7), as follows:

$$\frac{dz}{d\tau} = \frac{P+H}{\pi}E(\tau)e^{-i\beta\pi} \qquad on\ BCF_1F \cup AA_2F_2F \tag{15a}$$

$$\frac{V^2}{2g} + Im(z) = Const_2 \qquad on\ CF_1F \cup AA_2F_2F \tag{15b}$$

$$\int \frac{P+H}{\pi}E(\tau)e^{i\alpha\pi} \cdot d\tau = Ptg(\alpha\pi) + iP \qquad on\ BC \tag{15c}$$

where E, V and $V_{-\infty}$ are evaluated by equations (6b), (12) and (13), respectively. It is obvious that the boundary walls in the τ plane are two streamlines so that the top and lower boundaries in the z plane, mapped by equation (15a), unquestionably satisfy the streamline condition given by equation (7b). From equations (5) and (15a), it is easy to find out that the boundary condition described by (7d) is also satisfied automatically. Therefore, both of the conditions (7b) and (7d) need not be defined specially in the present model. Here we give the condition (15c) to fix the τ_{xb}, i.e. the location of the corner B in the τ plane. Apparently, we have confronted with an inverse problem in mathematics.

3. NUMERICAL METHOD

In the present model equation (15a) has to be numerically solved in an iterative way until the conditions (15b) and (15c) are satisfied. The authors have designed two numerical procedures, one of which is performed in the transformed plane and another in the physical plane. Here we choose only to show the first one which is considered to be superior. How this can be done in the physical plane will be reported on in Yang and Chen [11] [12]. The angle β appearing in equations (15a) and (6b) is virtually a given function of z depending on the geometry of the prescribed free surface. Unlike the existing Schwartz–Cristoffel methods, e.g. Woods[9] and Davis[10], with which the curved surface points and slopes are prescribed in the physical plane and the point locations in the transformed plane are determined as a part of the solution, here we will prescribe b in the τ plane and consider β to be determined as a function of b, for the reason which will be apparent later.

3.1. Discretized Equations

The discretized boundaries in the τ plane corresponding to the free surfaces are illustrated in Figure 3, where the lower boundary CF_1F is subdivided into n elements involving infinite one at the downstream and the top boundary AA_2F_2F into (ns−n+1) elements with two infinite ones at the upstream and the downstream. Therefore, equation (6b) can be written in a discrete form as

$$E(\tau_j) = exp\,[\,-\alpha ln|1-e^{-(\tau_{xb}-\tau_j)}| + \sum_{i=1}^{n-1}\int_{\beta_i}^{\beta_{i+1}} ln|1-e^{-(b-\tau_j)}|d\beta^L$$
$$+\int_{\beta_n}^{\frac{1}{2}} ln|1-e^{-(b-\tau_j)}|d\beta^L - \int_0^{\beta_{n+1}} ln|1-e^{-(b-\tau_j)}|d\beta^U$$
$$-\sum_{i=n+1}^{ns-1}\int_{\beta_i}^{\beta_{i+1}} ln|1-e^{-(b-\tau_j)}|d\beta^U - \int_{\beta_{ns}}^{\frac{1}{2}} ln|1-e^{-(b-\tau_j)}|d\beta^U\,] \tag{16}$$

in which τ_j represents the coordinate of finite node j in the τ plane, n is the number of the finite nodes by which the lower boundary CF has been subdivided and ns the number of all the finite nodes on both boundaries.

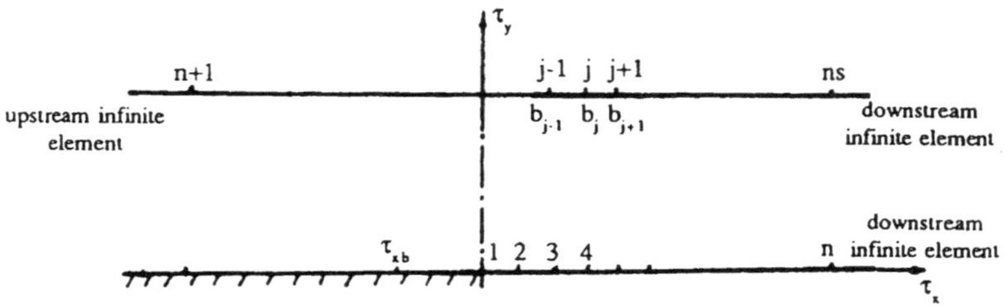

Figure 3. *Distribution of discrete elements or nodes*

Over a finite element, we approximate β with a linear polynomial in the dimensionless ξ coordinate system. Introducing the solution at the nodes results in

$$\beta = \beta_i + \triangle\beta_i\xi \qquad i = 1,2,\cdots,n-1,n+2,n+3,\cdots,ns-1 \tag{17}$$

where β_i represents the solution for β at the ith node, $\triangle\beta_i$ expressed by $(\beta_{i+1}-\beta_i)$ is the step of β, over the ith element, and ξ is given by

$$\xi = (b - b_i)/\triangle\tau_x \tag{18}$$

where and in the following subscript i refers to the ith node, $\triangle\tau_x$ is the element length which keeps same for all the finite ones, ξ is defined as local coordinate over an element which varies from zero to unity.

Over the upstream infinite element, i.e., the first one along the top boundary, where the β approaches zero at infinity and β_{n+1} at the (n+1)th node, the following linear polynomial expression for β is used

$$\beta = \beta_{n+1}\xi \tag{19}$$

where ξ is difined as

$$\xi = e^{b-b_{n+1}} \tag{20}$$

Over the downstream infinite elements, where the β's approach $\frac{1}{2}$ at infinity, higher order polynomial for β is neccesary to be developed since the β's vary much more greatly here than over the finite ones. Over the element along the lower boundary, i.e., F_1F, for example, we assume

$$\beta = C_0 + C_1\xi + C_2\xi^2 \tag{21}$$

where ξ is given by

$$\xi = 1 - e^{-(b-b_n)} \tag{22}$$

Because equation (21) is a quadratic function, three conditions must be imposed on it to evaluate the three constants C_0, C_1 and C_2. Introduce the requirement that $\beta\,|_{\xi=0} = \beta_n$, $\beta|_{\xi=1} = \frac{1}{2}$ and additional condition $\frac{d\beta}{db}|_{b=b_n^-} = \frac{d\beta}{db}|_{b=b_n^+}$ and note that $\frac{d\beta}{db}|_{b=b_n^-}$ and $\frac{d\beta}{db}|_{b=b_n^+}$ can be evaluated by equations (17) and (21), respectively. One can provide the solution for C_0, C_1, and C_2. Substitution of the result into equation (21) yields after some manipulation

$$\beta = \beta_n + \Delta\beta_n(B_n\xi + (1 - B_n)\xi^2) \tag{23}$$

where $\Delta\beta_n = \frac{1}{2} - \beta_n$, $\quad B_n = \frac{1}{\Delta\tau_x}\,\frac{\beta_n - \beta_{n-1}}{\Delta\beta_n}$

Similarly over the downstream infinite element along the top boundary, one has

$$\xi = 1 - e^{-(b-b_{ns})} \tag{24}$$

$$\beta = \beta_{ns} + \Delta\beta_{ns}(B_{ns}\xi + (1 - B_{ns})\xi^2) \tag{25}$$

where $\Delta\beta_{ns} = \frac{1}{2} - \beta_{ns}$, $\quad B_{ns} = \frac{1}{\Delta\tau_x} \cdot \frac{\beta_{ns} - \beta_{ns-1}}{\Delta\beta_{ns}}$

Obviously, it is assumed that the prescribed location of the n or ns point is far downstream and that the expression (23) or (25) is accurate enough to account for the variation of β from the n or ns location to infinity.

Substitution of equations (17), (19), (23) and (25) into (16) yields after some manipulation

$$E(\tau_j) = exp(-\alpha ln|1 - e^{-(\tau_{xb} - \tau_j)}| + \sum_{i=1}^{n} K_{i,j}\beta_i - \sum_{i=n+1}^{ns} K_{i,j}\beta_i + \frac{1}{2}K_p) \tag{26}$$

where $K_{i,j}$ is defined as the stiff matrix which is invariant during the iterative procedures. With use of equations (18), (20), (22) and (24), the expresions for $K_{i,j}$ are given as follows:

$$K_{i,j} = \int_0^1 ln|1 - S_{i-1,j}e^{-\xi\Delta\tau_x}|d\xi - \int_0^1 ln|1 - S_{i,j}e^{-\xi\Delta\tau_x}|d\xi \tag{27a}$$

$$i = 1,2,\cdots,n,n+2,n+3,ns$$

$$K_{n+1,j} = \int_0^1 ln|1 - S_{n+1,j}\xi|d\xi - \int_0^1 ln|1 - S_{n+1,j}e^{-\xi\Delta\tau_x}|d\xi \tag{27b}$$

$$K_p = \int_0^1 ln|1 - S_{n,j}\xi|(B_n + 2(1 - B_n)\xi)d\xi - \int_0^1 ln|1 - S_{ns,j}\xi|(B_{ns} + 2(1 - B_{ns})\xi)d\xi \tag{27c}$$

in which $S_{i,j} = e^{-(\tau_i - \tau_j)}$, i, j = 1, 2, ⋯, ns.

3.2. Calculation of the Free Surfaces and Discharge

The location of free surfaces can be determined by integrating equations (15a) and (26). Now the remaining problem in doing so is that τ_{xb} location in the transformed plane is unknown and must be solved as a part of the solution. For the prescribed τ_{xb}, condition (15c) is not satisfied generally. If subscribt c indicates computed value and the imaginary part is taken from the integration term in equation (15c), for a guessed τ_{xb} one has

$$P_c = \frac{P + H}{\pi} sin(\alpha\pi)\int_{\tau_{xb}}^{0} E(\tau)d\tau \tag{28}$$

through which the sloping angle of the weir BC will be found to be correct in the physical plane, but the weir height will be in error, i.e., $P_c \neq P$. The next guess on τ_{xb} is made as follows

$$\tau_{xb}^1 = \frac{P}{P_c}\tau_{xb}^0 \tag{29}$$

where upscribt 1 indicates the new value of τ_{xb} and upscribt zero the old one. The equations (28) and (29) are iteractively solved until the condition (15c) is satisfied to some prescribed accuracy. The numerical computation shows the fast convergence of the above procedure, which can be achieved in no more than five iterations typically.

Over the upstream infinite element, integrating equation (15a) and taking its imaginary part yield

$$\delta y_A = \frac{P+H}{\pi}\int_{-\infty}^{b_{n+1}} E(\tau)sin(-\beta\pi)d\tau \tag{30}$$

where δy_A is the height difference between point A_2 and upstream infinity. Note that β is rather small in this element, so that one has $\sin(\beta\pi)\approx\beta\pi$. With use of equations (19) and (20), equation (30) becomes

$$\delta y_A = -(P+H)\beta_{n+1}\int_0^1 E(b_{n+1}+ln\xi)d\xi \tag{31}$$

The above expression can be numerically integrated in a normal way and no special attention need to be paid to the singularity.

Integrate equation (15a) over downstream infinte elements along the lower and top boundaries respectively, and take their real parts. Substraction of the resulted expression along the top from the one along the lower yields with use of equations (22) and (24)

$$\delta x_F = \frac{P+H}{\pi}\int_0^1 [E(b_{ns}-ln(1-\xi))cos(\beta^U\pi)-E(b_n-ln(1-\xi))cos(\beta^L\pi)]\frac{d\xi}{1-\xi} \tag{32}$$

where δx_F is the horizontal length between the nth node (point F_1) and the nsth node (point F_2) in the physical z plane, β^U and β^L are described by (25) and (23), respectively.

Over a finite element, integrating equation (15a) using say trapezoidal integration formulae and splitting the result into real and imaginary parts yield

$$\Delta x_j = \frac{\Delta\tau_x}{2}\cdot\frac{P+H}{\pi}(E(\tau_j)cos(\beta_j\pi)+E(\tau_{j+1})cos(\beta_{j+1}\pi)) \tag{33a}$$

$$\Delta y_j = -\frac{\Delta\tau_x}{2}\cdot\frac{P+H}{\pi}(E(\tau_j)sin(\beta_j\pi)+E(\tau_{j+1})sin(\beta_{j+1}\pi)) \tag{33b}$$

where $\Delta x_j = x_{j+1}-x_j$, $\Delta y_j = y_{j+1}-y_j$, $j=1,2,\cdots,n-1,n+1,n+2,\cdots,ns-1$.

So far, during the iterations the profiles of free surfaces in the z plane can be determined point by point as follows:

$$x_1 = y_1 = 0$$
$$x_{j+1} = x_j + \Delta x_j$$
$$y_{j+1} = y_j + \Delta y_j \qquad j=1,2,\cdots,n-1 \tag{34}$$

for the lowes free surface, and

$$y_{n+1} = H + \delta y_A$$
$$y_{j+1} = y_j + \Delta y_j \qquad j=n+1,n+2,\cdots,ns-1 \tag{35}$$
$$x_{ns} = x_n + \delta x_F$$
$$x_j = x_{j+1} - \Delta x_j \qquad j=ns-1,ns-2,\cdots,n+1 \tag{36}$$

for the upper free surface.

As a result of the gravitational influence, an iterative procedure is unavoidable in all numerical methods. Using the present one, the numerical procedure is summed up as follows:

(1) Guess values for $\beta_j (j=2, 3, \cdots, ns)$ and τ_{xb}. Obviously, β varies from $-\alpha$ to $\frac{1}{2}$ along the lower boundary and from zero to $\frac{1}{2}$ along the upper.

(2) Evaluate the stiff matrix $K_{i,j}$ and K_F by equatons (26).

(3) Calculate the new value for τ_{xb} using the gussed β_j and τ_{xb} by equations (28) and (29)

(4) Calculate the $E_j's$ $(j=1, 2, \cdots, ns)$. Using the guessed β_j and new τ_{xb} from step (3) above, by equation (26), so that $V_{-\infty}$ and V_j can be evaluated by equations (13) and (12), respectively

(5) Calculate δy_A and δx_F by equations (31) and (32), respectively. The profiles of free surfaces can be determined using equations (34)–(36) for the guessed β_j and new τ_{xb} from step (3)

(6) Using the Bernoulli nonlinear condition (15b), we have

$$f_j(\beta_2,\beta_3,\cdots,\beta_{ns}) = H + \frac{V_{-\infty}^2}{2g} - y_j - \frac{V_j^2}{2g} = 0 \qquad j=2,3,\cdots,ns \tag{37}$$

Generally, the results for the prescribed $\beta_j's$ do not satisfy the above conditions, i.e., $f_j \neq 0$. Depending on the residues $f_j's$, new trial values $\beta_j^{1'}s$ for $\beta_j's$ are calcalated by employing the Newton–Raphson method, see, e.g., oden [13].

(7) A new iteration is performed by using the calculated new approximations of β_j from step (6) above. The iterative procedure from step (3) to step (6) is repeated until equations (47) are satisfied to some prescribed accuracy.

(8) Evaluate the discharge over the weir by equation (14a) or (14b)

One of the difficulties in solving the free surface flow problems under the effect of gravity, as pointed out in Section1, arises from the fact that the boundary is not known in advance. In the present method, the same fact above significantly makes it possible to perform the numerical solution in the transformed τ plane. In this case, the elements were taken to be equally spaced on both boundaries in the τ plane as illustrated in Figure 3, and the b_j locations are held fixed, so that the stiff matrix $K_{i,j}$ does not change as the locations of free surfaces are iterated. The method, therefore, does not require significantly more computer time than a case which only involves solid boundaries and the $b_j's$ are unkown. We know that the β at the first node is as same as the sloping angle of the weir, so that β_1 is out of the solution automatically. The Bernoulli equation like (37) between this point and upstream infinity has been used to solve $V_{-\infty}$, and therefore discharge over the weir.

4. NUMERICAL EXAMPLE

We take the flow over the sharp–crested weir with the 45° sloping angle and

P=H as an example. The practical calculation shows that the above procedure converges quite rapidly and appears to do so for all the cases, although no smoothing process has bee explored. The predicted profiles of the free surfaces are compared in Figure 4 with the cxperimental data from Ref. 〔14〕. Remarkably good agreement between them is observed. For example, the calculated results at the erest point a_0 (see Figuer 4), which is very important in the disign of spillway, are $(\frac{x_{a_0}}{H_s}, \frac{y_{a_0}}{H_s})_c = (0.215, 0.041)$, which are quite consistent with the experiment data of $(\frac{x_{a_0}}{H_s}, \frac{y_{a_0}}{H_s})_e = (0.221, 0.042)$ from Ref. 〔14〕. Therefore the relative errors are only (1.4%, 2.4%). In the above descriptions or Figuere 4, H_s is the total water head over the weir and is given as $H_s = H + V^2_{-\infty}/2g$, subscript c denotes the computed value and e the experimental one.

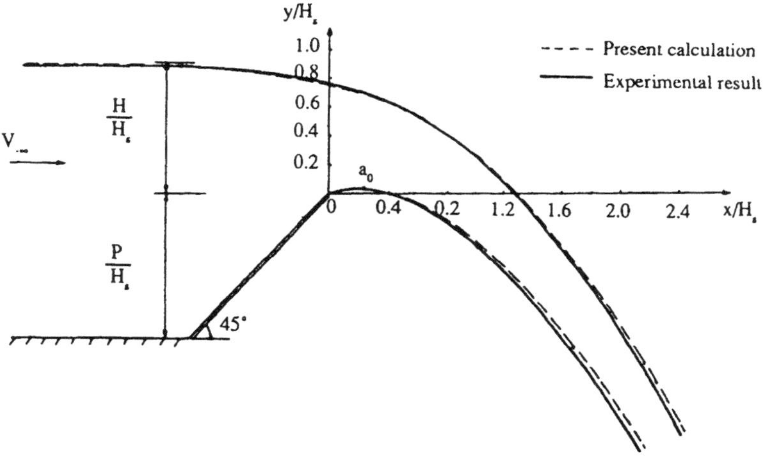

Figure 4. *Comparision of the predicted free surface profiles with the experimental results for* $\alpha = \frac{1}{4}$, $P = H$

If the discharge is desired, it can directly be obtained by equation (14a) or (14b). In this case, the result for the discharge coefficient is $\mu_c = 0.5$, where μ_c is the predicted discharge coefficient and it is evaluated from equations (13) and (14), so that

$$\mu_c = \frac{Q_c}{\sqrt{2gH}} = (P + H)\frac{E_1}{\sqrt{1 - E_1^2}}$$

where Q_c is the predicted discharge expressed by (14a). On the other hand, some the semi–therotical formulae for the discharge or discharge coefficient over the sharp–crested weir have been developed. The ВИИГ formula is a well known example and is normally described as

$$\mu_B = K(0.402 + 0.054\frac{H}{P})$$

where μ_B is the discharge coefficient from the ВИИГ and K is a modifying parametre due to the weir slope (K = 1.112 for $\alpha = \frac{1}{4}$). Substitution of P = H into the above equation yields $\mu_B = 0.507$, so that the relative error between it and the numerical one is only 1.2%.

5. CONCLUSION

Based on the extended Schwartz–Christoffel and infinite–strip transformations, the mathematical model and numerical method for the free surface gravity flow over a sharp–crested weir have been developed by the authors. Numerical calculations indicate that the method is very effective in dealing with multiple free surface problems strongly influenced by gravitation.

The numerical iteration is performed in the transformed plane only along a part of boundaries which corresponds to free surfaces in the physical plane, so that the method is provided with the merits which the boundary element method possesses and the unknows involved in the present method should be much less than in the FDM, the FEM, the FAM as well as the hodograph method discussed in section 1. The differences between the method and the BEM are that the present one requires less boundary condition, and allow for the development of infinite elements along the boundaries.

The use of the method can give the discharge or its coefficient directly from analysis. Instead, in most of other methods the discharge can not be found directly and has to be determined artificially, according to a series of the calculated free surfaces, see for example Ikegawa and Washigu [3], Aitchison [15].

On the basis of all evidence presented so far, it is obvious that the present method is superior to the existing numerical ones.

REFERENCE

1. Benjamin, T.B., On the Flow in Channels When Rigid Obstacles Are Placed in the Stream, *Journal of Fluid Mechanics*, Vol.1, PP.227–248, 1956
2. Southwell, R.V., and Vaisey, G., Relaxation Methods Applied to Engineering Problems: XII, Fluid Motions Characteriged by Streamlines, *Phylosophical Transactions of the Royal Society*, Lodon, Series A, Vol.240, PP.117–161, 1946
3. Ikagawa, M., and Washigu, K., Finite–Element Method Applied to Analysis of Flow Over a Spillway Crest, *Int. J. of Numerical Method in Engineering*, Vol.6, PP.179–189, 1973
4. Cheng, A. HD., et al, Boundary Calculation of Sluice and Spillway Flows, *J. of Hydraulic Division, ASME*, Vol.107, PP.1163–1178, 1981
5. Li, W., et al, Finite Analitical Solution of the Flow over Spillways, *J. of Engineering*

Mechanics, ASCE, Vol.115, 1989

6. Han, T.Y., and Chow, W.L, the Study of Sluice Gate and Sharp–Crested Weir Through Hodograph Transformations, *J. of Applied Mechanics, ASME,* Vol.48, PP.229–238, 1981
7. Trefethen, L.N., Numerical Computation of the Schwartz–Christoffel Transformation, *Stanford University Computer Science Report,* STAN–CS–79–710, 1979
8. Anderson, O.L., Caculation of Internal Viscous Flows in Axisymmetric Ducts at Moderate to High Reynolds Number, *Computer and Fluids,* 1979
9. Wood, L.C., The Theory of Subsonic Plane Flow, Cambridge University Press, 1961
10. Davis, R.T., Numerical Methods for Coordinate Generation Based on a Mapping Technique, in Computational Methods for Turbulent, Transonic, and Viscous Flows (Ed. Essers, J.A.), PP.1–44, HEMISPHERE PUBLISHING CORPORATION, Washigton, New York and Lodon, 1983
11. Yang, Z.F., and Chen, G.Q., The Numerical Method of Conformal Mappings for the Gravity Flow Involving Double Free Surfaces: I, Mathematical Model, *to appear in Chinese.*
12. Yang, Z.F., and Chen, G.Q., The Numerical Method of Conformal Mappings for the Gravity Flow Involving Double Free Surfaces: II, Numerical Method, *to appear in Chinese.*
13. Oden, J.T., Finite Elements of Non–Linear Continua, McGraw–Hill, New York, PP.279–283, 1967
14. Studies of Crests for Overall Dams, Bulletin 3, Part IV, *Boulder Canyon Project Final Reports,* Denver, Colorado, 1948
15. Aitchison, J.M., A Finite Element Solution for Critical Flow over a Weir, *Proceedings, 3rd Int. Conf. on Finite Elements in Flow Problems,* Banff, Alberta, V01.II, PP52–59, 1980

Two Open Boundary Conditions for Nonlinear Free Surface Flows

S. Jagannathan

The Glosten Associates, Inc., Seattle, Washington 98104, USA

ABSTRACT

The focus of this paper is the treatment of open boundaries in the simulation of two-dimensional nonlinear free surface flows. In a finite domain, a major concern is with the radiation condition to be applied at the open boundary. No general theoretical radiation conditions are known to exist. In this paper two open boundary conditions are investigated. The first is an extension of the Orlanski open boundary condition based on phase velocity at the open or truncation boundary. The second is based on energy flux equalization between the nonlinear inner domain and a linear outer domain. The nonlinear flow in the inner domain is solved numerically using a semi-Lagrangian procedure. The energy flux arriving at the open boundary is removed using a new open boundary condition which acts as a linear wave absorber. For the cases studied, the performance of these boundary conditions is found to be quite good.

INTRODUCTION

The use of linear hydrodynamic analysis has become quite commonplace and the results of such analysis are quite satisfactory for many engineering problems. Moreover, considering the essentially stochastic nature of the oceans, the use of linear analysis and hence superposition of solutions has led to the widespread use of spectral methods and impulse response function techniques. The domain equation for the velocity potential, assuming the fluid to be incompressible and the flow irrotational, is the Laplace equation which is linear. Nonlinearities are introduced into the problem through the boundary conditions at the free surface and on the body. In a linear solution, these boundary conditions are linearized leading to the requirements that the free surface elevation and the body remain in close proximity to their mean positions. This then precludes steep waves and large body motions. But these nonlinear effects are extremely important, particularly in the context of survivability at sea. There have been numerous cases of small vessels capsizing in rough seas. With the increasing number of offshore structures, appropriate methods of analysis are needed to evaluate the behavior of such structures under extreme conditions to ensure safety at sea. In recent years considerable progress has been made in addressing free surface and body nonlinearities directly in the time domain.[1,2]

A new era in nonlinear free surface calculations was initiated by Longuet-Higgins & Cokelet[2] in their simulation of unsteady two-dimensional waves. The waves are considered spatially periodic and the free surface is mapped into a closed contour. A Fredholm integral equation of the first kind is set up for the velocity potential and its normal for marker particles on the free surface. Hence at every instant in time, the velocity field for each marker particle is known through spatial differentiation. This permits the Lagrangian dynamic and kinematic free surface conditions to serve as evolution equations for the velocity potential and position for each marker particle. This then sets up the boundary value problem for the next instant in time.

This method is remarkably simple and effective and has the added advantage that the marker particles become concentrated near regions of sharp curvature. Vinje & Brevig[3] modified the method of Longuet-Higgins & Cokelet and by using Cauchy's theorem formulated the problem in terms of a Fredholm integral equation of the second kind for the velocity potential and the stream function. Subsequently Vinje & Brevig[4,5] analyzed a variety of body-fluid interaction problems using this method. This modified method is used as the basic method of solution in which the proposed boundary conditions are embedded. The mathematical formulation for this solution technique is described in detail in Jagannathan.[6]

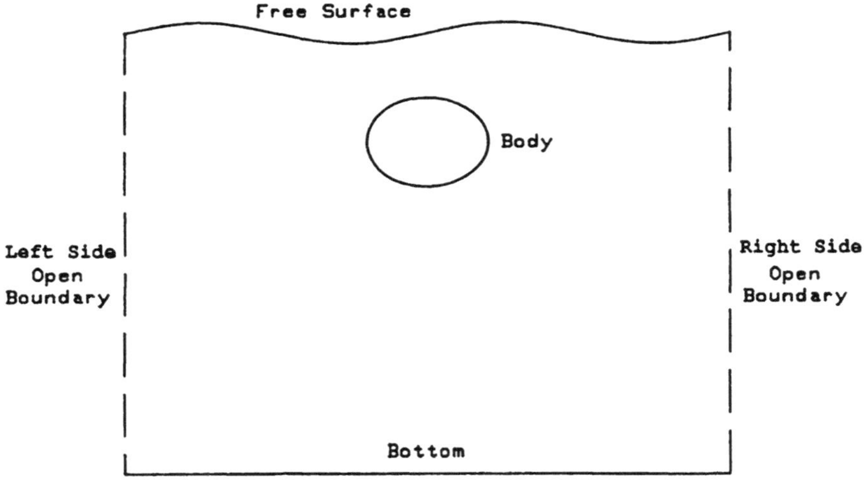

Figure 1. Definition sketch for the problem

Figure 1 shows a definition sketch for the problem, which is posed as an initial value problem concerning the forced motion of a two dimensional body in proximity to the free surface. The fluid is assumed to be inviscid, incompressible and the flow irrotational. The fluid domain consists of the free surface, the body, the bottom and two computational or 'open' boundaries. Exact boundary conditions are to be imposed on the free surface and on the body. For this two-dimensional problem, with potential theory being applicable (the fluid being considered inviscid, incompressible and the flow irrotational) the fluid motions are described using the complex velocity potential β. The contour of C of the domain is divided into elements with the variable of interest (β or β_t) being defined at nodal points. The part of the contour where the velocity potential ϕ is known is denoted as C_ϕ and where the stream function ψ is known it is denoted as C_ψ (and similarly for ϕ_t and ψ_t). When the body is in the interior of the domain, a 'branch cut' is taken out to the body to permit the latter to be considered part of the boundary for computational purposes. Details of the numerical solution procedure are given by Jagannathan.[6]

The process of simulation can be described in terms of two stages for each time-step: (a) solution of a well-posed boundary value problem for β (and β_t) at a fixed instant in time and (b) evolution of variables all over C to set up the boundary value problem for the next instant in time.

OPEN BOUNDARIES

The most difficult part in the solution of two dimensional nonlinear flow problems is the treatment of the open boundaries where a radiation condition needs to be applied. In the time domain, for an initial value problem, a radiation condition is actually not necessary. A sufficient condition is to state that in the far field, the disturbance is negligibly small. The implementation of such a condition is not practical (except for cases of non-propagative solutions at infinity using mapping techniques) for there would have to be a progressive domain expansion with time, leading to unreasonable computational costs. This consideration then implies the use of a truncated domain using open boundaries. Thus the advantage of an initial value formulation is not available.

Suitable radiation boundary conditions need to be furnished at these artificial boundaries to ensure a unique solution. Moreover, the open boundary conditions should be such that the solution so obtained will closely approximate the solution that would exist in the absence of the artificial boundary; i.e., the open boundary conditions should not distort the solution in the interior domain. For nonlinear free surface problems, no general open boundary conditions are known so far. Many individual approaches have emerged, such as those of Hedstrom[7] for nonlinear hyperbolic systems, Rudy and Strikwerda[8] for Navier-Stokes flows, and Engquist and Majda[9] for a class of wave equations. A review of open boundary conditions is provided by Jagannathan.[10]

To solve the boundary value problem at hand, either the velocity potential ϕ or the normal velocity ϕ_n (or the stream function ψ) should be known at the open boundary at each instant in time. In this paper two open boundary conditions are proposed, namely: (1) the Orlanski open boundary condition which defines ϕ at the boundary and (2) the energy flux equalization boundary condition which defines ϕ_n at the boundary.

ORLANSKI OPEN BOUNDARY CONDITION

For many physical problems the appropriate partial differential equation is hyperbolic in nature and the phenomena are characterized by wave propagation. Open boundary conditions for such problems are naturally based on the idea of wave absorption. For a time-harmonic problem given by $\phi = \bar{\phi} e^{-i\omega t}$ (where $\bar{\phi}$ is complex), the appropriate boundary condition is the well known Sommerfeld condition (see Yeung[1]).

$$\phi_x \mp ik\phi \rightarrow 0 \quad \text{as} \quad x \rightarrow \pm\infty \tag{1}$$

for two dimensions and k is the wave number corresponding to the frequency ν. In the time domain, if the solution is expected in the form $\phi(x, y, t) = \phi(x - ct, y)$, where c is the phase velocity, then the Sommerfeld condition above may be written as

$$\phi_t + c\,\phi_x = 0. \tag{2}$$

This condition applied numerically is often known as the Orlanski condition.

In many problems the value of the phase velocity to be used is not known. Kreiss[11] dealt with the problem of unknown phase velocity by assuming $c = \Delta x/\Delta t$, where Δx and Δt are the spatial and temporal grid sizes. Pearson[12] obtained the phase velocity from the linearized dispersion relationship using only the dominant wave. Orlanski[13] suggested that the phase velocity, instead of being a constant value, should be numerically determined from neighboring grid points. Han et al.[14] used a forward-time, upwind space variation of the Orlanski condition for application to stratified flows. They concluded that the open boundary condition performed remarkably well, even for these parabolic problems. Ertekin[15] used the Sommerfeld condition directly to study the behavior of long waves by using $c = \pm\sqrt{(gh)}$, where h is the water depth.

An extension to the Orlanski scheme is now proposed. In initial-value and unsteady flow problems the critical question is the value of the phase velocity to be used. With particular reference to the problem posed in Figure 1, the outgoing behavior of the velocity potential at the open boundaries is directly related to the free surface behavior near those boundaries. We then postulate the following:

(a) The value of the phase velocity at the open boundaries is the value of phase velocity on the free surface near those boundaries.

(b) The value of the phase velocity c at the free surface may be determined at any instant as

$$c = -\frac{\phi_t}{\phi_x} \tag{3}$$

The Orlanski condition is implemented in the simulation as follows: Consider a point on the *free surface* close to the open boundary. At some time $t = t_0$ let the complex potential $\beta = \phi + i\psi$ and the value ϕ_t be known at this point. The value of $u = \phi_x$ is obtained by numerical differentiation with neighboring free surface points. Hence, using Equation (3), the value of the phase velocity c is obtained. Now at a point on the *open* boundary ϕ_x is obtained by spatial differentiation and hence, using Equation (2), the value of ϕ_t and by integration the value ϕ are obtained. With this known value of ϕ on the open boundary and using the free surface, bottom and body boundary conditions, the boundary value problem is solved to obtain ψ on the open boundary. For the next instant in time, the free surface condition gives a new value of ϕ_t and β on the free surface and the process continues.

However, there could be problems in estimating c when $\phi_t, \phi_x \to 0$. To avoid this problem and to get a robust estimate of c, a 'numerical scanner' was implemented in the procedure. This scanner merely computes c according to Equation (3) at several points close to the open boundary but ignoring points where $\phi_t, \phi_x \to 0$. The average phase velocity c is then obtained and used for the open boundary condition in Equation (2). Further details of the numerical procedure can be obtained from Jagannathan.[6,10]

Application of Orlanski Condition

This approach to the implementation of an Orlanski boundary condition was tested using a small-amplitude progressive free surface wave of period 2π in a water depth 101.6 feet. At time $t = 0$, the wave configuration and its velocity potential were taken as initial conditions. The simulation was carried out with the Orlanski open boundary condition implemented at the port and starboard boundaries and it was found that this progressive wave could be simulated without distortion. Figure 2 shows the values of the phase velocity obtained using Equation (3) at free surface nodes 1, 8, 15, 23, 30, 37, 45 and 50 at three instants in time. It is to be noted that the numerically determined phase velocities over the length of the wave are quite close to the theoretical value of 32.2 ft/sec.

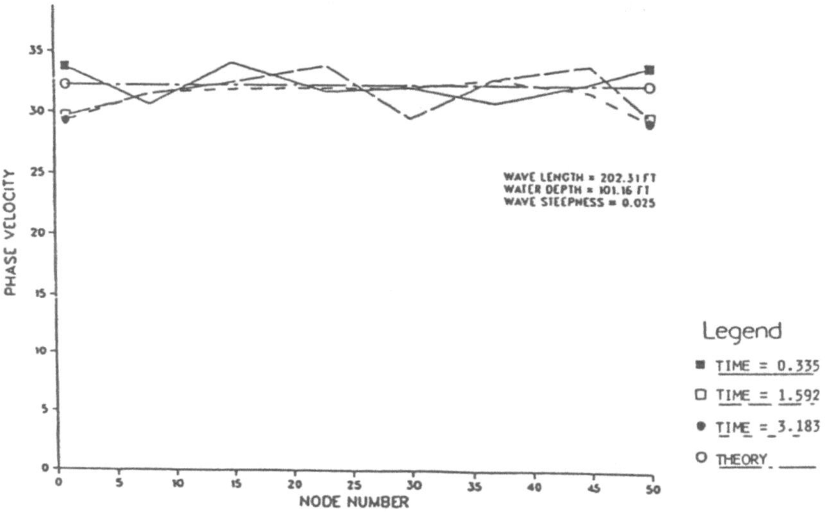

Figure 2. Phase velocity determination using the extended Orlanski condition for a small-amplitude progressive wave

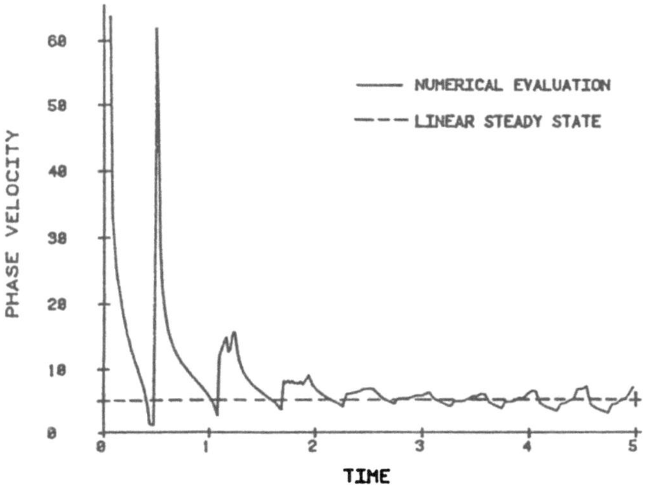

Figure 3. Phase velocity using the extended Orlanski condition at the right-hand side boundary

The forced motion of a submerged body was simulated using the Orlanski condition. The body was taken as a circular cylinder of radius 0.30 m (1.0') submerged with its axis horizontal, to a depth of (0.4 m) (1.3') at its center. The bottom was taken as uniformly flat at a depth of 1.60 m (5.25') and the body was oscillated sinusoidally in heave at a frequency of 6.261 rad/s. Both the open boundaries were considered and were located at a distance of 2.39 m (7.5') from the body.

Figure 3 shows the values of the numerically determined phase velocities near the right-hand side open boundary. Initially the phase velocities are high, corresponding to long waves generated as the body moves from rest. The numerical phase velocities converge to the linear steady state value after about two periods of oscillation. Figure 4 shows four free surface profiles about one period apart. Figure 5 provides a comparison of vertical force on the body obtained by simulation and the linear theory of Ogilvie.[16] The two compare very well and the simulation value reaches steady state after about two oscillations.

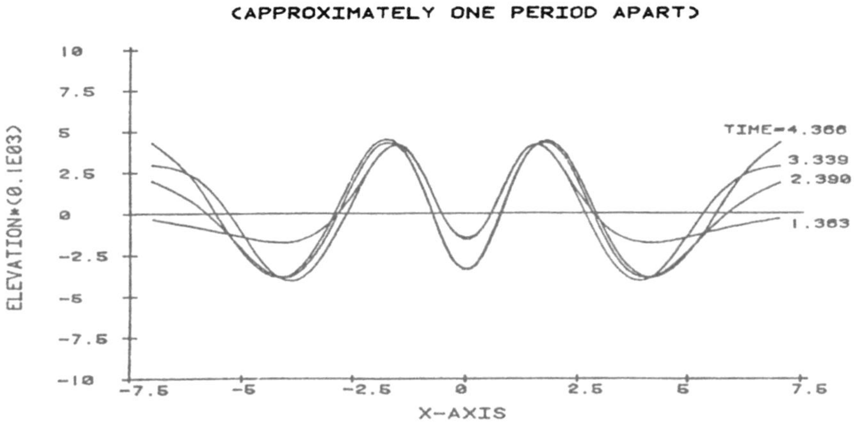

Figure 4. Free surface elevations for the forced body motion problem

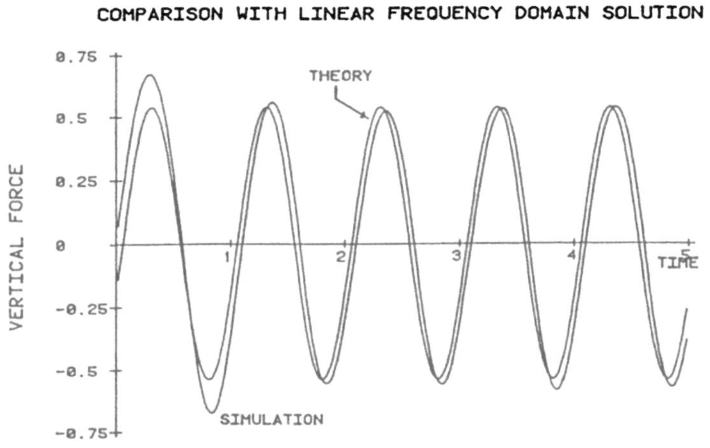

Figure 5. Comparison of vertical forces from simulation and frequency domain theory

ENERGY FLUX EQUALIZATION CONDITION

Energy Considerations

The problem of open boundaries may be posed in terms of the amount of energy that should be permitted to leave the computational domain through the open boundaries. In a two-dimensional horizontal-vertical plane, consider a finite domain Ω and its boundary $\Sigma = \Sigma_L \cup \Sigma_B \cup \Sigma_R \cup \Sigma_F$, where Σ_L is a vertical plane on the left side, Σ_B is the bottom boundary, Σ_F is the free surface boundary, and Σ_R is the open vertical boundary on the right side. Let free surface waves enter the domain through Σ_L and exit through Σ_R. Let E_Ω denote the total energy of fluid motion in the domain Ω. The energy theorem of John[17] (also see Yeung[1]) states

$$\dot{E}_\Omega = \rho\,(d/dt) \int_\Omega [1/2\,|\nabla\phi|^2 + gy]\, d\Omega \tag{4}$$

$$= \int_\Sigma [p\upsilon_n - \rho\phi_t(\phi_n - \upsilon_n)]\, d\Sigma \tag{5}$$

where υ_n is the normal velocity of any physical boundary. Let $\dot{E}_L$ be the rate at which energy enters the domain through Σ_L. In view of the boundary conditions at Σ_L, Σ_B and Σ_F, Equation (5) may be written as:

$$\dot{E}_L = \dot{E}_\Omega + \int_{\Sigma_R} [p\upsilon_n - \rho\phi_t(\phi_n - \upsilon_n)]\, d\Sigma \tag{6}$$

since $\upsilon_n = \phi_n = 0$ for Σ_B; $p = 0$, $\upsilon_n = \phi_n$ for Σ_F. The second term in Equation (6) is the energy flux from the open boundary Σ_R. This boundary may be perceived of either as a rigid movable boundary where $\phi_n = \upsilon_n$ or as an analytical open boundary without a physical boundary, i.e., $\upsilon_n = 0$. In any case, let E_R denote the energy flux at the open boundary Σ_R and hence we get:

$$\dot{E}_L(t) - \dot{E}_\Omega(t) = \dot{E}_R(t) \tag{7}$$

i.e.,

Rate of energy input into the system
- Rate of energy capture by the system
= Rate of energy flux from the system.

The integral of Equation (7) is also equally valid:

$$E_L(t) - E_\Omega(t) = E_R(t) \tag{8}$$

i.e.,

Total energy input into the system
- Total energy captured by the system
= Total energy flux from the system.

A choice of an open boundary condition would implicitly define the energy flux out of the open boundary and hence by Equation (7) would also determine the energy accumulation inside the fluid domain. Thus an improper open boundary condition would lead to improper accumulation of energy in the domain (i.e., through reflections) and lead to a progressive deterioration of the solution. Equations (7) and (8) may be used to monitor the accuracy of numerical solutions. In particular, for steady state problems, the average rate of energy input should equal the average rate of energy flux from the domain. It should however be noted that the occurrence of steady state does not necessarily imply that the solution is correct. The application of improper boundary conditions during the approach to steady state could clearly lead to a false steady state solution for nonlinear problems.

It is intuitively attractive to wonder whether an open boundary condition may be obtained using energy flux considerations. In Jagannathan,[10] this problem is approached from two viewpoints, namely that of energy flux maximization and energy flux equalization. In this paper we look at the energy equalization scheme where the optimal motion of the absorption boundary is sought based on the consideration that the incident energy flux from the inner domain be equal to energy flux radiated to the outer domain.

Energy Flux Equalization Hypothesis

Consider two-dimensional waves from the inner domain shown in Figure 6 to be incident upon a termination at x = 0. If the termination boundary were stationary then the waves would be reflected causing standing waves. We seek the optimal motion of the termination boundary which, based on the principle of equalization of energy flux, would minimize reflection and give a progressive wave solution.

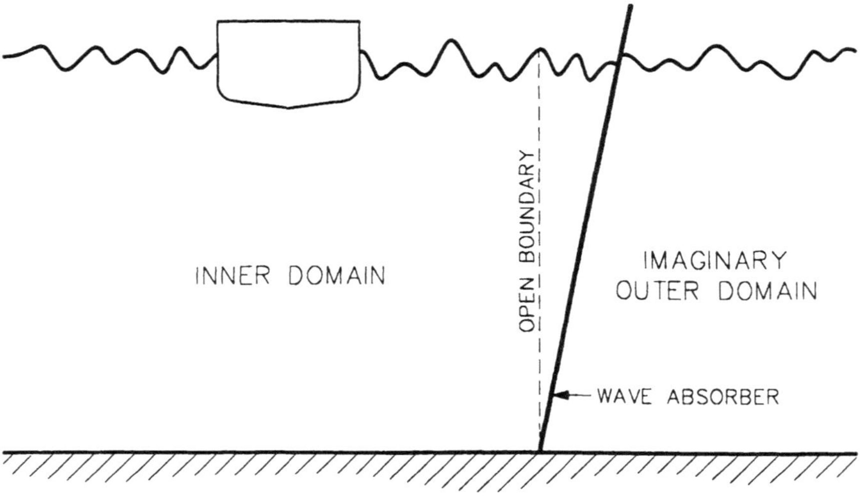

Figure 6. Definition sketch for energy flux equalization problem

Let us first consider that there is no termination boundary at x = 0. The incident two dimensional waves from the inner domain are allowed to pass unhindered to $x = \infty$. The normal velocity of the fluid particles at the plane x = 0 is then given by $\phi_x(y,t)$. Consider now that at x = 0 we impose a rigid termination boundary such that the normal velocity of this boundary $\dot{x}(y,t)$ equals the exact solution $\phi_x(y,t)$ of the incident wave in the absence of the termination boundary. Clearly the rest of the fluid would not 'know' that a rigid termination boundary has been imposed in the flow. All the incident energy is then 'absorbed' by the termination boundary and there is neither a reflection of progressive waves nor any local disturbance. A perfect absorber is possible if the exact behavior of the incident wave is known. For more general flows (linear or nonlinear) from the inner domain the exact solution at the open boundary is not known and hence the exact termination boundary solution to be used is also not known. Suppose that, due to this difficulty, we impose some arbitrary termination boundary, say a piston or hinge type in the simplest case. But how then is the motion of this boundary to be obtained? The proposition advanced here is that the boundary motion is to be such as to equalize the incident flux to the radiated flux at the termination boundary. In this paper this proposition is examined for problems with simple termination boundaries. In general the termination boundary can have infinite degrees of freedom. Analysis here is restricted to a one degree of freedom system.

Energy Flux Into Outer Domain

Clearly the forced motion of the termination boundary causes waves to propagate into the imaginary outer domain. For simplicity, the termination boundary is considered to have a single degree of freedom and linear free surface conditions are satisfied in the outer domain.

The radiation velocity potential is required to satisfy:

(i) $\nabla^2\phi_r = 0$ in Ω

(ii) $\phi_{ry} = 0$ on $y = -h$

(iii) Linearized free surface conditions[6]

(iv) $\phi_{rx}\ (y,t) = \dot{a}(t)\delta(y)$ at the termination boundary $x = 0$ (9)

where:

$\dot{a}(t)$ = time-dependent part of absorber velocity (to be determined)
$\delta(y)$ = absorber shape function (which is chosen)

In particular we have:

$\delta(y)$ = 1 for a 'piston' wave absorber
= $(y + h)/h$ for a 'hinged' wave absorber, hinged at $y = -h$

The pressure due to ϕ_r is

$$p = -\rho \frac{\partial \phi_r}{\partial t} \tag{10}$$

The hydrodynamic 'force' on the wave absorber is therefore:

$$-F_O(t) = \int_{-h}^{0} p\,\delta(y)\,dy \tag{11}$$

As shown in Jagannathan,[18] this can be written as:

$$-F_O(t) = M_h\ddot{a}(t) + \int_{-\infty}^{t} K(t-\tau)\,\dot{a}(\tau)\,d\tau \tag{12}$$

where

M_h = 'Genuine' added mass
$K(\tau)$ = Retardation function

M_h and $K(\tau)$ may be obtained by convolution from the frequency domain added mass and damping coefficients for the wave absorber, as detailed in Jagannathan.[18]

The work done by the absorber into the outer domain is due only to the radiation potential, which is given by

$$\dot{E}_O(t) = -\dot{a}(t)F_O(t). \tag{13}$$

Energy Flux from Inner Domain

Consider now the energy flux incident upon the termination boundary from the inner domain. If at any instant in time, $\phi(t)$ represents the velocity potential at the termination boundary, then we have

$$\phi_x(y,t) = \dot{a}(t)\delta(y)$$

The energy flux from the inner domain is given by

$$\dot{E}_I(t) = \int_{-h}^{\eta(t)} \phi_t \phi_x \, dy$$

$$= \int_{-h}^{\eta(t)} \phi_t \, \dot{a}(t) \, \delta(y) \, dy$$

$$= \dot{a}(t) \int_{-h}^{\eta(t)} \phi_t \, \delta(y) \, dy$$

$$= \dot{a}(t) \, F_I(t)$$

where

$$F_I(t) = \int_{-h}^{\eta(t)} \phi_t \, \delta(y) \, dy$$

= Hydrodynamic 'force' on wave absorber from the inner domain

Energy Flux Equalization

Equating the energy flux from the inner domain to the energy flux into the outer domain, we get:

$$\dot{E}_O(t) = \dot{E}_I(t) \tag{14}$$

i.e.,

$$-\dot{a}(t) \, F_O(t) = \dot{a}(t) \, F_I(t)$$

i.e.,

$$M_h \ddot{a}(t) + \int_{-\infty}^{t} K(t-\tau) \, \dot{a}(\tau) \, d\tau = F_I(t) \tag{15}$$

This is an integro-differential equation for the optimal motion $a(t)$ of the wave absorber which equalizes the incident energy flux to the radiated energy flux at every instant in time.

Note that in Equation (15) the left hand side denotes the hydrodynamic force due to a linear outer domain problem and the right hand side denotes the forcing function from a nonlinear inner domain problem. The extension to a nonlinear outer domain is possible using a Voltera kernel approach for the wave absorber, as elaborated in Jagannathan.[10,18]

Application of Energy Flux Condition

To investigate the efficiency of the energy flux equalization procedure a forced wave-maker problem was chosen. The left side boundary is considered a hinged type wave-maker with a forced sinusoidal motion of amplitude 0.005 foot at a frequency of 3 rad/s in water of depth 7 feet. The energy flux equalization procedure to be applied at the right side wave absorber is based on the velocity potential in the outer domain satisfying linearized free surface conditions. However, the forcing function derived at the open boundary (i.e., at the wave absorber) is based on the nonlinear inner domain where nonlinear free surface and body conditions are satisfied. Figures 7 and 8 show the added mass and damping function for the linear absorber, derived using two methods: the frequency domain solution, marked with a ■ and the impulse response solution, denoted by a ★. The two solutions are practically identical, indicating that the retardation function is accurate. The energy flux amplitude ratio L is 0.017 (see Jagannathan[10]) for this case and the absorption characteristics are clearly quite good. Our aim now is to incorporate this absorber into the nonlinear simulation procedure.

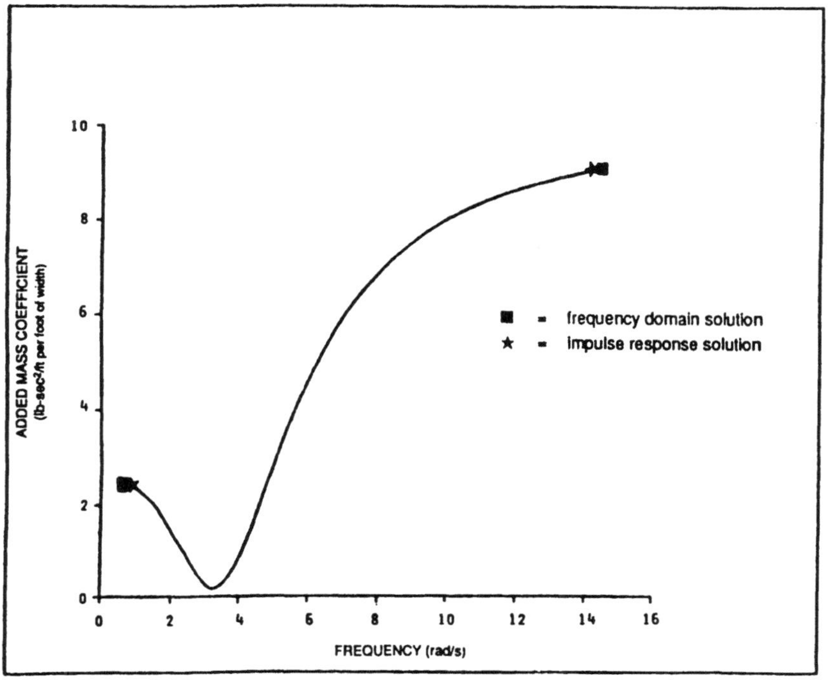

Figure 7. Added mass coefficient for wave absorber (type: hinged).

The length of the inner domain was chosen as 1.5 times the wavelength corresponding to the frequency of oscillation of the left side wave-maker. One hundred equally spaced nodes were used on the free surface, 28 on each side boundary, and 20 nodes at the bottom.

The wave-maker starts from rest. Figure 9 shows the energy input by the wave-maker into the domain. The optimal motion of the absorber is given in Figure 10. The amplitude of motion approximately equals the frequency domain solution but some slight peaks and troughs are noticeable. The energy absorbed corresponding to this

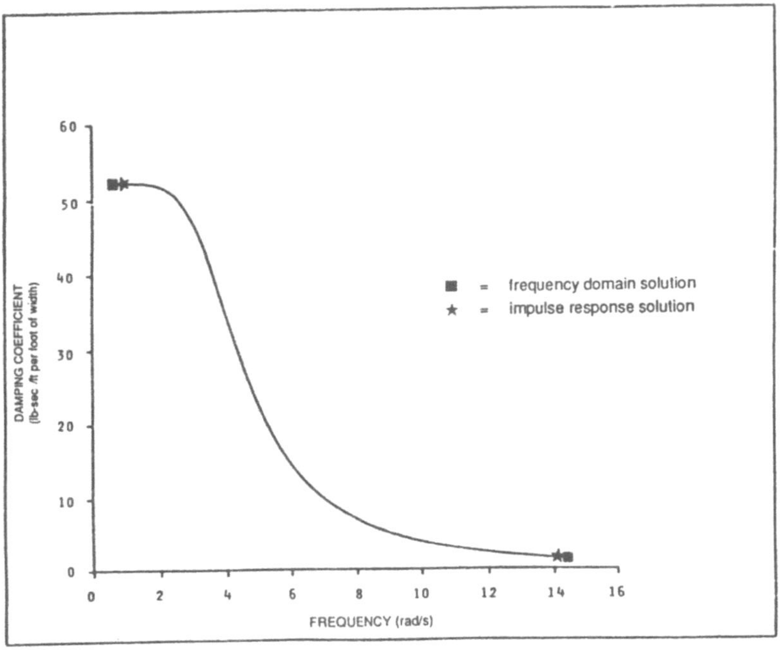

Figure 8. Damping coefficient for wave absorber (type: hinged)

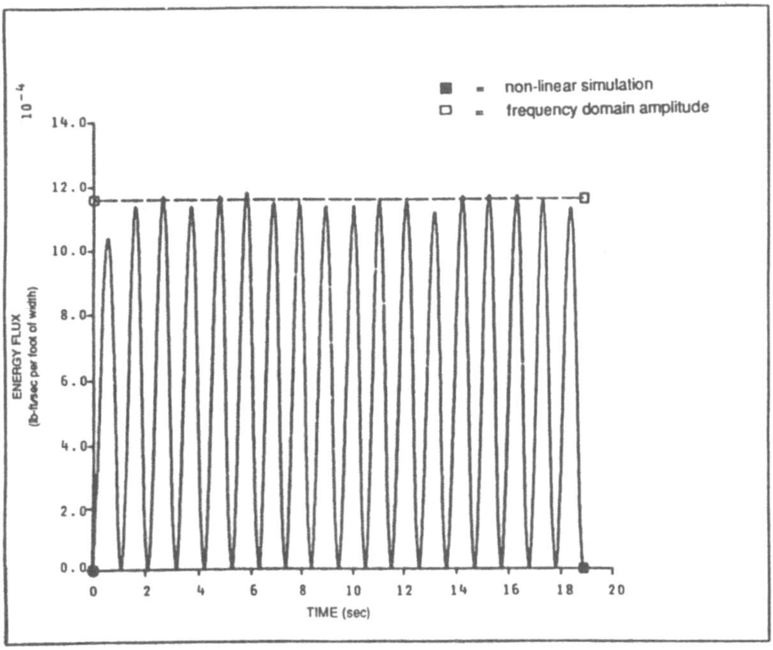

Figure 9. Energy input by left side wavemaker (type: hinged)

motion is shown in Figure 11. The comparison here with the frequency domain solution is fair with the occurrence of peaks and troughs more evident. This is because energy flux varies as the square of the motion and the errors are thereby made more evident. The error in the boundary motion is less than two percent and that in energy flux about four percent.

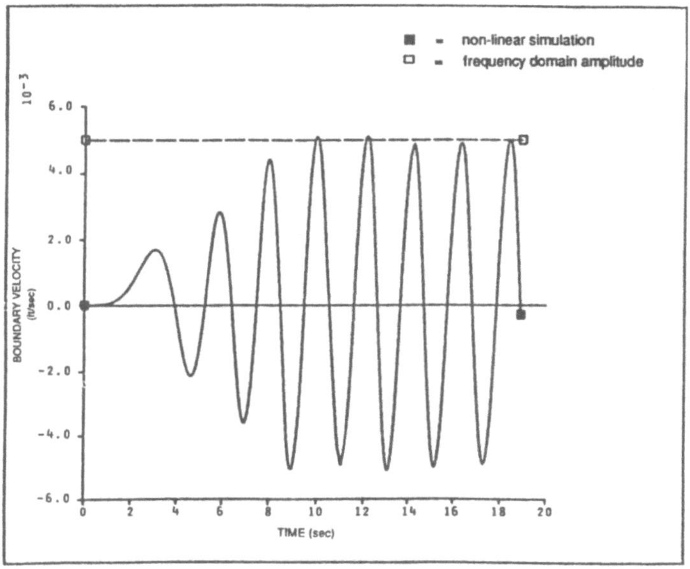

Figure 10. Optimal absorber boundary motion (type: hinged)

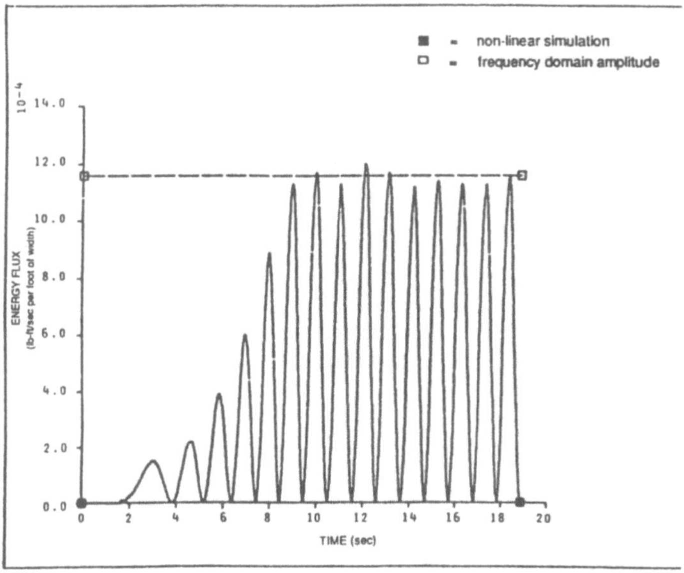

Figure 11. Energy absorbed for optimal boundary motion (type: hinged)

The numerical procedure was found to be sensitive to the method of integration as well as to the resolutions in space and time. The evolution of the free surface in time was solved simultaneously with the optimal motion of the absorber with a fourth order Runge Kutta algorithm. The time step used was one hundredth the oscillation period. The intersection of the absorber with the free surface gives rise to a logarithmic singularity which has to be treated properly as the free surface elevation at that intersection point contributes to the forcing function for the absorber motion. At the intersection point the velocity potential is obtained from the free surface condition, and the stream function from the absorber (i.e., the intersection point is fully determinate in terms of the boundary value problem). It is found that the vertical velocity of the intersection node is best approximated by backward spatial differentiation of the velocity potential along the absorber rather than backward differentiation along the free surface or complex differentiation around the intersection.

CONCLUSIONS

Of particular current interest is the treatment of the open boundary condition in the simulation of nonlinear free surface flows. For linear problems this is the radiation condition which is well known. For nonlinear problems, no general open boundary conditions are known so far. In this paper two open boundary conditions have been examined, namely the Orlanski condition and the energy flux equalization condition.

The Orlanski condition has been verified for the case of a freely moving gravity wave and an oscillating body. This condition appears quite non-reflective and is probably satisfactory for many practical applications. Tanaka and Nakamura[19] have implemented the Orlanski condition as described above and have provided a detailed evaluation of its performance and stability characteristics.

An energy flux equalization condition between an inner nonlinear computational domain and an external linear flow field has been proposed. The procedure for a linear single degree of freedom wave absorber has been detailed with the optimal solution obtained in the form of an integro-differential equation. The procedure allows extensions to multiple degrees of freedom and to nonlinear wave absorbers. The performance of this boundary condition is quite good for the nonlinear free surface problem that was investigated. For this initial value wave-maker problem the approach to the steady state is monotonic and the steady state is maintained with some low frequency oscillation. Further numerical experiments will be required to study the systematic improvements in this procedure associated with choice of shape functions and the degrees of freedom of the wave absorber.

It should be noted that for many three dimensional problems, the open boundary condition may be linearized based on the consideration of energy dissipation. In contrast, with two-dimensional nonlinear problems, there is no energy dissipation with distance and hence in principle the open boundary condition at any distance from the disturbance remains nonlinear.

ACKNOWLEDGEMENTS

This work forms a part of the author's Ph.D. dissertation and was done under the guidance of Prof. W.C. Webster, which is gratefully acknowledged. The author would like to thank Prof. Ronald Yeung for his advice and assistance. The financial support given by the Department of Naval Architecture and Offshore Engineering, U.C. Berkeley, is acknowledged with gratitude.

REFERENCES

1. Yeung, R.W. Numerical Methods in Free-Surface Flows, Amer. Rev. Fluid Mech., Vol. 14, pp. 395-442, 1982.
2. Longuet-Higgins, M.S. and Cokelet, E.D. The Deformation of Steep Surface Waves on Water, I. A Numerical Method of Computation, Proc. Roy. Soc. Lond. A, Vol. 350, pp. 1-26, 1976.
3. Vinje, T. and Brevig, P. Breaking Waves on Finite Water Depths, A Numerical Study, Ships in Rough Seas (SIS) Report, Division of Marine Hydrodynamics, Norwegian Institute of Technology, 1980.
4. Vinje, T. and Brevig, P. Nonlinear Two-Dimensional Ship Motion, Ships in Rough Seas (SIS) Report, Norwegian Institute of Technology, 1980.
5. Vinje, T. and Brevig, P. Nonlinear Ship Motions, Proc. 3rd Int. Conf. on Numerical Ship Hydrodynamics, Paris, 1981.
6. Jagannathan, S. Non-linear Free Surface Flows and an Application of the Orlanski Boundary Condition, Int. J. Numer. Methods Fluids, Vol. 8, No. 9, pp. 1051-1070, 1988.
7. Hedstrom, G.W. Nonreflecting Boundary Conditions for Nonlinear Hyperbolic Systems, J. Comput. Phys., Vol. 30, pp. 222-237, 1979.
8. Rudy, D.H. and Strikwerda, J.C. A Nonreflecting Outflow Boundary for Subsonic Navier-Stokes Calculations, J. Comput. Phys., Vol. 36, pp. 55-70, 1980.
9. Engquist, B. and Majda, A. Absorbing Boundary Conditions for the Numerical Simulation of Waves, Math. Comput., Vol. 31, No. 139, pp. 629-651, 1977.
10. Jagannathan, S. Nonlinear Free Surface Flows and Open Boundary Conditions, Ph.D. Dissertation, University of California, Berkeley, 1985.
11. Kreiss, H.O. Proceedings of a Symposium at Univ. of Wisconsin, Wiley, New York, 1966.
12. Pearson, R.A. J. Atmos. Sci., Vol. 31, 1974.
13. Orlanski, I. A Simple Boundary Condition for Unbounded Hyperbolic Flows, J. Comput. Phys., Vol. 21 pp. 251-269, 1976.
14. Han, T.Y., Meng, J.C.S. and Innis, G.E. An Open Boundary Condition for Incompressible Stratified Flows, J. Comput. Phys., Vol. 49, pp. 276-297, 1983.
15. Ertekin, R.C. Soliton Generation by Moving Disturbances in Shallow Waters: Theory, Computation and Experiment, Ph.D. Dissertation, University of California at Berkeley, 1984.
16. Ogilvie, T.F. First and Second Order Forces on a Cylinder Submerged Under a Free Surface, Vol. 16, No. 3, pp. 451-472, 1963.
17. John, F. On the Motion of Floating Bodies - I, Comm. Pure Appl. Math., Vol. 2, pp. 13-57, 1949.
18. Jagannathan, S. Energy Flux Equalization: An Open Boundary Condition for Non-Linear Free Surface Flows, submitted to Int. J. Numer. Methods Fluids.
19. Tanaka, Y. and Nakamura, T. Numerical Generation of Gravity Waves and Open Boundary Scheme, Offshore Mechanics and Arctic Engineering Symposium '91, Stavanger, Norway, 1991.

The Numerical Solution of Viscous Flows with a Free Surface

Y. Yuan, D.B. Ingham

Department of Applied Mathematical Studies, The University of Leeds, Leeds, LS2 9JT, England

ABSTRACT

In this paper we use the boundary element method to solve a very slow viscous flow with a free surface. An iterative scheme is presented and the numerical solutions are obtained for various non-dimensional fluid flow rate, q, and the non-dimensional surface tension, T.

1. INTRODUCTION

Viscous free surface flows are important in coating technology, polymer processing and other engineering applications as well as in certain areas of science. The free surface problem addressed in this paper is that of a fluid moving between two parallel plates which are in relative motion. At an arbitrary point the stationary plate turns through an angle of $\pi/2$ away from the moving plate. At this point the fluid forms a free surface which is extended downstream by the moving plate, see Fig. 1.

A numerical approach to this problem was first presented by Coyne and Elrod [1]. In their work a local perturbation technique was used but the shape of the free surface was restricted to assuming that the contact angle was fixed at $\pi/2$. Further numerical investigations of the problem have been performed by Saito and Scriven [2] and Carter [3]. Both of these works used the finite element technique to solve the problem which arises from slot coating flows and they allowed the contact angle to vary as required by the global solution on the free surface. However, they could not obtain results when the non-dimensional fluid flow rate through the plates was larger than 1/2.

Successful boundary element method (BEM) solutions of potential free surface flows were obtained by Liggett [4] and

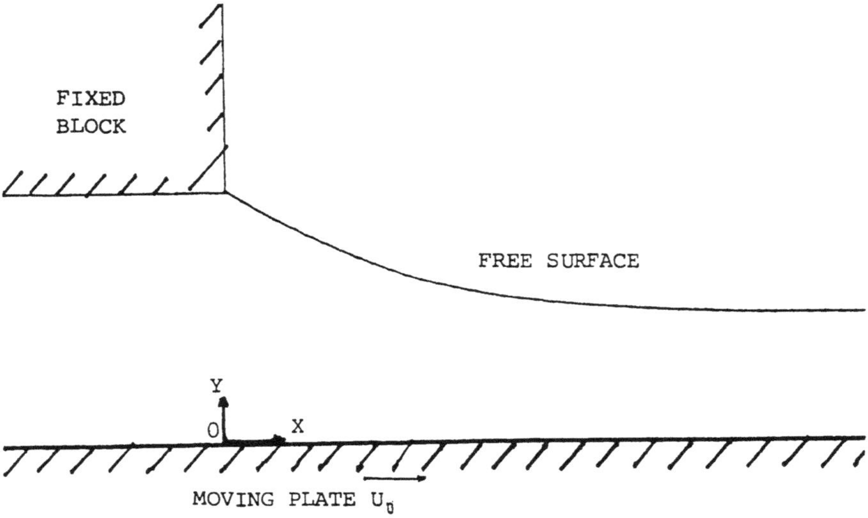

Fig. 1. The geometry of the problem.

Longuet-Higgins and Cokelet [5]. In these works the free surface location is obtained by an iterative process based on the satisfaction of the kinematic condition which provides an explicit relationship between known boundary element variables. However, for very slow viscous flows the available conditions provide no such explicit relationship. Ingham and Kelmanson [6] and Ritchie [7] dealt with the viscous flow using the constant BEM in which the viscous flow variables are stream function, velocity, vorticity and vorticity gradient. Both of these works ensured an analytical function as the fundamental shape of the free surface, then the problem reduced to determining one or more of the coefficients in the series solution. The disadvantage of this method is that it is difficult to extend to other free surface problems since the fundamental shape of the free surface depends on the problem itself. For example, the fundamental shapes which were introduced by Ingham and Kelmanson [6] and Ritchie [7] are not suitable to solve the sintering problem, see Kuiken [8].

In order to obtain more accurate solutions a linear BEM is presented in this paper in which the viscous flows, which are described by the biharmonic equation, are considered. Since the location of the free surface is unknown on the free surface three boundary conditions are required. Instead of using the fundamental shape method as described in [6] and [7] we parameterize the free surface as a collection of points which are joined by straight line segments and then use an iterative process to decide the location of the free surface. In this paper we have enforced the kinematic and the normal stress conditions and iterated on the shear stress condition in order to obtain the free surface. The reasons for this choice will be given later in this paper.

2. EQUATIONS AND BOUNDARY CONDITIONS

2.1 GOVERNING EQUATIONS

We consider the two-dimensional incompressible fluid flow between two parallel flat plates, as shown in Fig. 1. The infinite plate moves from left to right at a fixed speed, U_0. At a constant height, h_0, above the moving plate there is a stationary, semi-infinite block. The point on the moving plate directly below the edge of the stationary block is designated as the origin, O. The X-axis lies parallel to the moving plate and the Y-axis is at right angles to the moving plate with y increasing towards the stationary block. A free surface, Γ, forms at the edge of the stationary block and fluid is swept downstream by the moving plate where the free surface eventually runs parallel to the moving plate at a height, h_d, above the moving plate.

The following non-dimensional quantities are introduced

$$x = X/h_0 \qquad y = Y/h_0 \qquad \mathbf{u} = \mathbf{U}/U_0 \qquad p = P_0 h_0 / \rho\upsilon U_0$$

where υ is the kinematic viscosity of the fluid, ρ is the density, P_0 is the pressure of the fluid and $\mathbf{U}$ is the fluid velocity. For steady two-dimensional flow of an incompressible Newtonian fluid the Navier-Stokes equation and continuity equation reduce to

$$\nabla p = \nabla^2 \mathbf{u} \tag{2.1}$$

$$\nabla \cdot \mathbf{u} = 0 \tag{2.2}$$

when the Reynolds number, $Re = U_0 h_0 / \upsilon$, is assumed to be very small. On introducing the stream function ψ, the x- and y-components of velocity are then given by

$$u = \psi_y, \qquad v = -\psi_x \tag{2.3}$$

respectively. From equations (1.1-1.3) it may be shown, see Batchelor [9], that ψ satisfies the biharmonic equation

$$\nabla^4 \psi = 0 \tag{2.4}$$

On introducing the vorticity, ω, equation (2.4) may be written in the form

$$\left\{ \begin{array}{ll} \nabla^2 \psi = \omega & (2.5) \\ \nabla^2 \omega = 0 & (2.6) \end{array} \right.$$

2.2 BOUNDARY CONDITIONS

The fluid flow far upstream of the origin is constrained to flow between two parallel plates under the action of a pressure gradient and a moving plate, i.e. as $x \longrightarrow -\infty$ the the flow is a Couette-Poiseuille flow which means that the fluid velocity profile across the flow is parabolic. This results in the stream function being a cubic function of y. If the air resistance at the fluid / gas interface is neglected then downstream of the origin the fluid flow becomes plug flow, i.e. the fluid moves as a body with a uniform speed, U_0.

In order to determine the shape of free surface three boundary conditions are required, and these are

(i) No fluid crosses the interface.
(ii) There is zero shear stress on the free surface.
(iii) The normal stress is balanced by the pressure and the surface tension.

Let us now consider on the free surface the local right-handed coordinate system defined by the vectors **t** and **n**, where **t** denotes the tangent vector and **n** is the outward normal to the free surface vector. Then the boundary conditions on free surface may be expressed

$$\text{(i)} \qquad \psi = q \tag{2.7}$$

$$\text{(ii)} \qquad \psi'' - \psi_{tt} = 0 \tag{2.8}$$

$$\text{(iii)} \qquad - T\,k = - p + \sigma'' \tag{2.9}$$

where q ($= Q/h_0U_0$, where Q is the fluid flow rate) is a non-dimensional fluid flow rate, (′) denotes the normal derivative, T ($= T_0/h_0P_0$, where T_0 is the surface tension) is the non-dimensional surface tension, k is the local non-dimensional curvature of the free surface, p is the non-dimensional pressure and σ is the non-dimensional stress tensor, see Ingham and Kelmanson [6].

We may eliminate from the problem the pressure p. From equation (1.1), in terms of the local coordinate system (t,n) and considering the t-component, we find that

$$\partial p \,/\, \partial t = \nabla^2 u_t$$

or, in terms of the vorticity,

$$\partial p \,/\partial t = \nabla^2 (\partial\psi \,/\, \partial n) = \partial\omega \,/\, \partial n \tag{2.10}$$

Differentiating equation (2.9) with respect to t and substituting $\partial p/\partial t$ from equation (2.10) into the boundary condition (2.9) we obtain

$$2\,\frac{\partial^3\psi}{\partial n\,\partial t^2} + \frac{\partial\omega}{\partial n} + T\,\frac{dk}{dt} = 0 \tag{2.11}$$

Instead of the t-derivatives appearing in the boundary conditions (2.8) and (2.11) it is more convenient to have derivatives with respect to s, the arc length along the free surface. Coyne and Elrod [1] have shown that

$$\frac{\partial}{\partial t}\Big|_{\Gamma} = \frac{d}{ds}$$

$$\frac{\partial^2}{\partial t^2}\Big|_{\Gamma} = \frac{d}{ds^2} - k\,\frac{\partial}{\partial n}$$

which on substitution into the boundary condition (2.8) gives

$$\omega = -2\ k\ \psi' \tag{2.12}$$

Turning our attention now to the boundary condition (2.11) and considering a fixed point A on the free surface we observe that the derivative $\partial\psi/\partial n$ is taken with respect to a fixed local Cartesian coordinate system, which original point is at A. On the other hand ψ' denotes the outward normal derivative at each separate point of Γ. Let B be an arbitrary neighbourhood point of A on Γ, then we have

$$\partial\psi/\partial n\ |_B = \cos\theta\ \psi' + \sin\theta\ \partial\psi/\partial s = \cos\theta\ \psi'$$

where θ is the angle between the tangent to Γ at B and the tangent to Γ at A. It is clear that as B $\longrightarrow$ A we have $\theta \longrightarrow 0$ and $\partial\psi/\partial n|_B \longrightarrow \psi'|_A$. We can therefore now write

$$\frac{\partial^2}{\partial t^2}\frac{\partial\psi}{\partial n}\Big|_{\Gamma,A} = \frac{d^2}{ds^2}\frac{\partial\psi}{\partial n}\Big|_{\Gamma,A} - k\,\frac{\partial^2\psi}{\partial n^2}\Big|_{\Gamma,A}$$

$$= \frac{d^2}{ds^2}\left(\cos\theta\ \psi'\right)\Big|_{\Gamma,A} - \frac{k}{2}\,\omega\ \Big|_{\Gamma,A}$$

$$= \frac{d^2}{ds^2}\,\psi'\ \Big|_{\Gamma,A}$$

where we have used $d\theta/ds = k$ and the boundary conditions (2.8) and (2.12). The third boundary condition, (2.9), may now be expressed in the form

$$-T\,\frac{dk}{ds} = \omega' + 2\,\frac{d^2}{ds^2}\,\psi' \qquad \text{on } \Gamma \tag{2.13}$$

The full boundary conditions may now be written in the form

$$\psi = q, \qquad \psi_y = 0 \qquad \text{on } y=1,\ x \le 0 \tag{2.14}$$

$$\psi = 0, \qquad \psi_y = -1 \qquad \text{on } y=0,\ -\infty < y < \infty \tag{2.15}$$

$$\left.\begin{aligned} &\psi_y \longrightarrow 3\,(1-2q)y^2 + 2\,(3q-2)\,y + 1 \\ &\omega \longrightarrow 6\,(1-2q)y + 6q - 4 \end{aligned}\right\} \text{as } \begin{array}{l} x \longrightarrow -\infty \\ 0 \le y \le 1 \end{array} \tag{2.16}$$

$$\psi_y \longrightarrow 1, \qquad \omega \longrightarrow 0 \qquad \text{as } x \longrightarrow +\infty \quad 0 \le y \le q \tag{2.17}$$

$$\left.\begin{array}{ll} \psi = q & \omega = -2k\,\psi' \\ -T\,k_s = \omega' + 2\,\psi'_{ss} & \end{array}\right\} \text{ on } \Gamma \tag{2.18}$$

The conditions (2.14) and (2.15) give the no-slip velocity on the plate. The enforced upstream Couette-Poiseuille flow is given in (2.16) and the plug flow downstream is given by (2.17). The conditions (2.18) are specified on the free surface. Thus at each point on the boundary at least two conditions in terms of the BEM variables are known. We note that in order to balance the fluid flow rate in the upsteam and downstream we must have $0 \le y \le q$ downstream.

3. THE BOUNDARY ELEMENT METHOD

In order to solve the biharmonic equation (2.5) and (2.6) in the domain Ω we use the Boundary Element Method (BEM). For any $p = (x,y) \in \bar{\Omega}$ and $q = (x_0, y_0) \in \partial\Omega$, let

$$F_1(p,q) = \ln|p - q| \tag{3.1}$$

$$F_2(p,q) = |p - q|^2 (\ln|p - q| - 1) \tag{3.2}$$

where $|p - q| = \{ (x - x_0)^2 + (y - y_0)^2 \}^{1/2}$. Applying Green's second identity we have

$$\eta(p)\psi(x) = \int_{\partial\Omega} \psi(q)\,F_1'(p,q)\,ds(q) - \int_{\partial\Omega} \psi'(q)\,F_1(p,q)\,ds(q)$$

$$+ \frac{1}{4}\int_{\partial\Omega} \omega(q)\,F_2'(p,q)\,ds(q) - \frac{1}{4}\int_{\partial\Omega} \omega'(q)\,F_2(p,q)\,ds(q) \tag{3.3}$$

$$\eta(p)\omega(x) = \int_{\partial\Omega} \omega(q)\,F_1'(p,q)\,ds(q) - \int_{\partial\Omega} \omega'(q)\,F_1(p,q)\,ds(q) \tag{3.4}$$

where

(i) $ds(q)$ denotes differential increment of $\partial\Omega$ at q.

(ii) The prime (′) denotes differentiation with respect to the outward normal to Ω at q.

(iii) $$\eta(p) = \begin{cases} 2\pi & \text{when } x\in\Omega \\ \theta & \text{when } x\in\partial\Omega,\ \theta \text{ is the angle between the tangents to } \partial\Omega \text{ on either side of } x. \end{cases}$$

In order to obtain the numerical solution of the biharmonic equation from the integral equation (3.3) and (3.4) the simplest method to use is the classical BEM. However, for more accuracy we use the linear BEM in the present work. This is achieved by first subdividing the boundary $\partial\Omega$ into N segments $\partial\Omega_j$ where j = 1, 2,...,N. On each segment $\partial\Omega_j$, ψ, ψ', ω and ω' are approximated by linear functions such that

$$\psi = (1-\xi)\ \psi(q_{j-1}) + \xi\ \psi(q_j)$$

$$\psi' = (1-\xi)\ \psi'(q_{j-1}) + \xi\ \psi'(q_j)$$

$$\omega = (1-\xi)\ \omega(q_{j-1}) + \xi\ \omega(q_j)$$

$$\omega' = (1-\xi)\ \omega'(q_{j-1}) + \xi\ \omega'(q_j)$$

where q_{j-1} and q_j are the end points of the interval $\partial\Omega_j$, ξ is a linear function which increases from zero at q_{j-1} to unity at q_j. With these approximations the integral equations (3.3) and (3.4) become

$$\begin{aligned}\eta(p)\psi(p) = &\sum_{j=1}^{N} \{\psi_{j-1}\int_{\partial\Omega_j}(1-\xi)\ F_1'(p,q)ds(q) + \psi_j \int_{\partial\Omega_j}\xi\ F_1'(p,q)ds(q)\} \\ &- \sum_{j=1}^{N} \{\psi'_{j-1} \int_{\partial\Omega_j}(1-\xi)\ F_1(p,q)ds(q) + \psi'_j \int_{\partial\Omega_j}\xi\ F_1(p,q)d(q)\} + \\ &\frac{1}{4}\sum_{j=1}^{N} \{\omega_{j-1}\int_{\partial\Omega_j}(1-\xi)\ F_2'(p,q)ds(q) + \omega_j \int_{\partial\Omega_j}\xi\ F_2'(p,q)ds(q)\} - \\ &\frac{1}{4}\sum_{j=1}^{N} \{\omega'_{j-1} \int_{\partial\Omega_j}(1-\xi)\ F_2(p,q)ds(q) + \omega'_j \int_{\partial\Omega_j}\xi\ F_2(p,q)d(q)\}\end{aligned} \qquad (3.5)$$

$$\begin{aligned}\eta(p)\omega(p) = &\sum_{j=1}^{N} \{\omega_{j-1}\int_{\partial\Omega_j}(1-\xi)\ F_1'(p,q)ds(q) \\ &+ \omega_j \int_{\partial\Omega_j}\xi\ F'_1(p,q)ds(q)\} - \\ &\sum_{j=1}^{N} \{\omega'_{j-1} \int_{\partial\Omega_j}(1-\xi)\ F_1(p,q)ds(q) + \omega'_j \int_{\partial\Omega_j}\xi\ F_1(p,q)d(q)\}\end{aligned} \qquad (3.6)$$

If the segment $\partial\Omega_j$ is a straight line, then the integrals occurring in the equations (3.5) and (3.6) may be evaluated exactly, namely

$$\int_{\partial\Omega_j} F'_1(p,q)\, ds(q) = I_1 \tag{3.7}$$

$$\int_{\partial\Omega_j} F_1(p,q)\, ds(q) = J_1 \tag{3.8}$$

$$\int_{\partial\Omega_j} \xi\, F'_1(p,q)\, ds(q) = \frac{1}{h}\,(a \cos\beta\, I_1 + I_2) \tag{3.9}$$

$$\int_{\partial\Omega_j} \xi\, F_1(p,q)\, ds(q) = \frac{1}{h}\,(a \cos\beta\, J_1 + J_2) \tag{3.10}$$

$$\int_{\partial\Omega_j} F'_2(p,q)\, ds(q) = a\,(2J_1 - h)\sin\beta \tag{3.11}$$

$$\int_{\partial\Omega_j} F_2(p,q)\, ds(q) = \frac{1}{3}\,(h - a\cos\beta\,)^3(\,\ln b - \frac{4}{3}\,) + \frac{1}{3}\,(\,a\cos\beta\,)^3\,(\,\ln a - \frac{4}{3}\,) + (\,a\sin\beta\,)^2\,(J_1 - \frac{2}{3}\,h - \frac{1}{3}\,a\,\gamma\,\sin\beta\,) \tag{3.12}$$

$$\int_{\partial\Omega_j} \xi\, F_2(p,q)\, ds(q) = \frac{2}{h}\,a\sin\beta\,(a\cos\beta\, J_1 + J_2) - \frac{1}{2}\,a\,h\,\sin\beta \tag{3.13}$$

$$\int_{\partial\Omega_j} \xi\, F_2(p,q)\, ds(q) = \frac{1}{h}\,a\cos\beta \int_{\partial\Omega_j} F_2(p,q)\, ds(q) + \frac{1}{4h}\,(\,b^4\ln b - a^4\ln a\,) - \frac{5}{16h}\,(\,b^4 - a^4) \tag{3.14}$$

where

$$I_1 = \gamma$$

$$I_2 = a\sin\beta\,(\,\ln b - \ln a\,)$$

$$J_1 = a\cos\beta\,(\,\ln a - \ln b\,) - h + h\ln b + a\,\gamma\,\sin\beta$$

$$J_2 = \frac{1}{2}\,(\,b^2\ln b - a^2\ln a\,) - \frac{1}{4}\,(\,b^2 - a^2\,)$$

and a, b and h are the lengths of the lines joining p to q_{j-1}, p to q_j and q_{j-1} to q_j, respectively, and β and γ are the angles $q_j\, q_{j-1}\, p$ and $q_{j-1}\, p\, q_j$, respectively, see Fig. 2.

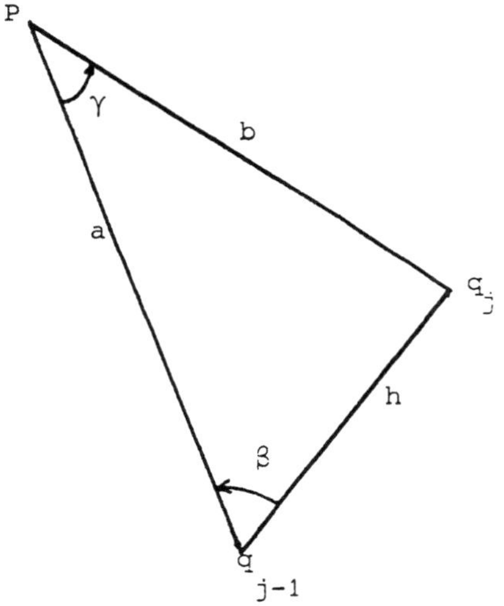

Fig. 2. Notation for the analytic evaluation of integrals on straight-line segment geometry.

4. NUMERICAL METHOD

4.1 CORNER TREATMENT

The successful application of the BEM for the biharmonic equation depends on the fact that two boundary conditions are required on rigid boundaries and three on the free surface. However, in our problem the number of equations could not be balanced against the number of unknowns, because there are some Dirichlet corners on the boundary at which there are two different normal derivatives, as shown in Fig. 3. Hence, we need a special treatment for the corner points and a number of papers address this problem. Lachat and Watson [10] suggested that the two gradients are equal, Ricardella [11] introduced the concept of the method of double notes, Alarcon *et al* [12] used of the relatively sophisticated additional relationships based on approximations to the local gradient of potential, and Ingham *et al* [13] introduced the method of determining derivatives and the method of element replacement. However, for a good accurate and easy to program approach we have used in this paper the method which was first introduced by Walker and Fenner [14]. They introduced an extra independent equation at the corner point which is

$$(\phi'_l - \phi'_r \cos\theta)\, \phi'_{n,l} + (\phi'_l \cos\theta - \phi'_r)\, \phi'_{n,r} = 0$$

and ϕ'_l and ϕ'_r may then be approximated by

$$\phi'_l = \frac{\phi(C) - \phi(A)}{|CA|} \qquad \phi'_r = \frac{\phi(B) - \phi(A)}{|BA|}$$

where ϕ is an arbitrary function, ϕ'_1 and ϕ'_r are the two tangent derivatives at the corner A, $\phi'_{n,1}$ and $\phi'_{n,r}$ are two normal derivatives at A and |CA| and |BA| are the lengths of the lines joining C to A and B to A, respectively, see Fig 3.

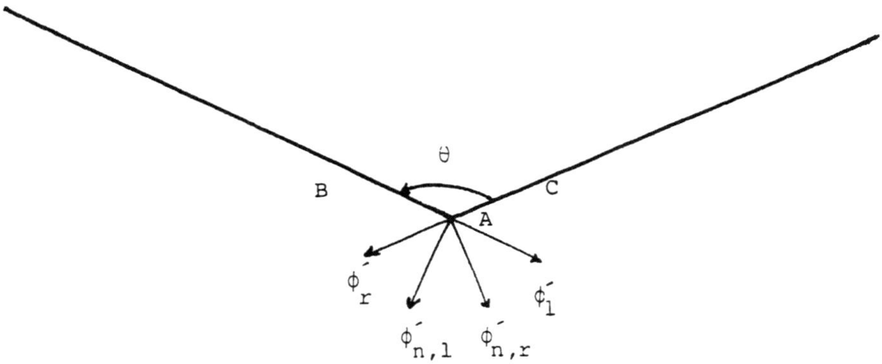

Fig. 3. The Dirichlet coner and the directional derivatives

4.2 THE ITERATIVE SCHEME

For any given combination of the non-dimensional fluid flow rate, q, and the non-dimensional surface tension, T, the problem reduces to finding the location of the free surface which satisfies all three boundary conditions (2.12). Once the location of the free surface is known then the stream function and vorticity may then be calculated everywhere within the fluid.

The fluid flow rate depends upon the pressure gradient, i.e. depends on the difference between the fluid pressure upstream, P_1, and the ambient pressure downstream, P_2. When $P_1 - P_2 = 0$ then $q = 1/2$, when $P_1 > P_2$ then $q > 1/2$ and when $P_1 < P_2$ then $q < 1/2$. If $q < 1/2$ then the contact angle, α_c, which is defined as the angle between the stationary plate and the free surface at the attachment point Q(0,1), may be less than $\pi/2$. This means that the free surface may have two possible values of y for a given value of x when $x < 0$, see Saito and Scriven [2]. On the other hand, when $q > 1$ the separation point of fluid on the stationary block may occur on $y > 1$, i.e the separation point may not be at Q(0,1) but at $Q_1(0,b)$, where $b > 1$ is an unknown function of q. It is very difficult to obtain an iterative process which will determine this separation point. Therefore in this paper in order to avoid having free surfaces, $y = \Gamma(x)$, which are either a multivalued function of x or involve an indeterminate separation point we have assumed that $1/2 < q < 1$.

We parameterize the free surface as a collection of points connected by N_1 straight line segments. Then solve the problem using two of the three boundary conditions and the third boundary condition is used to determine how the free surface should be moved in order to satisfy all three boundary conditions. We have three options in choosing the one boundary condition, (2.7), (2.12) or (2.13), as the iteration condition. The boundary condition (2.13) is the most complex condition and if used as the iteration condition then it is found to be difficult to obtain an iterative process which converges. The boundary condition (2.7) does not involve the local curvature of the free surface but the iterative process involves calculating the local curvature at each mesh point on the free surface and this may give rise to a large error. So in this paper we used the boundary condition (2.12) as the iteration condition and we have found that it is the most appropriate approach in order to ensure the iterative process converges most rapidly. The steps required to find free surface are detailed as follows:

Step 1. Specify a non-dimensional fluid flow rate, q, and a non-dimensional surface tension, T.

Step 2. Specify an initially free surface, $y = \Gamma(x)$.

Step 3. Solve the boundary value problem using the BEM. On the free surface the two boundary conditions are satisfied. Then we obtain all the values of ψ, ψ', ω and ω' on the boundary.

Step 4. Substitute the boundary values which are obtained in step 3 into the boundary condition (2.12) on the free surface, then at each mesh point we calculate the residual

$$R_{i,m} = \omega + 2\ k\ \omega'$$

where m denotes the mth iteration and $i = 1, 2, \ldots, N_1$, and the total residual is given by

$$R_m = \sum_{i=1}^{N1} |\ R_{i,m}\ |$$

Step 5. Consider the equation

$$k = (\ \alpha\ R_{i,m} - \omega\ )/\ 2\ \psi' \tag{4.1}$$

where k is the local curvature on free surface and $0 < \alpha < 1$ is the relaxation factor which depends on the total residual R_m. From the formula (4.1) we obtain the new curvature at all of the mesh points on free surface. Then consider the ordinary differential equation

$$\begin{cases} y'' - k\,(1 + y'^2)^{3/2} = 0 \\ y\,|_{x=0} = 1 \\ y \longrightarrow q \qquad \text{as } x \longrightarrow \infty \end{cases} \qquad (4.2)$$

which is solved numerically, and hence we obtained the next approximation to the free surface.

Step 6. The iteration is considered to have converged when both the residuals on the free surface and the change in position of the free surface between successive iterations are sufficiently small. That is

$$R < \varepsilon_1 \qquad (4.3)$$

and

$$\sum_{i=1}^{N1} |\, y_{i,m} - y_{i,m-1} \,| < \varepsilon_2 \qquad (4.4)$$

where $\varepsilon_1 > 0$ and $\varepsilon_2 > 0$ are two small preassigned constants. If both of the expressions (4.2) and (4.3) are satisfied then the process is complete and the free surface location is found, otherwise return to step 3.

5. NUMERICAL SOLUTION

The iterative scheme presented in section 4 requires some parameters, the non-dimensional fluid flow rate, q, the non-dimensional surface tension, T, the control parameters, ε_1 and ε_2, and the relaxation factor, α. As an example of the results obtained, we take these parameters as following

$$T = 1$$

$q = 0.6,\ 0.7,\ 0.75,\ 0.8$ and 0.9, respectively

$$\varepsilon_1 = 0.01, \qquad \varepsilon_2 = 0.0001$$

$$\alpha = \begin{cases} 0.1 & R_m \geq 1 \\ 0.2 & 1 > R_m \geq 0.5 \\ 0.3 & 0.5 > R_m \geq 0.1 \\ 0.5 & 0.1 > R_m \end{cases}$$

$$N_1 = 20$$

For the choice of the parameters ε_1, ε_2, α and N_1 we have considered many factors, including the need to obtain accurate solutions, the convergence of the iterative process, the convergence rate and the CPU time. We believe that the results presented in this paper are as accurate as the plotted solution. We have taken the fundamental shape as used by Ritchie [7] as the initial approximation to the free surface, which is given by

$$y = 1 - (1 - q) \tanh (x)$$

Fig. 4 shows the behaviour of the free surface profile as the fluid flow rate varies. These profiles indicate that the contact angle depends upon the fluid flow rate. The free surface profile is flatter at the higher flow rates than at the lower flow rates as one would expect from continuity. We have also investigated the dependence of the free surface profile upon the surface tension, and Fig. 5 shows the solution with q = 0.75 and T = 0, 0.5, 1 and 2, respectively. If T is very large then the solution domain in the x direction has to be very large and therefore no results are presented for T > 2 although, in principle, the method should be able to deal with such flows. These profiles show that the local curvature decreases as the surface tension increases, again as one would expect from physical consideration.

It is well known that the normal action of the surface tension on the free surface is equal to T k, see Batchelor [9]. Hence if the fluid flow rate is fixed, which means that the difference in the upstream fluid pressure and the downstream ambient pressure, $P_1 - P_2$, is specified, then in order to satisfy the boundary condition (2.13) the local curvature, k, has to decrease as the surface tension increases.

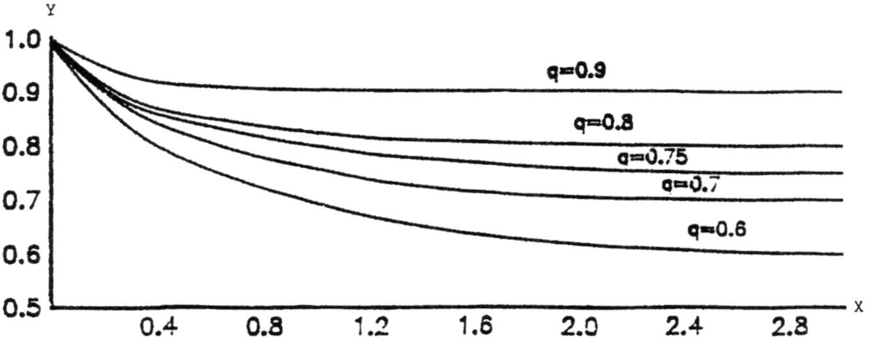

Fig. 4. The free surface profiles with various fluid flow rate, q.

On the other hand, if the surface tension is fixed, then the local curvature has to decrease as the fluid flow rate increases. Thus the free surface profile should be flatter at the higher flow rates than at the lower flow rates. The numerical solutions confirm this observation.

It is hoped to extend the method described in this paper to deal with the problems in which the fluid flow rate is no longer restricted such that $1/2 < q < 1$. On comparing this method with the method described by Ritchie, who used fundamental shapes, he was able to obtain solutions using only 10 to 20 iterations but in this work 100 iterations were required for convergence. Hence there is scope for further development of this technique in order to optimize the parameters which are used in the iterative scheme.

In conclusion, in this paper an iterative technique has been presented which is very effective in solving this free surface problem. Furthermore, this technique may be used with confidence in other very slow viscous flow problems in which free surfaces occur, e.g the cavitation bubble problem, the sintering problem, etc.

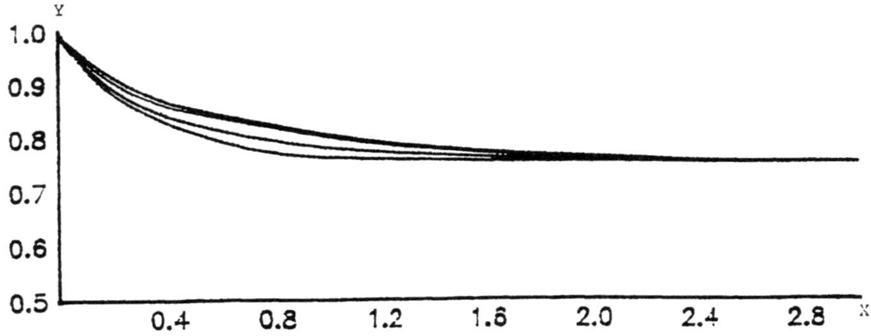

Fig. 5. The free surface profiles for various values of the surface tension, T. Reading from the bottom the curves correspond to T = 0, 0.5, 1.0 and 2.0, respectively.

ACKNOWLEDGEMENT

The authors would like to thank the Sino British Friendship Fellowship Scheme for the support of Y. Yuan.

REFERENCES

1. Coyne, J.C. and Elrod, H.G. Conditions for the Rupture of a Lubricating Film, Part 1: Theoretical Model, Trans. A.S.M.E. J. of Lub. Tech., Vol. 92, pp 451-456, 1970.

2. Saito, H. and Scriven, L.E. Study of Coating Flows by the Finite Element Method, J. Comp. Phys., Vol. 42, pp 53-76, 1981.
3. Carter, G.C. The Modelling and Analysis of Coating Processing, Ph.D. Thesis, Leeds University, 1985.
4. Liggett, J.A. Location of Free Surface in Porous Media, Trans. A.S.M.E. J. Hyd. Div., Vol. 104, pp 353-365, 1977.
5. Louguet-Higgins, M.S. and Cokelet, E.D. The Deformation of Steep Surface Waves on Water, part 1: A Numerical Method of Computation, Proc. R. Soc. London, Vol. A350, pp 1-26, 1976.
6. Ingham, D.B. and Kelmanson, M.A. Boundary Integral Equation, Analyses of Singular, and Biharmonic Problems, Lecture Notes in Eng., Vol. 7, Springer-Verlag,Berlin Heidberg and New York, 1984.
7. Ritchie, J.A. The Boundary Element Method in Lubrication Analysis, Ph.D. Thesis, Leeds University, 1989.
8. Kuiken, H.K. Viscous Sintering: The Surface-Tension-Driven Flow of a Liquid from Under the Influence of Curvature Gradients at its Surface, J. Fluid Mechanics, Vol. 14, pp 503-515, 1990.
9. Batchelor, G.K. An Introduction to Fluid Dynamics, Cambridge University Press, 1967.
10. Lachat, J.C. and Watson, J.O. Effective Numerical Treatment of BIE, Int. J. Num. Meth. in Eng., Vol. 10, pp 991-1005, 1976.
11. Riccardella, P. An Implementation of the Boundary Integral Techniques for Plane Problems in Elasticity and Elasto-Plasticity, Ph.D. Thesis, Carnegie-Mellon University, Pittsburgh,,1973.
12. Alarcon, E., Martin, A. and Paris, F. Boundary Element in Potential and Elasticity Theory, Comp. and Stru., Vol.10, pp 351-3622, 1979.
13. Ingham, D.B., Ritchie, J.A. and Taylor, C.M. The Linear Boundary Element Solution of Laplace's Equation with Dirichlet Boundary Conditions, Mathematics and Computer Modelling, 1991. To appear.
14. Walker, S.P. and Fenner, R.T. The Treatment of Corners in BIE Analysis of Potential Problems, Int. J. for Num. Meth. in Eng., Vol. 28, pp 2569-2581, 1989.

Implementing the Boundary Element Method for 2-D Viscous Sintering

G.A.L. van de Vorst, R.M.M. Mattheij
Department of Mathematics, University of Technology, P.O. Box 513, 5600 MB Eindhoven, The Netherlands

Abstract

By viscous sintering it is meant processes in which a granular compact is heated to a temperature at which the viscosity of the material under consideration becomes low enough for surface tension to cause the powder particles to deform and coalesce. Here a two-dimensional model is considered. The governing (Stokes) equations describe the deformation of a viscous fluid region under the influence of the curvature of the outer boundary. A boundary element method is applied to solve the equations for an arbitrarily initial-shaped fluid region. The numerical problems that can arise in computing the curvature, in particular at places where a *cusp* is arising, are discussed. A number of numerical examples is shown for *simply connected* regions which transform themselves into circles as time increases.

1 Introduction

Sintering is the process of bringing a powder of metals, ionic crystals, or glasses to temperatures near to their melting point so that, due to the high(er) mobility, the compact densifies, thereby releasing the surface energy of the powder particles. The driving force arises from the excess free energy of the surface of the powder over that of the solid material.

There are a number of physical principles which can be held responsible for the sintering phenomena; for a review see for example Exner [2]. We are mainly interested in the case of sintering when the material transport can be modeled as a viscous Newtonian volume flow, driven solely by surface tension (viscous sintering). This gives us a simple model of what is known as the sol-gel technique, which can for example be used to produce high-quality glasses. In this technique a glassy aerogel is heated to a temperature at which the viscosity of the glass becomes low enough for surface tension acting on the interior surface of the gel to cause the gel to collapse into a dense homogeneous material.

It is impossible to give a deterministic description of the flow within such a complex sintering geometry as an aerogel. We shall therefore investigate simple geometries; to start with in 2-D only, aiming at eventually deriving constitutive laws of the effects obtained.

A classical problem in sintering literature is the two-dimensional initial-stage unit model, the sintering of two cylinders, which has been solved analytically by Hopper [4]. The sintering of an infinite line of cylinders was simulated numerically by Ross *et al.* [9]; they were also the first to perform the simulation using a finite element method. Jagota and Dawson [5] have reported some results obtained with the finite element method for the sintering of two spheres and an infinite line of cylinders (i.e. 3-dimensional axisymmetric problems). Recently, Kuiken [6] applied a boundary element method to solve viscous sintering problems for bodies with rather smooth boundaries.

Here we present another way of implementing a boundary element method for viscous sintering problems. Our aim is to develop a code which tells us how a fluid with an arbitrarily shaped region transforms itself through time, driven only by the surface tension. The motivation to use a boundary element method rather than a finite element method is given by the fact that for computing the shape of the movement of the fluid region in time, only the velocity of the boundary curve has to be obtained. Another reason is that remeshing a boundary curve is much easier than remeshing a full 2-dimensional grid, as has to be done in a finite element method.

This paper is built up as follows: Firstly, we shall rewrite the viscous sintering problem, which is described by a set of partial differential equations (Stokes), into a set of integral equations. These integral equations are solved by the boundary element method, as proposed by Brebbia [1], in which *linear* elements are employed. Next, we shall discuss the numerical problems that arise in computing the curvature, at places on the boundary where a cusp is arising. We have solved this problem by using a specially tailored remeshing scheme. Finally, we shall give a number of numerical examples to demonstrate the usefulness of our method.

2 Problem formulation

The viscous sintering problem is modeled by a 2-dimensional viscous incompressible Newtonian fluid flow, see also Kuiken [6]. We denote the dimensionless *velocity* of the fluid by $\mathbf{v}$ and the *pressure* by p. The region of flow is defined by a closed curve Γ and the interior area denoted by Ω. Furthermore we assume the region of flow is a *simply connected* surface.

A viscous incompressible Newtonian fluid can be described by the Stokes' equations, i.e. in dimensionless form (Batchelor [3]),

$$\begin{aligned} \Delta \mathbf{v} - \operatorname{grad} p &= 0 \\ \operatorname{div} \mathbf{v} &= 0, \end{aligned} \tag{1}$$

with stress tensor

$$\mathcal{T}_{ij} = -\delta_{ij} p + \left(\frac{\partial v_i}{\partial x_j} + \frac{\partial v_j}{\partial x_i} \right). \tag{2}$$

On the boundary Γ the normal component of the stress tensor is proportional to the local curvature κ (=div $\mathbf{n}$) of the boundary. This condition can also be expressed as

$$\mathcal{T}\mathbf{n} = \kappa \mathbf{n}, \tag{3}$$

where $\mathbf{n}$ is the outward unit normal vector of Γ.

The equations (1)-(3) described above do not ensure a unique solution $\mathbf{v}$. It can be seen, cf [10], that a superposition of an arbitrary rigid-body *translation* or an arbitrary rigid-body *rotation* upon any particular solution of these equations, is also a solution of equation (1) and will not alter the stress field at the boundary. Thus in total we need to add three extra conditions to ensure that the velocity field obtained is unique. In order to get rid of the translation freedom, we formulate the problem to be stationary at a (reference) point in the fluid, say $\mathbf{x}^r$. With regard to this reference point the velocity of the boundary points are computed. The most natural choice for this reference point is the centre of mass: the point where the gravity forces would grip the body, thus:

$$\mathbf{v}(\mathbf{x}^r) = 0. \tag{4}$$

Furthermore we assume the tangential component of the velocity at the boundary is zero, i.e.

$$\int_\Gamma (\mathbf{v}, \tau)\, d\Gamma = 0, \tag{5}$$

where τ is the tangential vector of the boundary Γ. Combining this with Stokes formula it follows from equation (5) that the flow in Ω is *irrotational.*

The problem defined by the equations (1)-(5) has a unique solution. If we were to solve these equations for a fixed boundary Γ we would find, in general, a non-zero flow field $\mathbf{v}$ on Γ. Thus in time (t, i.e. the dimensionless time) the boundary is moving. The displacement of the boundary can be found from the derived velocity field $\mathbf{v}$ in the following way,

$$\frac{d\mathbf{x}}{dt} = \mathbf{v}(\mathbf{x}) \qquad (\mathbf{x} \in \Gamma), \tag{6}$$

subjected to an initial boundary Γ_0 at $t = t_0$. It is worth noting that equation (6) says that the *material* points of the boundary are moving in the direction of the *characteristic* curves.

We remark that we are only interested in the movement of the region Ω. Hence only the velocity at the boundary is required, from which the shape and motion of the body can be calculated directly. Therefore we shall represent the solution in terms of boundary distributions of the single- and double-layer potentials for the Stokes' equation. The integral equation that can be derived for the Stokes' equation at a point **x**, say, when the boundary is sufficiently "smooth" (see also [10]), reads in matrix notation

$$\mathcal{C}\mathbf{v}(\mathbf{x}) + \int_\Gamma \mathcal{Q}\mathbf{v}\, d\Gamma_y = \int_\Gamma \mathcal{U}\mathbf{b}\, d\Gamma_y. \tag{7}$$

Here $\mathcal{C}$,$\mathcal{Q}(\mathbf{x},\mathbf{y})$ and $\mathcal{U}(\mathbf{x},\mathbf{y})$ are 2×2 matrices with coefficients c_{ij}, q_{ij} and u_{ij} respectively:

$$c_{ij} = \begin{cases} \delta_{ij} & \mathbf{x} \in \Omega \\ \frac{1}{2}\delta_{ij} & \mathbf{x} \in \Gamma, \end{cases} \tag{8}$$

$$q_{ij} = \frac{r_i r_j}{\pi R^4} r_k n_k, \tag{9}$$

$$u_{ij} = \frac{1}{4\pi}\Big[\, \delta_{ij} \log \frac{1}{R} + \frac{r_i r_j}{R^2} \Big], \tag{10}$$

where $r_i = x_i - y_i$ and $R = \sqrt{r_1^2 + r_2^2}$, and the vector **b**

$$\mathbf{b} = \kappa\, \mathbf{n}. \tag{11}$$

Many authors attribute the analysis and the integral equation (7) to Ladyzhenskaya [7] (1963), but actually it was Lorentz [8] who derived this formulation essentially, back in 1896.

In order to make the integral equation (7) uniquely solvable, we have to add the conditions (4) and (5). Since the reference point $\mathbf{x}^r$ is taken in Ω, we obtain from equation (7)

$$\int_\Gamma \mathcal{Q}^r \mathbf{v}\, d\Gamma_y = \int_\Gamma \mathcal{U}^r \mathbf{b}\, d\Gamma_y. \tag{12}$$

Here $\mathcal{Q}^r = \mathcal{Q}(\mathbf{x}^r,\mathbf{y})$ and $\mathcal{U}^r = \mathcal{U}(\mathbf{x}^r,\mathbf{y})$. From the equations (5), (7) and (12) we derive the following two integral equations which have to be solved:

$$\mathcal{C}\mathbf{v}(\mathbf{x}) + \int_\Gamma (\mathcal{Q} - \mathcal{Q}^r)\mathbf{v}\, d\Gamma_y = \int_\Gamma (\mathcal{U} - \mathcal{U}^r)\mathbf{b}\, d\Gamma_y, \tag{13}$$

$$\int_\Gamma (\mathbf{v}, \tau)\, d\Gamma = 0. \tag{14}$$

3 The Boundary Element Method

The problem is ideally suited to be solved by a boundary element method (Brebbia [1]). Therefore the boundary Γ will be discretized into a sequence of N elements and the velocity and surface tension are written in terms of their values at a sequence of nodal points. From the discretized form of equation (13) for every nodal point, together with the discretized form of equation (14), we derive a system of (2N+1) linear algebraic equations with 2N unknowns.

From this system we obtain the approximate velocity at time $t = t_k$ at the nodal boundary points. The displacement of the boundary at time $t = t_{k+1} = t_k + \Delta t$ can then be obtained by discretizing equation (6). Here we will use a simple Forward Euler discretization scheme, i.e.

$$\mathbf{x}(t_{k+1}) = \mathbf{x}(t_k) + \Delta t \mathbf{v}(\mathbf{x}(t_k)). \tag{15}$$

As mentioned in section 2, equation (15) is an approximation of the trajectories of the material boundary points. These trajectories may not intersect each other, which gives restrictions for the time step Δt. The integral equations, subjected to the derived (new) boundary, are solved again which give the velocity at $t = t_{k+1}$, etc...

After discretization of the boundary Γ into a polygon, we define the polynomial functions $\tilde{\mathbf{v}}$ and $\tilde{\mathbf{b}}$, cf (11), which apply at a typical element 'j',

$$\tilde{\mathbf{v}} = \Phi v^j \qquad \text{and} \qquad \tilde{\mathbf{b}} = \Phi b^j, \tag{16}$$

where v^j and b^j are the velocity and surface tension of the boundary nodal point $\mathbf{x}^j$. The interpolation function Φ is a $2 \times 2\text{M}$ matrix of shape functions:

$$\Phi = \begin{bmatrix} \phi_1 & 0 & \phi_2 & 0 & \dots & \phi_M & 0 \\ 0 & \phi_1 & 0 & \phi_2 & \dots & 0 & \phi_M \end{bmatrix}. \tag{17}$$

The functions ϕ_l are the standard finite-element-type polynomials, see also Brebbia [1]. The number M is equal to the degree of the polynomial approximation plus one. After substituting the approximation polynomial (16) into equation (13) with a discretized boundary, we obtain the following equation for an *arbitrary* nodal point i:

$$\mathcal{C}^i v^i + \sum_{j=1}^{N} \Big(\int_{\Gamma_j} (\mathcal{Q} - \mathcal{Q}^r) \Phi \, d\Gamma_y \Big) v^j = \sum_{j=1}^{N} \Big(\int_{\Gamma_j} (\mathcal{U} - \mathcal{U}^r) \Phi \, d\Gamma_y \Big) b^j. \tag{18}$$

For constant elements, i.e. where $\tilde{\mathbf{v}}$, $\tilde{\mathbf{b}}$ are assumed to be constant over each element, we note that $\mathcal{C}^i = 0.5\mathcal{I}$ for a "smooth" boundary. $\mathcal{C}^i$ will be for higher-order elements in general a 2×2 matrix.

Consider the following types of integrals, in matrix notation, which have to be evaluated for every element Γ_j and every nodal point $\mathbf{x}^i$,

$$\begin{aligned} \hat{H}^{ij} &= \int_{\Gamma_j} \mathcal{Q} \Phi \, d\Gamma_y, & H^j_r &= \int_{\Gamma_j} \mathcal{Q}^r \Phi \, d\Gamma_y, \\ G^{ij} &= \int_{\Gamma_j} \mathcal{U} \Phi \, d\Gamma_y, & G^j_r &= \int_{\Gamma_j} \mathcal{U}^r \Phi \, d\Gamma_y, \\ H^{ij} &= \begin{cases} \hat{H}^{ij} & i \neq j \\ \hat{H}^{ij} + \mathcal{C}^i & i = j. \end{cases} \end{aligned} \tag{19}$$

We obtain for equation (18)

$$\sum_{j=1}^{N}\left(H^{ij}-H_{\tau}^{j}\right)v^{j} = \sum_{j=1}^{N}\left(G^{ij}-G_{\tau}^{j}\right)b^{j}. \tag{20}$$

If we now let i vary from 1 to N, and note that the right-handside vector, say $\mathbf{F}$, of equation (20) is known we derive the following system:

$$\mathcal{H}\mathbf{v} = \mathbf{F}, \tag{21}$$

where $\mathcal{H} = H^{ij}$.
The system (21) can be solved uniquely when the discretized form of the extra relation (14) is added.

In what follows we consider *linear* approximations of $\mathbf{v}$ and $\mathbf{b}$ over an element, i.e. we apply linear boundary elements only. For more details we refer to [10].

4 Computation of curvature and mesh

The driving force of the boundary movement is a tension that depends on the curvature of the boundary. Therefore it is important that a good approximation is used to determine this curvature. Especially when in a certain point of the boundary a *cusp* arises. Here the curvature becomes unbounded, thus the approximate curvature can have large errors.

The curvature at a boundary point, say $\mathbf{x}^2$ is approximated by fitting a quadratic polynomial through this point and its neighbours $\mathbf{x}^1$ and $\mathbf{x}^3$, i.e.

$$\tilde{\mathbf{x}}(s) = \mathbf{x}^1\phi_1(s)+\mathbf{x}^2\phi_2(s)+\mathbf{x}^3\phi_3(s), \tag{22}$$

where $\phi_1(s)=\frac{1}{2}s(s-1)$; $\phi_2(s)=1-s^2$; $\phi_2(s)=\frac{1}{2}s(s+1)$ and $-1\leq s\leq 1$. For the approximate curvature at the nodal point $\mathbf{x}^2=\tilde{\mathbf{x}}(0)$, we obtain:

$$\begin{aligned}\kappa(\mathbf{x}^2) &= \frac{(x_2)_s(x_1)_{ss} - (x_1)_s(x_2)_{ss}}{\left[((x_1)_s)^2 + ((x_2)_s)^2\right]^{\frac{3}{2}}} \\ &\doteq \frac{4\left[(x_2^3-x_2^1)(x_1^1-2x_1^2+x_1^3) - (x_1^3-x_1^1)(x_2^1-2x_2^2+x_2^3)\right]}{\left[(x_1^3-x_1^1)^2 + (x_2^3-x_2^1)^2\right]^{\frac{3}{2}}}.\end{aligned} \tag{23}$$

This way we can derive the curvature at all nodal points. The curvature at any point of an element is obtained from the polynomial fit (16) through the approximate curvature (23) of these nodal points.

As mentioned above, when approximating the curvature, problems are arising if a cusp occurs. Especially when we make an extra refinement of the mesh in the neighbourhood of such a point, larger errors can be made when computing the curvature there using equation (23). This can be seen as follows: we can assume

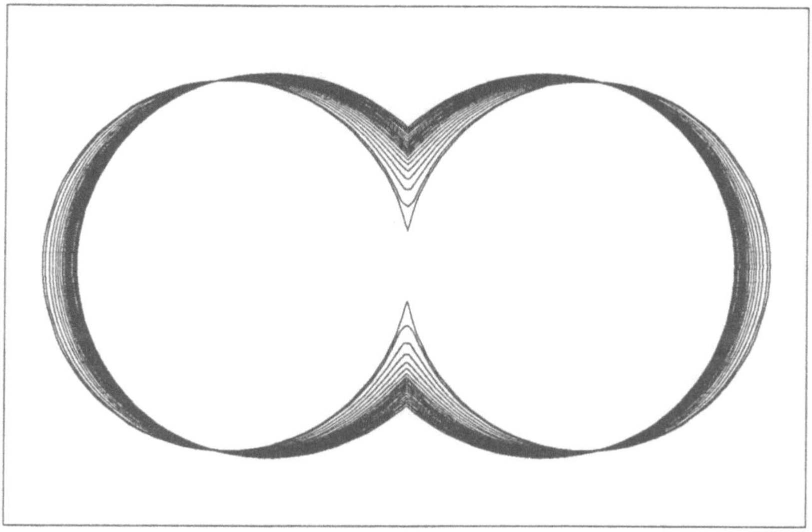

(a)

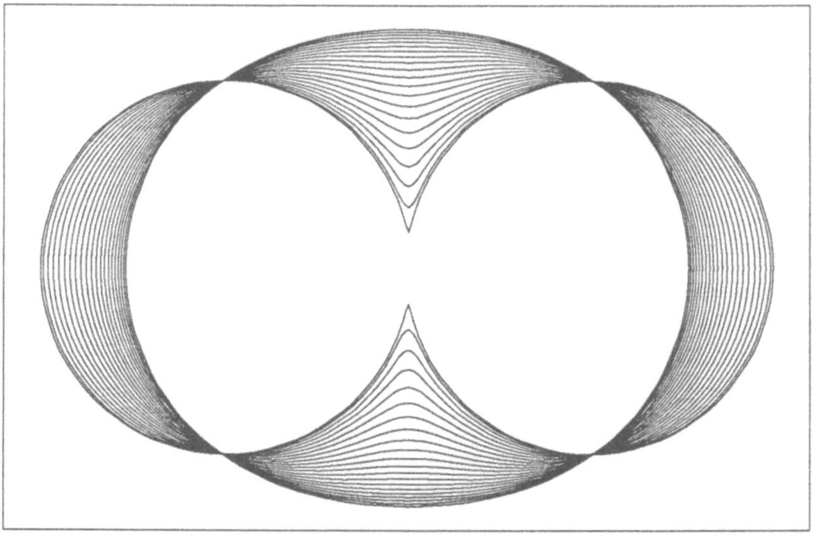

(b)

Figure 1: Two sintering cylinders with equal diameters. The boundary curves refer to values of time t=0.0(0.1)2.0. Figure (a) is obtained when the initial mesh is used throughout the simulation. In figure (b) is done a mesh verification at each time step.

that the *spatial* discretization error, i.e. the error that is made by the approximation of the boundary by a polygon, is smaller than the *time* discretization error, i.e. the error that arises from the time stepping scheme (15). This Forward Euler scheme gives a global time discretization error $\mathcal{O}(\Delta t)$. Furthermore, the time step Δt can not be made very small, because then the computing time would become prohibitively large. Thus we can say that the computed boundary deviates $\mathcal{O}(\Delta t)$ from the exact curve. When approximating the curvature at points where the mesh is fine, we loose some accuracy because some points are necessarily very close to each other. The quotient of the approximate curvature of equation (23) will then be of $\mathcal{O}(\Delta t)$ over $\mathcal{O}(\Delta t)$. Hence the computed curvature can deviate considerably from the exact curvature. Thus the mesh has to be checked every time step in such a way that the collocation points do not get to close to each other. On the other hand, it seems reasonable that the collocation points have to lie close to each other at places on the boundary where the curvature is large; since there we are expecting large variations of the velocity field $\mathbf{v}$ of the boundary. These two conflicting aspects have to be wrought together in an algorithm that takes care of the mesh adaptation and verification.

The implementation of the latter algorithm is done in the following way. Assume that a suitable initial mesh is given. If at a certain time step Δt is satisfying the conditions, which will follow shortly, nothing is done. If this is not the case, the mesh has to be updated locally.

We introduce two positive constants h_{min} and h_{max}, which are given bounds for the distance between two successive collocation points, i.e.

$$h_{min} \leq \delta_i \leq h_{max}, \tag{24}$$

where δ_i denotes the actual distance between two such points: $\delta_i = \|\mathbf{x}^{i+1} - \mathbf{x}^i\|$. The value of h_{min} is taken of $\mathcal{O}(\Delta t)$. Furthermore, only when a "cusp" arises at a collocation point, say $\mathbf{x}^j$, the distance between the two neighbours of $\mathbf{x}^j$ must be of order Δt:

$$\|\mathbf{x}^{j-1} - \mathbf{x}^{j+1}\| \geq h_{min}. \tag{25}$$

Otherwise, we lose some digits when computing the curvature at this point using equation (23). Thus the approximate curvature will certainly have large errors then. The restriction (25) also limits the magnitude of the curvature. From equation (23) we derive that the maximum value of the curvature is $\mathcal{O}(\Delta t^{-1})$. The requirement of a fine mesh at places where the curvature is large, can be fulfilled through verifying that the straight line between two contiguous nodal points is not deviating too much (in a relative sense), from that part of the boundary which lies between those points; i.e. we look for a kind of equidistribution of the curvature over the elements. In [10] we used the following implementation for this criterion:

$$\max\left(\mathrm{h}_i^i, \mathrm{h}_i^{i+1}\right) \leq \epsilon_c \delta_i, \tag{26}$$

where

$$\mathrm{h}_i^k = \frac{1}{|\kappa(\mathbf{x}^k)|}\left(1 - \sqrt{1 - \frac{1}{4}\delta_i^2[\kappa(\mathbf{x}^k)]^2}\right), \quad k = i, i+1 \tag{27}$$

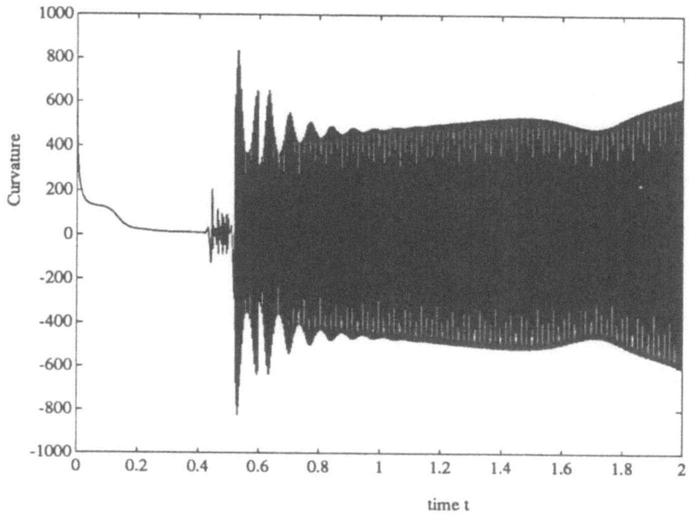

Figure 2: The curvature of the neck point. The same mesh is used throughout the simulation. Large oscillations develop when the trajectories of the nodal points in the neck region come close together.

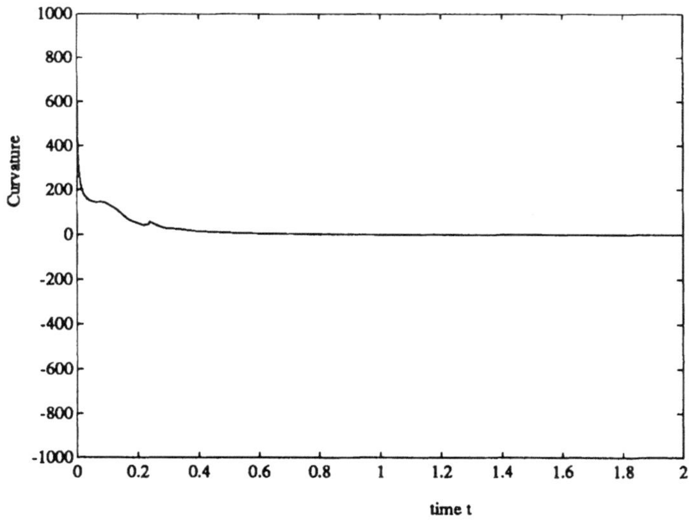

Figure 3: The curvature of the neck point. The mesh is "updated" each time step.

and ϵ_c is a certain threshold value. At least we want the collocation points to lie fairly uniformly distributed on the boundary, i.e.

$$\epsilon_s \leq \frac{\delta_i}{\delta_{i-1}} \leq \frac{1}{\epsilon_s}, \tag{28}$$

where ϵ_s is a given parameter, greater than 1.

In decreasing order of importance of the conditions which have to be fulfilled are: (i) the distance condition (24) and (25), (ii) the curvature criterion (26) and at least, (iii), the smooth variation of the mesh (28). We assume the initial mesh satisfies those conditions. When, after some steps, one of the conditions is not fulfilled any more at a certain nodal point, we update the mesh locally. We try to *shift* this badly behaving point along the neighbouring elements. This is done in such a way that the mesh points do not get too close for those neighbouring collocation points. If this is somehow not possible, then an extra collocation point is introduced and put into the mesh, using a locally quadratic interpolation; or otherwise this badly behaving point is removed. Only when this point is a *cusp*, than the point may **not** be shifted along an element or removed. In this case, we only shift or remove the two neighbouring collocation points. In order to ensure that the curvature does not change in the cusp, the cusp is shifted slightly in the normal direction; hereby care is taken that the distance between the "new" and the "old" cusp point is smaller than the discretization error.

5 Numerical results

In this section we show a number of results for some simply connected surfaces. Since, as we said in section 1, the driving force for sintering arises from the excess of free surface energy, a 2-dimensional viscous fluid region Ω transforms itselfs into a circle when the time t is going to infinity. This is because the length of the boundary curve attains its minimum then. Thus a circle must be the result of the simulation considered. Also, the fluid is assumed to be incompressible. In 2-dimensions this means that the *total* surface of the moving fluid region must be constant in time.

To demonstrate the usefulness of the mesh distribution algorithm, described in section 4, we consider the case of two *sintering* spheres in 2-dimensions, fairly shortly after they have made contact. In 2-dimensions these are two circles and in 3-dimensions this models two infinity long cylinders. In figure 1 the numerical results obtained at different time steps for the coalescence of the two circles are shown. As an initial time step we chose $\Delta t = 0.002$. During the simulation this time step is automatically updated using the formula we had derived in [10]. The centre of mass is taken as the reference point, $\mathbf{x}^r$, which is chosen to be the origin.

In figure 1^a we show the results obtained without mesh verification. Thus the trajectories of the nodal points of the initial mesh are followed. When a mesh verification is applied at each time step (as proposed in section 4) we derive

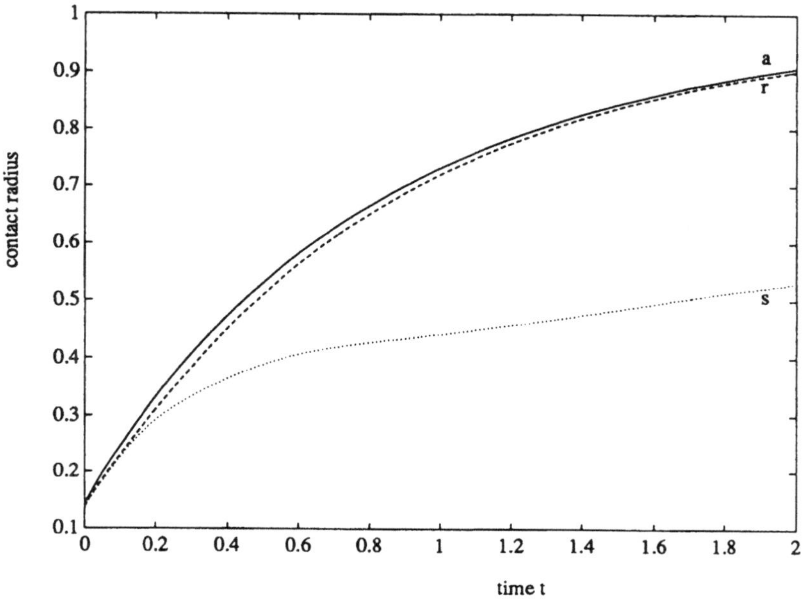

Figure 4: The contact radius of the sintering cylinders, obtained by both numerical simulations compared with the analytical solution (**a**), i.e. the same mesh (**s**) and when the mesh is updated (**r**) at each time step.

the shapes as shown in figure 1^b. The shapes of both graphs are plotted at the same time points. When comparing these graphs, it can be seen that the shape evolution of both curves, completely differ already at an early stage, especially near the contact region of both circles. (In sintering literature, this contact region is usually denoted as the *neck* region, see Exner [2]).

The reason that the shape evolutions, develop so differently is due to an oscillation in the approximation of the curvature from the boundary in the neck region. In figure 2 we show the curvature of the point at the boundary where both circles are making contact. Here, the mesh is not changed throughout the numerical process. As can be seen, a very rapid oscillation is occuring in the approximate curvature, at a time when the trajectories of the nodal points come very close to each other. This oscillation is caused, as is mentioned in section 4, by the loss of accurate digits when computing the approximate curvature. These oscillations of the curvature do not disturb the movement of the boundary, in the sense of a brake-down of our numerical algorithm. The fact that these oscillations are bounded is due to the "natural" damping of such viscous fluid. Kuiken [6] has shown that a small disturbance of the boundary is damped out rapidly. In figure 3 we again give the curvature of this contact point. Now a mesh verification is done at every time step. Here, the oscillations in the (approximate) curvature have disappeared.

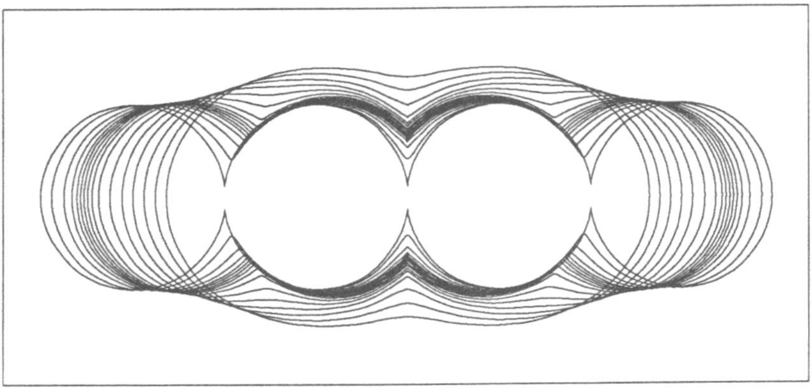

(a)

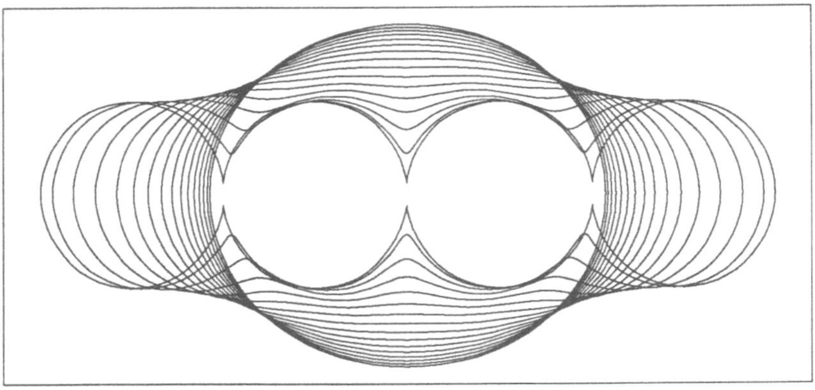

(b)

Figure 5: Four sintering cylinders with equal diameters. The boundary curves refer to time values t=0.0(0.2)3.0. Figure (*a*) is obtained when the mesh is kept constant used during the simulation. In figure (*b*) a mesh verification is done at each time step.

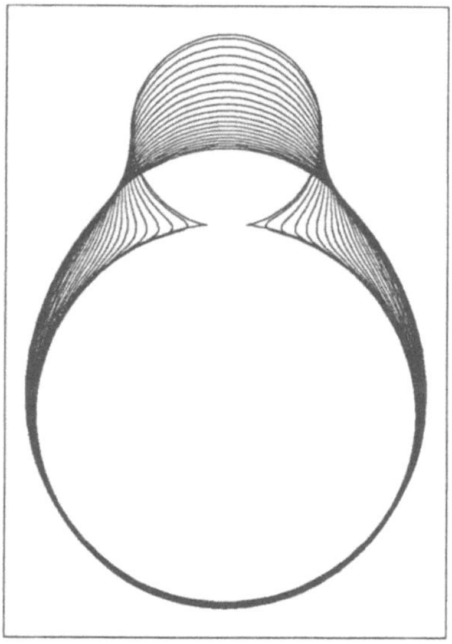

Figure 6: The transformation of two sintering cylinders with different diameters. The diameters of the cylinders are taken to be 0.5 and 1, respectively. The boundary curves refer to values of time t=0.0(0.1)2.0.

The development during sintering of the contact radius, i.e. the neck, is also of physical interest. This contact radius is a measure of how "strong" a sintering compact already is. When this contact radius is small, a smaller force is necessary to brake the contact between both cylinders than at later stages of the sintering process, when the contact radius is bigger. In figure 4 the contact radius is shown, which is obtained in both numerical simulations. In this graph we have also plotted the analytical solution for the sintering of two circles, which has been derived by Hopper [4]. As can be seen the numerical solution, matches well with the exact solution when a mesh verification is done.

Another example of two different numerical simulations for a same initial shape is shown in figure 5. Here we simulate the sintering of four equal cylinders which are lying in a row. Again, this example illustrates the unpredictability of the shape evolution, especially when more complex geometries are taken. The shapes obtained by the mesh verification algorithm is the one, which is expected in physical reality.

In figure 6 we show another example of a shape evolution at different time points, of a fluid region with a "near" cusp. The initial shape consists of two cylinders of arbitrary diameters. The diameters of the cylinders is taken 0.5 and 1, respectively. Near the region of contact of the cylinders we have to deal with a

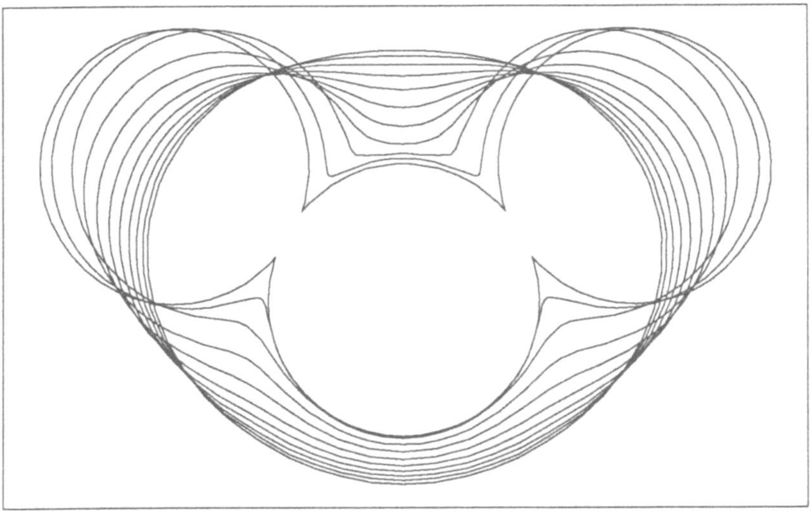

Figure 7: The transformation of three sintering cylinders with a angle between the mid points of the initial cylinders of 120 degrees. The boundary curves refer to values of time t=0.0(0.2)2.0.

large varying curvature. Again, the centre of mass (which is lying somewhere on the y axis) is taken as reference point. Our remeshing algorithm did work well.

At last, we consider the geometry of three equal cylinders, which are making an angle with each other. For the angle between the mid points of the circles we took 120 degrees. In figure 7 we show the obtained shape evolution when the mesh verification algorithm is used.

In future, we plan to investigate multiply connected regions, i.e. viscous fluid regions with gas bubbles.

ACKNOWLEDGMENT

This research was supported by the Technology Foundation (STW).

References

[1] Brebbia C.A. and Dominguez J., Boundary Elements An Introductory Course, Computational Mechanics Publications, Southampton, 1989.

[2] Exner H.E., Grundlagen von Sintervorgängen, Gebrüder Borntraeger, Berlin-Stuttgart, 1978.

[3] Happel J. and Brenner H., Low Reynolds Number Hydrodynamics with special applications to particulate media, Prentice-Hall, London, 1965.

[4] Hopper R.W., Plane Stokes flow driven by capillarity on a free surface, J. Fluid Mech., Vol. 213, pp. 349-375, 1990.

[5] Jagota A. and Dawson P.R., Simulation of the Viscous Sintering of Two Particles, J. Am. Ceram. Soc., Vol. 73, pp. 173-177, 1990.

[6] Kuiken H.K., Viscous sintering: the surface-tension-driven flow of a liquid form under the influence of curvature gradients at its surface, J. Fluid Mech., Vol. 214, pp. 503-515, 1990.

[7] Ladyzhenskaya O.A., The Mathematical Theory of Viscous Incompressible Flow, Gordon and Beach., New York-London, 1963.

[8] Lorentz H.A., Eene Algemeene Stelling omtrent de Beweging eener Vloeistof met Wrijving en eenige daaruit afgeleide Gevolgen, Verslag Kon. Acad. v. Wetensch. Amsterdam, Vol.5, pp. 168-175, 1896.

[9] Ross J.W., Miller W.A. and Weatherly G.C., Dynamic Computer Simulation of Viscous Flow Sintering Kinetics, J. Appl. Phys., Vol. 52, pp. 3884-88, 1981.

[10] Vorst G.A.L. van de, Mattheij R.M.M., Kuiken H.K., A boundary element solution for 2-dimensional viscous sintering, J. Comput. Phys. (submitted), 1990.

SECTION 5: MATHEMATICAL PROBLEMS AND COMPUTATIONAL TECHNIQUES

A Computational Method for Variational Inequalities

M.A. Noor

Mathematics Department, College of Science,
P.O. Box 2455, Riyadh 11451, Saudi Arabia

ABSTRACT

It is well known that a wide class of unrelated free and moving boundary value problems arising in various branches of mathematical and engineering sciences can be studied in the general and unified framework of variational inequalities. In this paper, we use the auxiliary principle to prove not only the existence of a solution of nonlinear variational inequalities, but also to suggest a novel and innovative iterative algorithm for computing the approximate solution. We also study its convergence criteria. Finally we discuss many special cases, which can be derived from our main result.

INTRODUCTION

Variation inequality theory has emerged as an interesting and important branch of applicable mathematics, which provides a number of nice and deep results. This theory has become a rich source of inspiration and motivation to study a large number of problems arising in pure and applied sciences in a general and unified framework. In recent years, variational inequalities have been generalized and extended in various directions. The variety of problems to which variational inequality techniques may be applied is impressive and amply representative for the richness of the field. One of the charms of this theory is that the location of the free boundary boundary contact area) becomes an instrinsic part of the solution and no special techniques are needed to locate it. We also remark that the theory so far developed is only applicable for moving and free boundary value problems of even order. In 1988, Noor [1] introduced and studied a new class of variational inequalities, which enabled him to study the constrained boundary value problems of both odd and even orders.

There is already a substantial number of numerical methods for finding the approximate solution of variational inequalities, see [2,3,4,5] and the references therein. Among the most effective numerical technique is the projection method and its variant forms. It is worth mentioning that the scope of the projection algorithm is quite limited due to the fact that it is very difficult to find the projection of the space into the convex set except in very simple cases. Secondly, the projection method cannot be applied for other classes of variational inequalities of type (1). These facts motivated Noor [6,7] to develop alternative methods to study the existence of solution of some classes of variational inequalities.

Inspired and motivated by the recent research work going on this fascinating and challenging field, we consider and study a new class of variational inequalities. Using the auxiliary principle technique of Noor [6,7], we suggest and analyze an innovative iterative algorithm for finding the approximate solution of variational inequalities. The proposed algorithm is quite general, flexible and unifying one. Extension of this technique for quasi variational inequalities is an open problem, which is another direction for further research work in this field.

FORMULATION AND BASIC RESULTS

Let H be a real Hilbert space with its dual H', whose inner product and norm are denoted by $\langle .,. \rangle$ and $\|.\|$ respectively. We identify the Hilbert space H with its dual space H'. Let K be a closed convex set in H.

Given continuous operators, $T,A,g:H \longrightarrow H$, we consider the problem of finding $u \in H$ such that $g(u) \in K$ and

$$\langle Tu, g(v) - g(u)\rangle + \phi(v) - \phi(u) \geq \langle A(u), g(v) - g(u)\rangle, \tag{1}$$

for all $g(v) \in K$, where $\phi(v)$ is a proper, convex and lower semicontinuous functional on H. The inequality is known as the general nonlinear variational inequality.

<u>Special Cases</u>

I. If $g(u) = u \in K$, then problem (1) is equivalent to finding $u \in K$ such that

$$\langle Tu, v-u\rangle + \phi(v) - \phi(u) \geq \langle A(u), v-u\rangle, \text{ for all } v \in K \tag{2}$$

The problem (2) is due to Noor [7]. For mathematical and physical formulation, see Noor [7,8].

II. If $g(u) = u \in K$ and $\phi(u)$ is a zero function, then problem (1) becomes: Find $u \in K$ such that

$$\langle Tu, v-u\rangle \geq \langle A(u), v-u\rangle, \quad \text{for all } v\varepsilon K. \tag{3}$$

Inequalities of the type (3) are known as mildly nonlinear variational inequalities which were introduced and considered by Noor [8] in the theory of constrained mildly nonlinear elliptic boundary value problems.

III. If $A(u) = 0$ and $g(u) = u\varepsilon K$, then problem (1) is equivalent to finding $u\varepsilon K$ such that

$$\langle Tu, v-u\rangle + \phi(v) - \phi(u) \geq 0, \quad \text{for all } v\varepsilon K, \tag{4}$$

which is known as the variational inequalities of the mixed type, see Kikuchi and Oden [3].

IV. If $\phi(u) = 0$, then problem (1) reduce to: Find $u\varepsilon H$ such that $g(u)\varepsilon K$ and

$$\langle Tu, g(v)-g(u)\rangle \geq \langle A(u), g(v)-g(u)\rangle \text{ for all } g(u)\varepsilon K \tag{5}$$

The problem (5) is originally due to Noor [9,10].

V. If $A(u) \equiv 0$, $\phi(u) \equiv 0$ and $g = I$, the identity mapping, then problem (1) is equivalent to finding $u\varepsilon K$ such that

$$\langle Tu, v-u\rangle \geq 0, \quad \text{for all } v\varepsilon K. \tag{6}$$

which is known as variational inequality. The problem (6) is originally due to Lions and Stampacchia [11].

VI. If $K^* = \{u\varepsilon H, (u,v) \geq 0, \text{ for all } v\varepsilon K\}$ is a polar cone of the convex cone K in H and $\phi(u) = 0$, then problem (1) is equivalent to finding $u\varepsilon H$ such that

$$g(u)\varepsilon K, \ (Tu-A(u))\varepsilon K^*, \ \langle Tu-A(u), g(u)\rangle = 0, \tag{7}$$

which is known as the general mildly nonlinear complementarity problem. The problem (7) appears to be new one and includes many known linear and nonlinear complementarity problems as special cases.

VII. If $K = H$ and $\phi(u) \equiv 0$, then problem (1) redcues to finding $u\varepsilon H$ such that

$$\langle Tu-A(u), g(v)\rangle = 0, \quad \text{for all } g(v)\varepsilon H, \tag{8}$$

which is known as the weak formulation of the odd order boundary value problems, see Ciarlet, Schultz and Varga [12].

It is clear that problems (1) - (8) are special cases of the general nonlinear variational inequality (1) introduced in this paper. In brief, we conclude that the problem (1) is a most general and unifying one, which enables us to study a number of moving and free boundary problems of both odd and even orders. This is one of the main motivations of this paper.

We also need the following concepts.

Definition: An operator $T:H \longrightarrow H$ is said to be:

(a) Strongly monotone, if there exists a constant $\alpha>0$ such that

$$\langle Tu-Tv,\ u-v\rangle \geq \alpha\|u-v\|^2, \quad \text{for all } u,\ v\in H$$

(b) Lipschitz continuous, if there exists a constant $\beta>0$ such that

$$\|Tu-Tv\| \leq \beta\|u-v\|, \quad \text{for all } u,v\in H$$

It is clear that if β exists, then so does α and $\alpha\leq\beta$.

GENERAL ALGORITHM

In this section, we consider the auxiliary principle and suggest an iterative algorithm for finding the approximate solution of variational inequality of type (1).

Auxiliary Principle

For given $u\in H$, we consider the auxiliary variational inequality problem of finding $w\in H$ such that $g(w)\in K$ and

$$\langle w,v-w\rangle+\rho\phi(v)-\rho\phi(w) \geq \langle u,v-w\rangle-\rho\langle Tu,g(v)-g(u)\rangle +\rho\langle A(u),g(v)-g(w)\rangle \tag{9}$$

for all $u\in H$ such that $g(v)\in K$ and some positive constant ρ.

We note that if the operator g is linear, then problem (9) is equivalent to finding the minimum of the functional $F(w)$ on K, where

$$F[w] = \tfrac{1}{2}\langle w,w\rangle + \rho\langle Tu-A(u),g(w)\rangle+\langle u,w\rangle+\rho\phi(w), \tag{10}$$

which is a differentiable functional associated with the auxiliary variational principle.

It is obvious that if $w=u$, then w is the solution of the problem (1). Based on these observations, we now suggest the following general algorithm for computing the approximate solution of variational inequalities.

Iterative Algorithm

(i) At n=0, start with some initial w_0.

(ii) At step n, solve the auxiliary problem (9) with $u=w_n$. Let w_{n+1} denote the solution of the problem (9).

(iii) If, for given $\varepsilon>0$, $\|w_{n+1} - w_n\| \leq \varepsilon$, stop. Otherwise repeat (ii).

CONVERGENCE ANALYSIS

In this section, we consider the convergence properties of the iterative algorithm, which is the main motivation of the following result.

THEOREM
Let w_n and w be the solutions of (9) related to $u_n, u \in H$ respectively. Let the continuous operator $T, g: H \longrightarrow H$ be both strongly monotone and Lipschitz continuous. If the operator A is Lipschitz continuous, then

$$w_n \longrightarrow w \quad \text{in } H,$$

for $\rho < 2\dfrac{\alpha - (k + \gamma\delta)}{\beta^2 - (k + \delta\gamma)^2}$, $\gamma\delta < (\alpha - k)$ and $\rho < 1/(k + \gamma\delta)$.

PROOF
Since w_n and w are solutions of (9) related to $u_n, u \in K$ respectively. In order to prove the convergence of the approximate solution to the exact solution w of (9), it is enough to show that for $\rho>0$ well chosen,

$$\|w_n - w\| \leq \theta\|u_n - u\|,$$

with $0<\theta<1$. Now taking $v=w$ (respectively w_n) in (9) related to u_n (respectively u), we have

$$\langle w_n, w-w_n\rangle + \rho\phi(w) - \rho\phi(w_n) \geq \langle u_n, w-w_n\rangle - \rho\langle Tu_n, g(w)-g(w_n)\rangle$$
$$+ \rho\langle A(u_n), g(w)-g(w_n)\rangle$$

and

$$\langle w, w_n-w\rangle + \rho\phi(w_n) - \rho\phi(w) \geq \langle u, w_n-w\rangle - \rho\langle Tu, g(w_n)-g(w)\rangle$$
$$+ \rho\langle A(u), g(w_n) - g(w)\rangle$$

Adding these inequalities, we have

$$\langle w_n - w, w_n - w\rangle \leq \langle u_n - u, w_n - w\rangle - \rho\langle Tu_n - Tu, g(w_n)-g(w)\rangle + \rho\langle A(u_n) - A(u), g(w_n) - g(w)\rangle,$$

$$= \langle u_n - u - \rho(Tu_n - Tu), w_n - w\rangle + \rho\langle Tu_n - Tu, w_n - w - (g(w_n) - g(w))\rangle + \rho\langle A(u_n) - A(u), g(w_n) - g(w)\rangle,$$

from which it follows that

$$\|w_n - w\|^2 \leq \|u_n - u - \rho(Tu_n - Tu)\| \, \|w_n - w\| + \rho\|A(u_n) - A(u)\| \, \|g(w_n) - g(u)\| + \rho\|w_n - w - (g(w_n) - g(w))\| \, \|Tu_n - Tu\| \quad (11)$$

Since T, g are both strongly monotone and Lipschitz continuous, so by using the technique of Noor [13], we have

$$\|u_n - u - \rho(Tu_n - Tu)\|^2 \leq (1 - 2\alpha\rho + \beta^2\rho^2)\|u_n - u\|^2 \quad (12)$$

and

$$\|w_n - w - (g(w_n) - g(w))\|^2 \leq (1 - 2\sigma + \delta^2)\|w_n - w\|^2 \quad (13)$$

From (11), (12), and (13), we obtain by using the Lipschitz continuity of T, g and A,

$$\|w_n - w\| \leq \left\{\sqrt{1 - 2\alpha\rho + \beta^2\rho^2} + \rho\beta\sqrt{1 - 2\sigma + \delta^2} + \rho\gamma\delta\right\}\|u_n = u\|$$

$$= \left\{t(\rho) + \rho(k + \gamma\delta)\right\} \|u_n - u\|$$

$$= \theta\|u_n - u\|,$$

where

$$k = \beta\sqrt{1 - 2\sigma + \delta^2}, \quad t(\rho) = \sqrt{1 - 2\alpha\rho + \beta^2\rho^2}$$

and $\theta = t(\rho) + \rho(k + \gamma\delta)$

We have to show that $\theta<1$. Now $t(\rho)$ assumes its minimum value for $\bar{\rho} = \alpha/\beta^2$ with $t(\bar{\rho}) \sqrt{1 - \alpha^2/\beta^2}$. For $\rho = \bar{\rho}$, $t(\rho) + \rho(k+\gamma\delta)<1$ implies that $\rho < \frac{1}{k + \gamma\delta}$. Thus it follows that $\theta<1$ for all ρ with $\rho<2(\alpha-(k+\gamma\delta)/(\beta^2 - (k+\delta\gamma)^2)$, $\rho<1(k+\gamma\delta)$ and $\gamma\delta<(\alpha-k)$.

Since $\theta<1$, so the approximate solution w_n converges to w, the exact solution of (9) strongly in H, the required result.

CONCLUSION

In this paper, we have considered and studied an auxiliary variational inequality problem, which includes the previously known ones as special cases. It is shown that the auxiliary problem can be used to suggest an innovative and novel iterative algorithm for various classes of variational inequalities. The auxiliary problem is quite general and flexible. By appropriate choice of the auxiliary problem, one may be able to select a suitable method to solve the variational inequalities and related optimization problems. The auxiliary problem and algorithm suggested in this paper may be extended for the multivalued operators involving variational inequalities. Development and improvement of an implementable algorithm for various classes of variational and quasi variational inequalities deserve further research efforts.

REFERENCES

1. Noor, M.A., General Variational inequalities, Applied Math. Letters, Vol. 1, pp. 119-122, 1988.

2. Glowinski, R., Lions, L. and Temolieres, R., Numerical Analysis of Variational Inequalities, North-Holland, Amsterdam, 1982.

3. Kikuchi, N. and Oden, J.T., Contact Problems in Elasticity, Soc. Ind. Appl. Math., Philadelphia, 1988.

4. Crank, J., Free and Moving Boundary Problems, Clarendon Press, Oxford, U.K., 1984.

5. Baiocchi, C. and Capelo, A., Variational and Quasi Variational Inequalities, J. Wiley and Sons, New York, 1984.

6. Noor, M.A., General Nonlinear Variational Inequalities, J. Math. Anal. Appl., Vol. 126, pp. 78-84, 1987.

7. Noor, M.A., Variational Inequalities Related with a Signorini Problem, C.R. Math. Rep. Acad. Sci. Canada, Vol.7, 267-272, 1985.

8. Noor, M.A., On Variational Inequalities, Ph.D. Thesis, Brunel University, U.K. 1975.

9. Noor, M.A., On Nonlinear Variational Inequalities, Int. J. Math. & Math. Sci. in Press.

10. Noor, M.A., An Iterative Algorithm for Variational Inequalities, J. Math. Anal. Appl. in Press.

11. Lions, J. and Stampacchia, G., Variational Inequalities, Comm. Pure Appl. Math. Vol. 20, pp. 493-519, 1967.

12. Ciarlet, P.G., Schultz, M.H. and Varga, R.S., Numerical Methods of High Order Accuracy for Nonlinear Boundary Value Problems, V. Monotone Operator Theory, Numerische Math., Vol. 13, pp. 51-77, 1969.

13. Noor, M.A., An Iterative Scheme for a Class of Quasi Variational Inequalities, J. Math. Anal. Appl. Vol. 110, pp. 463-468, 1985.

Optimal Shape Design of a Simple Harbour Model

S. Lyle, P.H. Soan

School of Mathematics, Kingston Polytechnic, Kingston upon Thames, Surrey, KT1 2EE, United Kingdom

INTRODUCTION

A variety of different numerical methods have now been developed to solve optimal shape design problems [4, 7] most of which involve a somewhat arbitrary combination of a method for solving partial differential equations (PDEs) with an optimisation technique. These are linked in an iterative scheme whereby the PDE solver is applied over several different shaped regions, each being successively chosen by the optimiser in such a way as to tend towards the optimum domain. However, the theoretical analysis of the convergence properties of such methods is not yet very far advanced so that they cannot always be guaranteed to reach a solution.

An alternative approach is to solve the state equation and the optimisation problem simultaneously. One way that this may be achieved is to use a transformation technique [6] whereby a mapping is sought, in terms of unknown parameters, which maps the unknown optimal shape onto a known domain. If such a mapping can be found then the optimal shape design problem is transformed into the problem of finding those values of the mapping parameters which optimise the cost functional, subject to the state equations and any other constraints, over a known, fixed, domain. A set of necesssary conditions for a solution to this problem may be deduced, usually as a system of nonlinear PDEs, and then solved by an appropriate numerical technique. Or, alternatively, the optimisation problem may be solved approximately by some direct method such as Ritz's Method.

In this work the transformation technique is considered in the context of a particular linear/quadratic shape optimisation problem. Specifically, the problem of choosing the shape of a harbour basin to minimise the average wave height inside is explored. The nonlinear nature of optimal shape design problems means that, after a numerical discretisation, a nonlinear system of algebraic equations will be produced. Solving such a system on a computer is a very time consuming process, so the mathematical model used to describe the harbour is, of necessity, fairly simple and the resolution available is low. As a consequence of this the results obtained for the optimal harbour shape cannot be expected to provide the definitive solution. Nevertheless, it is felt that the general features of the shapes obtained should be valid and reflect the situation in the real world. Furthermore, the feasibility of applying this method to such problems is demonstrated and it may be shown that the method will converge to the correct solution if the resolution is sufficiently fine. Thus it should be possible, given greater computing power, to make the model more sophisticated to include more effects encountered in reality.

THE HARBOUR MODEL

For the model considered here the harbour, when viewed in plan form, is taken to occupy the region S bounded by the curve $C = C_1 \cup C_2$ (see Figure 1). The straight line C_1 represents the harbour mouth which is of known size and C_2 is the unknown shape of the harbour walls which is to be optimised. It is assumed that the unknown optimum shape of the harbour will be symmetric about the y-axis. The harbour is regarded as having an area S_0 which is fixed i.e. the curve C_2 is only permitted to take forms that will give the harbour an area equal to S_0. Inside the harbour the sea-bed is assumed to be flat and the water shallow i.e. the time averaged depth of water, h, is constant across S and is small compared with the other dimensions of the problem.

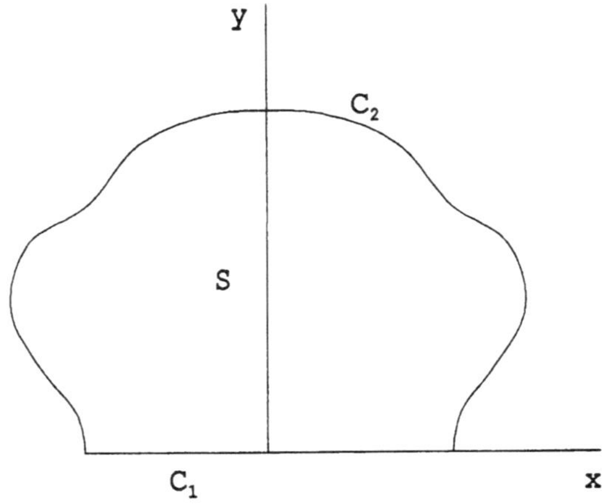

Figure 1. Plan view of the harbour.

The flow conditions are assumed to be inviscid, incompressible and irrotational and disturbances are taken to have wavelengths which are long compared with the average depth of the water. Under these conditions where gravity, acting vertically downwards, is assumed to be the only restoring force acting on the water the fluid flow is described by the linearised shallow water equations which may be written [5]:

$$\frac{\partial u}{\partial t} = -g\frac{\partial \zeta}{\partial x}, \qquad \frac{\partial v}{\partial t} = -g\frac{\partial \zeta}{\partial y} \tag{1}$$

$$\frac{\partial \zeta}{\partial t} + h\left(\frac{\partial u}{\partial x} + \frac{\partial v}{\partial y}\right) = 0 \tag{2}$$

where g is the acceleration due to gravity, u and v are the x and y components of velocity respectively, and $\zeta(x,y)$ is the perturbed height of the water surface above the constant average depth h. In this formulation (1) are the equations of motion and (2) is the equation of continuity.

No attempt is made to simulate explicitly the conditions in the sea outside the harbour. However, it is assumed that the conditions in the external ocean produce oscillations of a single frequency ω at the harbour mouth. Only disturbances with this frequency ω are then sought inside the harbour. Thus the variables u, v and ζ are taken to be simple harmonic and may be written as $u = U(x,y)\cos\omega t$, $v = V(x,y)\cos\omega t$ and $\zeta = z(x,y)\sin\omega t$. In this way time may be eliminated from the equations of fluid motion (1)-(2) which now become:

$$\omega U = g\, z_x\,, \qquad \omega V = g\, z_y \tag{3}$$

$$\omega z + h\,(U_x + V_y) = 0. \tag{4}$$

Combining these gives

$$z_{xx} + z_{yy} + k^2 z = 0 \tag{5}$$

where $k^2 = \omega^2/gh$, which is the two dimensional version of the reduced wave or Helmholtz equation.

The variable z gives the spatial variation of the height of the water surface above (or below) the mean depth when the energy of the water is all stored as potential. A measure of the (mean square) average deviation from the mean depth at such a time is thus given by

$$\int\!\!\int_S \frac{1}{2}\,\gamma_0 z^2\; dx\, dy \tag{6}$$

where γ_0 is a weighting factor which is chosen here to be constant.

Boundary conditions for the problem are chosen to be

$$\frac{\partial z}{\partial n} = 0 \qquad \text{on } C_2 \tag{7}$$

$$\frac{\partial z}{\partial n} = \frac{a_0}{g} \qquad \text{on } C_1 \tag{8}$$

where a_0 is a non-zero constant and $\partial/\partial n$ denotes the outward normal derivative. The condition on C_2 is the standard reflective boundary condition indicating that no energy is transmitted out through the harbour walls. The condition on C_1 indicates that there is a fluid velocity normal to this boundary.

The linearity of the equation (5) and the boundary conditions (7)-(8) mean that the size of the surface elevation disturbance z may be scaled to any desired magnitude and the only change that this necessitates is in the value of a_0.

FORMULATION OF THE OPTIMISATION PROBLEM

The aim is to choose the unknown boundary curve C_2 (Figure 1) in such a way that the average wave height within the harbour takes its lowest possible value. The integral (6) is taken to represent the average wave height across the area of the harbour. Thus the optimisation problem may be stated as follows:

$$\underset{z,C_2}{\text{Minimise}} \iint_S \frac{1}{2} \gamma_0 z^2 \, dx \, dy \tag{9}$$

subject to

$$z_{xx} + z_{yy} + k^2 z = 0 \qquad \text{in } S \tag{10}$$

$$\frac{\partial z}{\partial n} = 0 \qquad \text{on } C_2 \tag{11}$$

$$\frac{\partial z}{\partial n} = \frac{a_0}{g} \qquad \text{on } C_1 \tag{12}$$

$$\iint_S dx \, dy = S_0 \tag{13}$$

where (13) is the constant harbour area constraint. It should be noted that it is assumed that z possesses continuous second partial derivatives with respect to x and y throughout the region S, so that all of the quantities appearing in equations (10)-(13) exist and are well-defined.

TRANSFORMATION TO A KNOWN DOMAIN

In theory, the solution to the optimisation problem specified above can be found by deriving and solving a set of necessary conditions for an extremum. In practice, however, this means solving a system of partial differential equations over an unknown domain. Without further information on the nature of the unknown boundary curve C_2 this is not an easy task. To overcome this difficulty the region of interest is mapped onto a known fixed domain. Although this transformation makes the actual partial differential equations more complicated, nevertheless the problem is more tractable in this form. The transformation is accomplished through the following steps.

First the harbour problem is transformed from its previous Cartesian formulation onto plane polar co-ordinates. The unknown boundary C_2 can then be represented by the equation $r = \ell(\theta)$, where ℓ is an unknown function of angle θ, assumed to be continuous, with continuous first and second derivatives, single valued and greater than zero. Choosing this representation for C_2 places restrictions on the set of admissible boundaries. Henceforth, only smooth boundaries, such that every point on the boundary is in direct line of sight with the origin, will be considered.

It should be noted that this problem is symmetric about the line $\theta = \pi/2$ and so for reasons of computational efficiency only half of the harbour ($0 \leq \theta \leq \pi/2$) need be considered. This introduces a new boundary segment C_3 along the line $\theta = \pi/2$, and from the symmetry requirement the boundary condition on this line must be $\partial z/\partial n = 0$.

In order to remove any possible singularity at the origin in polar co-ordinates, and to avoid any problems with multiple values of θ there, a simple strategy is to exclude the origin, and a very small neighbourhood around it, from the domain specified for the harbour. This is achieved by setting the boundary of the harbour to include a new curve, C_4, given by $r = \epsilon \ell(\theta)$ $0 \le \theta \le \pi/2$, where $\epsilon \ll 1$, around the origin linking C_1 with C_3. The zero normal derivative boundary condition $\partial z/\partial n = 0$ is imposed on C_4. If the value of ϵ is kept very small (say 1.0×10^{-6}) then it is anticipated that the perturbation that this causes to the original optimisation problem will not be great. The modified region used for the calculations is shown in Figure 2.

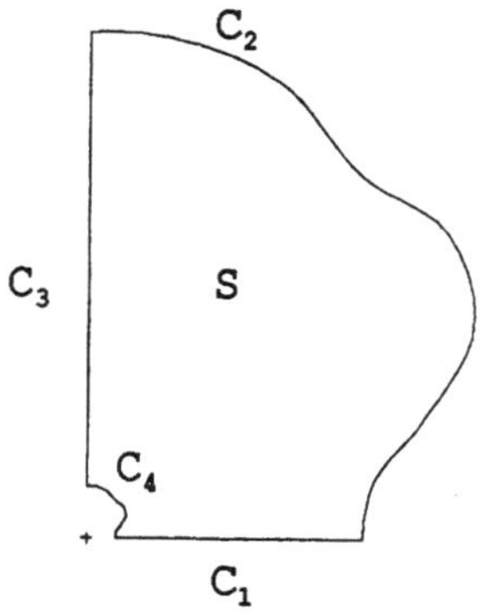

Figure 2. Plan view of calculation domain.

The modified constrained optimisation problem with z now a function of r and θ may be written

$$\underset{z,\ell(\theta)}{\text{Minimise}} \int_0^{\pi/2} \int_{\epsilon\ell(\theta)}^{\ell(\theta)} \frac{1}{2} \gamma_0 z^2 \; r \; dr \; d\theta \qquad (14)$$

subject to

$$z_{rr} + \frac{1}{r} z_r + \frac{1}{r^2} z_{\theta\theta} + k^2 z = 0 \quad \text{in } S \qquad (15)$$

$$\frac{\partial z}{\partial n} = \frac{1}{(\ell^2 + \ell'^2)^{\frac{1}{2}}} \left\{ \ell z_r - \frac{\ell'}{\epsilon \ell} z_\theta \right\} = 0 \quad \text{on } C_4 \qquad (16)$$

$$\frac{\partial z}{\partial n} = \frac{1}{r} z_\theta = 0 \quad \text{on } C_3 \qquad (17)$$

$$\frac{\partial z}{\partial n} = \frac{1}{(\ell^2 + \ell'^2)^{\frac{1}{2}}} \left\{ \ell z_r - \frac{\ell'}{\ell} z_\theta \right\} = 0 \quad \text{on } C_2 \qquad (18)$$

$$\frac{\partial z}{\partial n} = -\frac{1}{r} z_\theta = \frac{a_\ell}{g} \quad \text{on } C_1 \qquad (19)$$

and

$$\int_0^{\pi/2} \int_{\epsilon\ell(\theta)}^{\ell(\theta)} r\ dr\ d\theta = \frac{S_\ell}{2}. \tag{20}$$

The optimisation problem, albeit slightly modified, is now well defined in polar co-ordinates. However, the domain of the problem, or, more specifically, the range of integration with respect to r is still unknown.

Transforming again, to new co-ordinates (R,Θ) given by

$$R = \frac{r}{\ell(\theta)} \quad , \quad \Theta = \theta \tag{21}$$

the previously unknown boundary C_2 can now be represented by the equation $R = 1$, while C_4 is represented by the equation $R = \epsilon$. The domain of the optimisation problem, the interior of the harbour S, is now the set of points (R,Θ) such that $\epsilon \leq R \leq 1$ and $0 \leq \Theta \leq \pi/2$.

If the transformation (21) maps $z(r,\theta)$ onto $Z(R,\Theta)$ and, for consistency, the constants γ_0 and a_0 are renamed Γ_0 and A_0, respectively, then the optimisation problem (14) is now transformed to

$$\underset{Z,\ell}{\text{Minimise}} \int_0^{\pi/2} \int_\epsilon^1 \frac{1}{2} \Gamma_0 Z^2\ R\ell^2\ dR\ d\Theta \tag{22}$$

while the constraint equation (15) becomes

$$\frac{1}{\ell^2}\left[1 + \left(\frac{\ell'}{\ell}\right)^2\right] Z_{RR} + \frac{1}{R\ell^2}\left[1 - \frac{\ell''}{\ell}\right] Z_R - \frac{\ell'}{R\ell^3}\left[Z_{\Theta R} + Z_{R\Theta}\right]$$

$$+ \frac{1}{R^2\ell^2} Z_{\Theta\Theta} + k^2 Z = 0 \tag{23}$$

the boundary conditions (16)-(19) are now

$$\frac{\partial Z}{\partial n} = \frac{1}{(\ell^2 + \ell'^2)^{\frac{1}{2}}} \left\{\left(1 + \frac{\ell'^2}{\ell^2}\right) Z_R - \frac{\ell'}{\epsilon\ell} Z_\Theta \right\} = 0 \quad \text{on } C_4 \tag{24}$$

$$\frac{\partial Z}{\partial n} = -\frac{\ell'}{\ell} Z_R + \frac{1}{R\ell} Z_\Theta = 0 \quad \text{on } C_3 \tag{25}$$

$$\frac{\partial Z}{\partial n} = \frac{1}{(\ell^2 + \ell'^2)^{\frac{1}{2}}} \left\{\left(1 + \frac{\ell'^2}{\ell^2}\right) Z_R - \frac{\ell'}{\ell} Z_\Theta \right\} = 0 \quad \text{on } C_2 \tag{26}$$

$$\frac{\partial Z}{\partial n} = \frac{\ell'}{\ell} Z_R - \frac{1}{R\ell} Z_\Theta = \frac{A_\ell}{g} \quad \text{on } C_1 \tag{27}$$

respectively, while the constant area constraint (20) may now be written

$$\int_0^{\pi/2} \int_\epsilon^1 R\ell^2\ dR\ d\theta = \frac{S_\ell}{2}. \tag{28}$$

In order to convert the constrained optimisation problem (22)-(28) into an unconstrained optimisation problem a Lagrange multiplier approach is adopted. The constraint equations (23) and (28) are adjoined to the cost functional using a Lagrange multiplier function $\Lambda = \Lambda(R,\Theta)$ (assumed to possess continuous second partial derivatives with respect to both R and Θ) and a constant Λ_0 to form the Lagrangian L given by

$$L = \int_0^{\pi/2} \int_\epsilon^1 \Big\{ \tfrac{1}{2}\Gamma_0 Z^2 + \Lambda\Big\{\frac{1}{\ell^2}\Big[1 + \Big(\frac{\ell'}{\ell}\Big)^2\Big]Z_{RR} + \frac{1}{R\ell^2}\Big[1 - \frac{\ell''}{\ell}\Big]Z_R$$

$$- \frac{\ell'}{R\ell^3}\Big[Z_{\Theta R} + Z_{R\Theta}\Big] + \frac{1}{R^2\ell^2}Z_{\Theta\Theta} + k^2 Z\Big\} \Big\} R\ell^2 \; dR\, d\theta$$

$$+ \Lambda_0\Big\{ \int_0^{\pi/2} \int_\epsilon^1 R\ell^2 \; dR\, d\theta - \frac{S_\ell}{2} \Big\}. \tag{29}$$

Integrating by parts, to remove the second derivatives, and then applying the boundary conditions (24)-(27) leads to

$$L = \int_0^{\pi/2} \int_\epsilon^1 \Big\{\frac{1}{2}\Gamma_0 Z^2 + k^2\Lambda Z - \frac{1}{\ell^2}\Big[1 + \frac{\ell'^2}{\ell^2}\Big]Z_R\Lambda_R + \frac{\ell'}{R\ell^3}\Big[Z_\Theta\Lambda_R + Z_R\Lambda_\Theta\Big]$$

$$- \frac{1}{R^2\ell^2}Z_\Theta\Lambda_\Theta \Big\} R\ell^2 \; dR\, d\theta + \int_\epsilon^1 \Big\{ \Lambda\ell\frac{A_\ell}{g}\Big|_{\Theta=0} \Big\} dR$$

$$+ \Lambda_0\Big\{ \int_0^{\pi/2} \int_\epsilon^1 R\ell^2 \; dR\, d\theta - \frac{S_\ell}{2} \Big\}. \tag{30}$$

The optimisation problem may now be stated:
Find the optimum values Z^*, ℓ^*, Λ^* and Λ_0^* of Z, ℓ, Λ and Λ_0 such that

$$L[Z^*,\ell^*,\Lambda^*,\Lambda_0^*] = \inf_{Z,\ell}\ \sup_{\Lambda,\Lambda_0} L[Z,\ell,\Lambda,\Lambda_0] \tag{31}$$

over all admissible functions Z, ℓ, and Λ, and all $\Lambda_0 \in \mathbf{R}$.
It can be shown [8] that the necessary conditions for a solution to this optimisation problem are equivalent to the necessary conditions for a solution to the problem in cartesian co-ordinates so that the nature of the problem has not been altered by the transformation.

APPLICATION OF THE FINITE ELEMENT METHOD

The finite element method [1, 9] is used to find an approximate solution to the problem. The region over which the problem is now defined i.e. $\{ (R,\Theta) \mid \epsilon \le R \le 1,\ 0 \le \Theta \le \pi/2 \}$ is divided up into a set of non-overlapping elements by the superimposition of a radial mesh with M × N elements (Figure 3). The mesh is constructed by dividing the region in the R direction into M intervals, the first of length $h_R = 1/M-\epsilon$ the rest of length $h_R = 1/M$, and dividing the region in the Θ direction into N equal intervals each of length $h_\Theta = \pi/2N$.

This uniform mesh is used for most of the calculations, however, for some later examples modifications are made to the mesh. In particular, different schemes to refine the mesh close to $\Theta=0$ are considered.

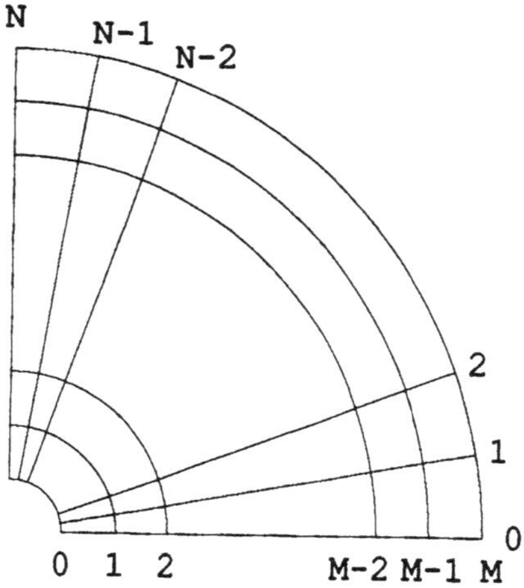

Figure 3. The uniform radial finite element mesh (the size of ϵ is exaggerated for clarity).

Since Z and Λ are functions of both R and Θ while ℓ is a function of Θ alone, at least two separate sets of basis functions will be required. The two dimensional basis functions are denoted by $w_i(R,\Theta)$ for i=0,1,...,m and the one dimensional basis functions are denoted by $\psi_j(\Theta)$ for j=0,1,...,n. Finite linear combinations of these basis functions are then used to give approximations to the variables of interest in this problem i.e. Z, ℓ and Λ:

$$Z(R,\Theta) = \sum_{i=0}^{m} \alpha_i w_i(R,\Theta)$$

$$\ell(\Theta) = \sum_{j=0}^{n} \beta_j \psi_j(\Theta) \qquad (32)$$

$$\Lambda(R,\Theta) = \sum_{i=0}^{m} \gamma_i w_i(R,\Theta).$$

The same set of basis functions $w_i(R,\Theta)$ is used to represent both Z and Λ.

If the approximations (32) are substituted into the Lagrangian (30) this means that solutions $\bar{Z}$, $\bar{\ell}$, $\bar{\Lambda}$ and $\bar{\Lambda}_0$ are now sought in finite dimensional spaces and are such that

$$L[\,\bar{Z},\bar{\ell},\bar{\Lambda},\bar{\Lambda}_0\,] = \inf_{\substack{Z\epsilon A_m \\ \ell\epsilon B_n}} \sup_{\substack{\Lambda\epsilon A_m \\ \Lambda_0\epsilon \mathbf{R}}} L[\,Z,\ell,\Lambda,\Lambda_0\,] \qquad (33)$$

where

$$A_m = \{\ a \mid a = \sum_{i=0}^{m} a_i w_i(R,\Theta),\ \epsilon \le R \le 1,\ 0 \le \Theta \le \frac{\pi}{2},\ a \in \overline{H^1(S) \cap N(Z^*)}\ \}$$

$$B_n = \{\ b \mid b = \sum_{j=0}^{n} b_j \psi_j(\Theta),\ 0 \le \Theta \le \frac{\pi}{2},\ b \in \overline{H^1(S) \cap N(\ell^*)}\ \}$$

$$C_m = \{\ c \mid c = \sum_{i=0}^{m} c_i w_i(R,\Theta),\ \epsilon \le R \le 1,\ 0 \le \Theta \le \frac{\pi}{2},\ c \in \overline{H^1(S) \cap N(\Lambda^*)}\ \}$$

and **R** is the set of real numbers. Here $H^1(S)$ is the space of all functions $u(R,\Theta)$ which are, together with their first partial derivatives square integrable over the harbour region S, N(.) denotes a neighbourhood and the 'overbar' denotes closure. The problem has now become that of finding the extremum of the Lagrangian, L, with respect to the coefficients α_i, β_j, γ_i and also Λ_0 ($i=0,\dots,m$; $j=0,\dots,n$). A necessary condition for such an extremum is that the partial derivatives of the Lagrangian with respect to each of the coefficients α_i ($i=0,\dots,m$), β_j ($j=0,\dots,n$), γ_i ($i=0,\dots,m$) and Λ_0 should be zero. This gives the following set of Ritz-Galerkin equations:

$$\frac{\partial L}{\partial \alpha_p} = \int_0^{\pi/2} \int_\epsilon^1 \Big\{ \Big[\Gamma_0 ZR\ell^2 + k^2 \Lambda R\ell^2 \Big] w_p + \Big[-R\Big(1 + \frac{\ell'^2}{\ell^2}\Big)\Lambda_R + \frac{\ell'}{\ell}\Lambda_\Theta \Big] w_{pR} + \Big[\frac{\ell'}{\ell}\Lambda_R - \frac{1}{R}\Lambda_\Theta \Big] w_{p\Theta} \Big\}\, dR\, d\Theta = 0 \qquad p=0,\dots,m \tag{34}$$

$$\frac{\partial L}{\partial \beta_q} = \int_0^{\pi/2} \int_\epsilon^1 \Big\{ \Big[\Gamma_0 Z^2 R\ell + 2k^2 \Lambda ZR\ell + 2R\frac{\ell'^2}{\ell^3} Z_R \Lambda_R - \frac{\ell'}{\ell^2} Z_\Theta \Lambda_R - \frac{\ell'}{\ell^2} Z_R \Lambda_\Theta + 2\Lambda_0 R\ell \Big] \psi_q + \Big[-2R\frac{\ell'}{\ell^2} Z_R \Lambda_R + \frac{1}{\ell} Z_\Theta \Lambda_R + \frac{1}{\ell} Z_R \Lambda_\Theta \Big] \psi_q{}' \Big\}\, dR\, d\Theta + \int_\epsilon^1 \Lambda \frac{A_\ell}{g}\, \psi_q \Big|_{\Theta=0} dR = 0 \qquad q=0,\dots,n \tag{35}$$

$$\frac{\partial L}{\partial \gamma_p} = \int_0^{\pi/2} \int_\epsilon^1 \Big\{ \Big[k^2 ZR\ell^2 \Big] w_p + \Big[-R\Big(1 + \frac{\ell'^2}{\ell^2}\Big) Z_R + \frac{\ell'}{\ell} Z_\Theta \Big] w_{pR} + \Big[\frac{\ell'}{\ell} Z_R - \frac{1}{R} Z_\Theta \Big] w_{p\Theta} \Big\}\, dR\, d\Theta + \int_\epsilon^1 \frac{A_\ell}{g} \ell w_p \Big|_{\Theta=0} dR = 0 \quad p=0,\dots,m \tag{36}$$

$$\frac{\partial L}{\partial \Lambda_0} = \int_0^{\pi/2} \int_\epsilon^1 R\ell^2\, dR\, d\Theta - \frac{S_\ell}{2} = 0. \tag{37}$$

The above integrals may be evaluated analytically but this is quite tedious so numerical methods are employed. For the double integrals a four point Gauss-Legendre quadrature formula for integration over each element is chosen. This Gauss-Legendre quadrature rule will integrate exactly any polynomial term $R^p\Theta^q$, where $p \leq 3$ and $q \leq 3$, over an element [1]. Unfortunately the nonlinearity in the current problem means that, even with linear and bilinear elements, polynomial terms of higher order in R and Θ will occur and, as a result, the integration will not be exact.

For the boundary integrals in equations (35) and (36) two numerical methods have been used. For most calculations a 5-point composite Simpson's Rule formula has been applied to each element on the boundary. Some later calculations have used the one dimensional 2-point Gauss-Legendre quadrature formula to evaluate the boundary integrals over each element on the boundary. The Gauss-Legendre formula is preferable to Simpson's Rule because it has comparable accuracy, both formulae integrate cubics exactly, with less function evaluations required. Test calculations have shown that the above two formulae give virtually identical results.

After the numerical integration the system of equations (34)-(37) becomes a set of nonlinear algebraic equations of the form

$$\mathbf{f}(\mathbf{x}) = \mathbf{0} \tag{38}$$

where **x**, the vector of unknowns, contains the coefficients α_i, β_j and γ_i from the approximations (32) together with Λ_0. The system (38) contains $2(m+1)+(n+1)+1 = 2m+n+4$ equations in $2m+n+4$ unknowns. However, one equation and one unknown may be deleted because $\ell(0) = 1$ is already given and hence β_1 is determined. Thus, finally, a set of $2m+n+3$ nonlinear algebraic equations in $2m+n+3$ unknowns is obtained.

This set of equations is then solved using the Harwell Library Subroutine NS03AD [3]. This routine uses a modification of the Marquardt Algorithm [2] - a combination of Newton's Method and the Method of Steepest Descent which is designed to ensure convergence even from a starting point far from the actual solution and yet have a fast rate of convergence as the solution is approached. It has been found that when the number of equations becomes quite large (i.e. a few hundred) even this routine requires an initial point fairly close to the actual solution for the system considered here.

The finite element method, used as a direct method of optimisation, as outlined above, leads to the approximate solution $(\bar{Z}, \bar{\ell}, \bar{\Lambda}, \bar{\Lambda}_0)$. The convergence of this approximation to the exact solution as the finite element mesh is refined is established in the following theorem.

Theorem. Let $(\bar{Z}, \bar{\ell}, \bar{\Lambda}, \bar{\Lambda}_0) \in A_m \times B_n \times C_m \times \mathbf{R}$ be the solution of the necessary conditions (34)-(37) above. Then there exist non-negative constants C_1, C_2 and C_3, independent of m and n, such that

$$\| Z^* - \bar{Z} \|_2 + \| \ell^* - \bar{\ell} \|_2 \leq C_1 \left[\left\{ \|\epsilon(\hat{Z})\|_{2,1} + \|\epsilon(\hat{\ell})\|_{2,1} \right\}^{\frac{1}{2}} + \right.$$

$$\left. \left\{ \|\epsilon(\hat{Z})\|_{2,1} + \|\epsilon(\hat{\ell})\|_{2,1} \right\} + \left\{ \|\epsilon(\hat{\Lambda})\|_{2,1} + \|\epsilon(\hat{\Lambda}_0)\|_{2,1} \right\} \right] \tag{39}$$

and

$$L^* - C_2 \left\{ \| \epsilon(\hat{\Lambda}) \|_{2,1} + \| \epsilon(\hat{\Lambda}_0) \|_{2,1} \right\}^2 \leq \bar{L} \leq$$

$$L^* + C_3 \left[\left\{ \| \epsilon(\hat{Z}) \|_{2,1} + \| \epsilon(\hat{\ell}) \|_{2,1} \right\} + \left\{ \| \epsilon(\hat{Z}) \|_{2,1} + \| \epsilon(\hat{\ell}) \|_{2,1} \right\}^2 \right] \tag{40}$$

where

$$\epsilon(\hat{Z}) = \hat{Z} - Z^*, \qquad \epsilon(\hat{\ell}) = \hat{\ell} - \ell^*$$

$$L^* = L[Z^*, \ell^*, \Lambda^*, \Lambda_0^*], \qquad \bar{L} = L[\bar{Z}, \bar{\ell}, \bar{\Lambda}, \bar{\Lambda}_0]$$

and $\hat{Z}$ is the unique best approximation to Z^* in A_m, $\hat{\ell}$ is the unique best approximation to ℓ^* in B_n, and $\hat{\Lambda}$ is the unique best approximation to Λ^* in C_m. Here $\|.\|$ denotes a Sobolev norm. If it is assumed that A_m, B_n and C_m are good approximating spaces in the sense that $\hat{Z}$, $\hat{\ell}$ and $\hat{\Lambda}$ converge to Z^*, ℓ^* and Λ^* as $m \to \infty$, $n \to \infty$ then it follows that

$$\| Z^* - \bar{Z} \|_2 \to 0, \qquad \| \ell^* - \bar{\ell} \|_2 \to 0, \qquad | L^* - \bar{L} | \to 0$$

as $m,n \to \infty$ and the convergence is demonstrated. A proof of this theorem is given in [8] - the proof in fact requires that the approximating spaces A_m, B_n and C_m be restricted to contain only functions which are, together with their first and second partial derivatives, square integrable. This would restrict the basis functions to Hermite polynomials which preserve the continuity of the first derivatives of the approximations (32). However, the absence of any second derivatives in the Ritz-Galerkin equations (34)-(37) suggests that this constraint may be unnecessarily severe, and calculations have been performed with approximations which do not have the theoretically required continuity.

RESULTS

Results for the optimal harbour shape are presented in Figures 4-7. On these diagrams each asterisk represents a value of $\ell(\theta)$, a point on the boundary. The figures illustrate half of the harbour with the line $\Theta = 0$ (the harbour mouth) along the horizontal axis and Θ increasing anti-clockwise to $\Theta = \pi/2$ along the vertical axis (the axis of symmetry). Calculations of the optimal harbour shape have been performed with uniform meshes using the following finite element approximations:
(i) $\ell(\theta)$ linear, Z and Λ bilinear
(ii) $\ell(\theta)$ quadratic, Z and Λ bilinear
(iii) $\ell(\theta)$ quadratic, Z and Λ biquadratic
(iv) $\ell(\theta)$ (Hermite) cubic, Z and Λ (Hermite) bicubic.
The results presented were all obtained using the following values for the parameters of the model: $\Gamma_0 = 1.0$, $k = 1.0$, $A_0/g = 2.0$, $\epsilon = 1.0 \times 10^{-6}$, $S_0 = (1 - \epsilon^2)\pi/2$ and $\ell(0) = 1.0$ (half width of harbour mouth). Choosing these values means that the model represents the action of long waves in a small harbour with a wide entrance. Further details of these and other related calculations may be found in [8].

As the mesh resolution is increased the results for the optimal harbour shape obtained from calculations with approximations (i)-(iii) above all have similar general features - a typical example is shown in Figure 4. The boundary is smooth for values of Θ greater than about $\pi/8$ - the smoothness extends across the axis of symmetry where clearly $\ell' = 0.0$. The appearence of a smooth boundary without any smoothing criteria being explicitly imposed is a very encouraging feature of these results. However, the perturbation close to $\Theta = 0$ is clearly not desirable. This perturbation may be removed by local mesh refinements close to the 'corner' of the harbour at $(R,\Theta) = (1,0)$ where there is a discontinuity in the normal derivative boundary condition. The simplest of all mesh refinement strategies - i.e. considering an initially uniform mesh and adding extra Θ mesh lines close to $\Theta = 0$ and extra R mesh lines close to $R = 1$ - has been found to be effective. The results obtained from a calculation with quadratic and biquadratic approximations, (iii) above, are illustrated in Figure 5 - the mesh is shown in Figure 5(a) and the optimal harbour shape in Figure 5(b).

Since Hermite cubics explicitly approximate first derivatives the smoothing condition $\ell' = 0.0$ at $\Theta = \pi/2$ is explicitly imposed for calculations with approximations of type (iv). The results for the optimal harbour shape obtained from calculations with Hermite cubic and bicubic approximations, as anticipated, show a greater uniformity of smoothness than those with linear or quadratic elements - Hermite approximations have continuous first derivatives. Figure 6 shows a typical result from a Hermite cubic calculation with a uniform mesh. The boundary is generally smooth throughout although there is a noticeable perturbation at about $\Theta = \pi/8$. In a calculation with 2 extra mesh lines (one at $\Theta = \pi/128$ and one at $\Theta = \pi/64$) added onto the uniform mesh this perturbation is removed and a uniformly smooth boundary is produced (Figure 7).

All of the harbour shapes produced by the aforementioned calculations share similar general features. The optimal harbour shape appears to be close to part of an ellipse with the major axis parallel to the harbour entrance. In fact, in these calculations with a wide entrance to the harbour, the optimal boundary shape is not too far removed from a semi-circle. Indeed the difference in the values of the cost functional (6) between the optimal shape geometry and the semi-circular geometry is only of the order of 1%-2% for the values of the model parameters used here. This indicates that the problem has a 'shallow minimum' and also suggests that from a practical standpoint a semi-circle is a good sub-optimal solution. Test calculations [8] performed with values of the parameter k between 1.0 and 2.0 reveal that the optimal harbour shape does not appear to be sensitive to changes in the wavelength of the oscillations within the harbour (at least within a fairly narrow range). On the other hand it has been found that altering the width of the harbour mouth relative to the other dimensions does significantly influence the final optimal harbour shape.

The value of the current calculations is at present limited by a number of factors, most connected with restrictions on the available computing power. Currently the number of elements that may be used limits the investigation to wavelengths which are of a similar magnitude to the dimensions of the harbour. Also it is currently impossible to include any of the sea external to the harbour in the calculations - if this were possible it would permit the imposition of a more realistic boundary condition on the harbour entrance (the flow conditions could be solved inside and outside the harbour and then matched across the mouth). However, notwithstanding these limitations the method presented here does appear to provide a way of obtaining the general features of the optimal shape of a harbour once certain parameters have been specified. Furthermore, with greater computing power it should be possible to significantly improve the realism of the harbour model without greatly increasing the complexity of the optimal shape design problem.

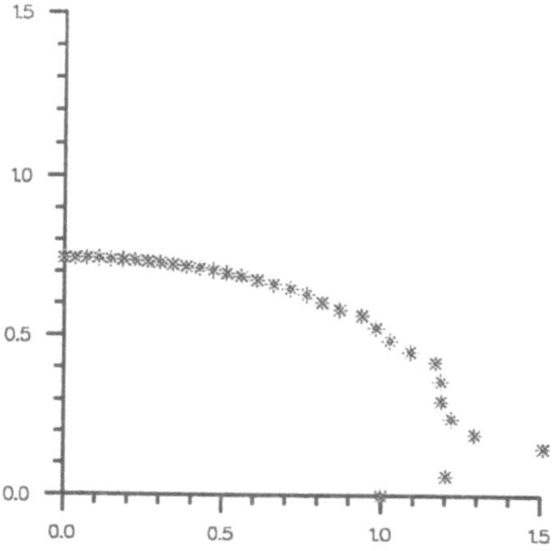

Figure 4. Harbour shape. Quadratic boundary ℓ - biquadratic state Z and adjoint Λ - 4 × 16 elements (uniform mesh).

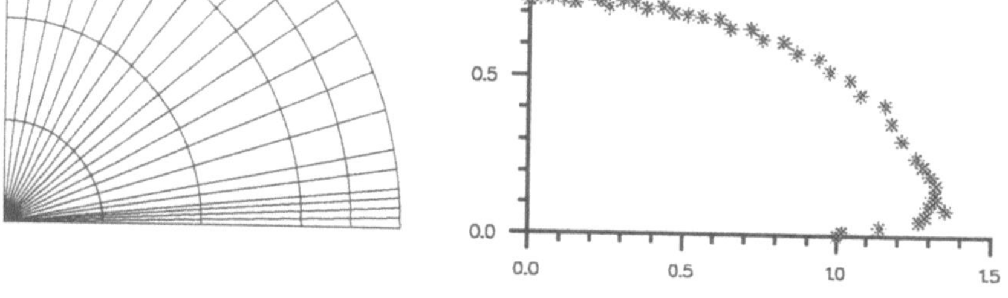

(a) Mesh (b) Harbour shape

Figure 5. Quadratic boundary ℓ - biquadratic state Z and adjoint Λ - 5 × 20 elements.

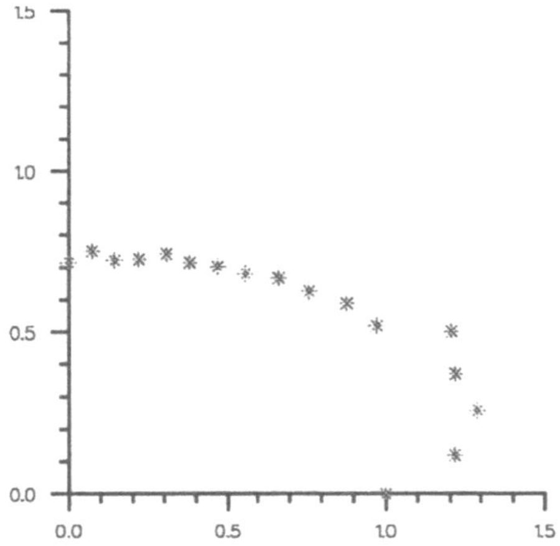

Figure 6. Harbour shape. Cubic boundary ℓ - bicubic state Z and adjoint Λ - 4 × 16 elements (uniform mesh).

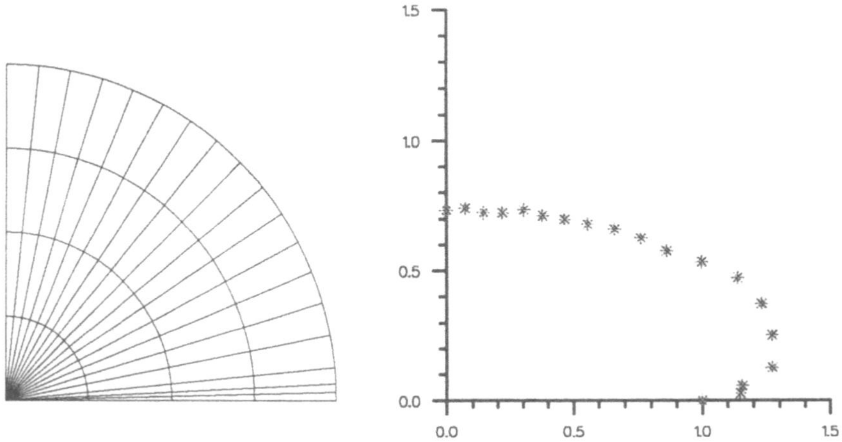

(a) Mesh (b) Harbour shape

Figure 7. Cubic boundary ℓ - bicubic state Z and adjoint Λ - 4 × 18 elements.

ACKNOWLEDGEMENT

A research grant from Kingston Polytechnic to support this work is gratefully acknowledged.

REFERENCES

1. Burnett, D. S. : Finite Element Analysis, Addison-Wesley, 1987.

2. Fletcher R : A Modified Marquardt Subroutine for Non-Linear Least Squares, AERE Report R6799, 1971.

3. Harwell : Harwell Subroutine Library Manual, Seventh Edition, 1987.

4. Haslinger, J. and Neittaanmäki, P. : Finite Element Approximation For Optimal Shape Design: Theory and Applications, John Wiley and Sons, 1988.

5. Landau, L. D. and Lifshitz, E. M. : Fluid Mechanics, (A Course of Theoretical Physics - Vol. 6), Pergamon Press, 1959.

6. Lyle, S. and Nichols, N. K. : On the Finite Element Solution of Optimal Design Problems Using Transformation Techniques; in The Mathematics of Finite Elements and Applications IV (Ed. Whiteman, J. R.), pp 183-191, Academic Press, 1982.

7. Pironneau, O. : Optimal Shape Design for Elliptic Systems, Springer-Verlag, 1984.

8. Soan, P. H. : Transformation Techniques in Optimal Design Problems with Application to Harbour Shapes, Ph.D Thesis, Kingston Polytechnic, 1990.

9. Wait, R. and Mitchell, A. R. : Finite Element Analysis and Applications, John Wiley and Sons, 1985.

Bend Theory of Stream Meanders on a Smooth Plate

K. Mizumura, T. Nakagawa

Department of Civil Engineering, Kanazawa Institute of Technology, 7-1, Ohgigaoka, Nonoichi, PO Kanazawa-South, Ishikawa 921 Japan

ABSTRACT

To obtain the condition of stable meandering on a smooth plate, the bend theory on water rivulets including the effect of the surface tension is derived. The criterion from the linear stability theory shows the relationship between the meander wavelength of water rivulets and flow properties. As a result, the derived equation is found to be in good agreement with the observed data of stable meandering on a smooth plate.

INTRODUCTION

A window pane speckled with rain drops offers an enigma in the physics of fluids. When the rain runs down the window pane in a stream, why does it normally meander instead of going straight? Such a stream will be, hereafter, refered to as a rivulet (Photo 1), which flows on a smooth hydrophobic surface driven by the component of the gravity along the surface (Nakagawa & Scott[11]). Rivulets are also seen on automobile windshields or on the walls of showers. The perimeter of the contact area between the rivulet and the surface is called the triple-phase line because of the conjunction of water, air, and solid. The angle at which the water touches the solid at the triple-phase line is called the contact angle. It is measured between the solid and the tangent to the water surface and, conventionally, it is the angle which includes the liquid phase.

In the early part of the 19th century, Thomas Young (1805) stated that the size of the contact angle for a liquid drop is set by the tendency toward equilibrium of the three tensions pulling on the triple-phase line. Obviously a triple-phase line is in equilibrium over a range of contact angles rather than at a single value. The liquid is observed to advance over the surface of a solid body only when the contact angle exceeds some upper limit and it recedes only when contact angle drops below some lower limit. After either event the contact angle is again within the range of values that makes for stability.

The upper limit is called the advancing contact angle and the lower one the receding contact angle. The ability of the triple-phase lines of a rivulet to be in equilibrium at different values of the contact angle is refered to as the contact-angle hysteresis. It is the hysteresis that enables a liquid drop to cling to an inclined surface. Why a triple-phase line displays hysteresis is a subject of current research.

Since water particles in the rivulet are pulled downward by the gravity, it may be natural to expect the locus to be straight down. Yet in many experiments(e.g. Tanner[15]; Kern[8][9]; Gorycki[5]; Nakagawa[10]; Sieber[13]; Culkin & Davis[2]; Davic & Tinker[4]), the rivulet meanders, thus developing bends that connect relatively straight stretches slanting to one side or the other. Furthermore, rivulets display a large variety of intriguing instability phenomena. Kern[8,9], for example, sees the break-up of rivulets into drops, rivulet meandering, and turbulent rivulet. It is suggested by Davis[3] that droplet formation is due to capillary instability [12] where the surface tension on the liquid-air interface causes capillary pressure gradient that enhances small interfacial corrugations. Large amplitude surface waves on the rivulet are also apparent in many situations. Nakagawa & Scott[11] have found that the rivulet forms a stable meander if the volume flow rate is between an upper and lower limit. If the flow rate exceeds the upper limit, the meander is movable and changes constantly. If the flow rate becomes smaller than the lower limit, the rivulet breaks up into drops that slide separately down the surface. A plausible mechanism of rivulet meanders on a smooth hydrophobic surface has been proposed by many researchers. Progress of the mathematical theory on the rivulet meanders is, however, relatively slow. The existing theory is essentially limited to a straight rivulet solution (Towell & Rothfeld[16]; Allen & Biggin[1]). Towell et al.[16] have treated a cylindrical static meniscus, straight parallel static contact lines and undirectional steady parallel flow down the smooth solid surface. Allen et al.[1] have extended the zero-order solution of Towell et al. to the first-order solution, and calculated the velocity distribution within the rivulet. On the one hand, Davis[3] has investigated capillary instabilities on a small static rivulet on a horizontal surface by the linear stability theory. The triple-phase lines treated in the theory are fixed, moving but having fixed contact angles which are smooth functions of triple-phase line speeds. No results have also been obtained by him for a system that displays contact-angle hysteresis, for this phenomena, although physically common, is outside of the realm of the linear stability theory. Weiland & Davis[17] have studied long wave instabilities for flat rivulets on a vertical wall, in corporating the same triple-phase line condition into their analysis. They result in a suggestion that contact-angle hysteresis would lead to absolute stability of rivulets against small-amplitude long waves.

All the above analyses are unsatisfactory due to the fact that rivulet meandering is not allowed. They treat only the velocity distribution in an idealized ribulet or the capillary and long wave instabilities. In the present analysis the bend theory due to Ikeda et al.[6,7] is used to develop a formal stability theory of sinuous rivulets. The stability criterion is described in terms of the growth-rate of lateral bend amplitude. The linear thory herein yields a relation for initial characteristic bend wavelength of the rivulets on the inclined smooth hydrophobic surface, and has shown good agreement with the present data as well as Sieber's data[13].

EXPERIMENTS

Fig.1 shows a schematic diagram of the experimental arrangement. The optically smooth plexiglas plate was 100 cm long and 60 cm wide, and the water tank allowed a constant discharge rate of water through a vinyl tube of 10 cm inner diameter. The tube mouth was arranged on the centerline of the plate, at a point 20 cm down the slope from the upper edge. The experimental variables were the discharge rate, measured by weighing the water discharged during a given time, and the surface slope, measured with a protractor. The maximum discharge rate for each surface slope is an amount above which the meandering pattern becomes unstable, and the surface slope was varied from 5° to 35° in 5° step. The form of the water rivulet was determined by tracing the plan locus of its central axis, using ink, onto the transparent plate, and then the wavelength, to be discussed, is obtained with it. The Reynolds number of the flow within the vinyl pipe was in the range 100 to 650, well below the critical value, and the flow was observed to be laminar. During the course of the experiment it was found that the stream is very sensitive to the condition of the plate surface. For example, the stream would tend to follow parts that had already been wetted. Thus, a dry surface was carefully prepared for each experimental run, wiping the surface with soft hygroscopic tissues, and then leaving the surface for 30 minutes prior to the next run.

The relationship between wavelength of meandering λ and discharge Q obtained from the experiments is plotted in Fig.2. It is expressed as the wavelength is approximately proportional to $Q^{0.710}$. The correlation between them is pretty good. The relationship between Froude number F_r and wave number of meandering k is described in Fig.3. The correlation between two variables is not as good as the previous relationship. As a result, through the experiments the followings are obtained:
(1) Normal rivulet width is almost constant as shown in Photo 1.
(2) Wavelengths of meandering are almost constant along each rivulet.
(3) The computed mean water depth of rivulet H, by using normal rivulet width b, discharge Q, and resistance law, is much smaller than normal rivulet width.
During this computation the shape of the cross section of rivulet is assumed to be a rectangular for simplicity.

GOVERNING EQUATIONS

Since the rivulet flow is found to be very shallow by the experiments, the equations of motion for a sinuous rivulet on a uniform slope (Fig.1) are based on the St Venant equations of shallow water flow. To simplify them, the following assumptions are made:
(1) Normal rivulet width is constant. This is considered to be reasonable by the experiments (Photo 1).
(2) The centerline radius of curvature is much larger than the normal rivulet width. Therefore, the effect of the curvature radius of the rivulet in along-stream direction is neglected.
(3) The classical quasi-steady assumption holds. It is considered that the centerline movement of the water rivulet is very slow in comparison with the time change of flow variables.
Including surface tension effect, the simplified equations are written as follows:

$$\frac{\partial hu^2}{\partial s} + \frac{\partial huv}{\partial n} + 2chuv + hg\frac{\partial h}{\partial s} - hgI + \frac{\tau_s}{\rho} + \frac{hT}{\rho}\frac{\partial}{\partial s}\left(\frac{1}{r}\right) = 0 \tag{1a}$$

$$\frac{\partial huv}{\partial s} + \frac{\partial hv^2}{\partial n} - chu^2 + hg\frac{\partial h}{\partial n} + \frac{\tau_n}{\rho} + \frac{hT}{\rho}\frac{\partial}{\partial n}\left(\frac{1}{r}\right) = 0 \tag{1b}$$

$$\frac{\partial hu}{\partial s} + \frac{\partial hv}{\partial n} + chv = 0 \tag{1c}$$

in which u and v = depth-averaged velocity components in the s and n directions, respectively; c = centerline curvature; h = water depth; g = the gravitational acceleration; τ_s and τ_n = the bed stresses in the s and n directions, respectively; T = surface-tension force acting on the water rivulet; ρ = water density; r = curvature radius of water rivulet in across-stream direction; I = bottom slope. The bed stresses are expressed with the use of a friction factor C_f:

$$\tau_s = \rho C_f \bar{U} u$$

$$\tau_n = \rho C_f \bar{U} v$$

in which $\bar{U} = \sqrt{u^2 + v^2}$. Local flow is defined in summation of unperturbed quantity plus a primed one denoting a perturbation induced by channel curvature as follows:

$$\begin{aligned} u &= U + u' \\ v &= v' \\ h &= H + h' \\ c &= c' \\ r &= R + r' \end{aligned} \tag{2}$$

in which U = reach-averaged tangential velocity. Since the normal rivulet width is much narrower than the meandering wavelength of the rivulet, the perturbation of the tangential velocity u' is considered to be much larger than the normal velocity v' in the perturbation induced by the channel curvature. For the straight water rivulet, the contact angles on the north and the south sides of the rivulet are denoted by $\bar{\beta}_N$ and $\bar{\beta}_S$, respectively as shown in Fig.5. When the centerline of the rivulet is deviated y from the initial location, the contact angles on the both sides are given by

$$\begin{aligned} \bar{\beta}_N &= \bar{\beta}_N + \Delta\beta_N \\ \bar{\beta}_S &= \bar{\beta}_S + \Delta\bar{\beta}_S \end{aligned} \tag{3}$$

in which $\Delta\beta_N$ and $\Delta\beta_S$ = the deviations of the contact angles from $\bar{\beta}_N$ and $\bar{\beta}_S$, respectively. Further, to simplify the effect of the surface tension in Eq.(1b), we assume that the contact angle is constant for the straight water rivulet and the deviation $\Delta\beta_N$ and $\Delta\beta_S$ are the functions of the deviation from the straight stream y as follows:

$$\bar{\beta}_N = \bar{\beta}_S = \beta_0 \tag{4}$$

$$\Delta\beta_N = -\Delta\beta_S = \beta'(y) \tag{5}$$

Thus, Eq.(3) is transformed into Eq.(6).

$$\begin{aligned} \beta_N &= \beta_0 + \beta'(y) \\ \beta_S &= \beta_0 - \beta'(y) \end{aligned} \tag{6}$$

For zeroth order, Eqs.(1a), (1b), and (1c) yield

$$C_f U^2 = gHI \tag{7a}$$

$$\frac{\partial}{\partial n}\left(gH + \frac{T}{\rho R} \right) = 0 \tag{7b}$$

$$\frac{\partial}{\partial s}(UH) = 0 \tag{7c}$$

Eq.(7b) expresses the basic state of the straight water rivulet. For first order, Eq.(1a), (1b), and (1c) yield

$$(U^2+gH)\frac{\partial h'}{\partial s} + 2HU\frac{\partial u'}{\partial s} + \frac{\partial HUv'}{\partial n} - gIh' + C_f Uu' - \frac{TH}{\rho R^2}\frac{\partial r'}{\partial s} = 0 \tag{8a}$$

$$-cHU^2 + gH\frac{\partial h'}{\partial n} + gh'\frac{\partial H}{\partial n} - \frac{TH}{\rho R^2}\frac{\partial r'}{\partial n} = 0 \tag{8b}$$

$$H\left(\frac{\partial u'}{\partial s} + \frac{\partial v'}{\partial n} \right) + U\frac{\partial h'}{\partial s} = 0 \tag{8c}$$

Refering Fig.6, we obtain

$$\sin \beta_N = \frac{b}{r} \tag{9}$$

However,

$$\beta_N = \beta_0 + \beta'(y) \tag{10a}$$

$$r = R + r' \tag{10b}$$

For the straight water rivulet,

$$\sin \beta_0 = \frac{b}{R} \tag{11}$$

The substitution of Eqs.(10a), (10b), and (11) into Eq.(9) and the neglection of small terms give

$$r' = -\frac{\beta'(y)R^2 \cos \beta_0}{b} \tag{12}$$

Eq.(8b) is integrated to yield

$$h'(n=b) = \frac{c'U^2b}{g} - \frac{T\cos\beta_0\ \beta'(y)}{g\rho b} \tag{13}$$

In order to obtain the first order solution, the flow velocity U and the water depth H of the zeroth order are assumed to be constant, since the set of equations (7a), (7b), and (7c) are not analytically solved to get the zeroth order solution. This corresponds that the cross section is assumed to be a rectangular. If the water rivulet is straight, $\beta'(y) = 0$. When the water rivulet is deviated y from the initial location (straight stream) and y is very small, $\beta'(y)$ is assumed to be expressed as:

$$\beta'(y) \simeq k_* y \tag{14}$$

in which k_* = a constant value. An equation for $u_b' = u'(n=b)$ is derived by substituting Eq.(13) into Eq.(8) as:

$$(U^2+gH)\frac{\partial}{\partial s}\{\frac{c'U^2b}{g} - \frac{T\beta'(y)\cos\beta_0}{g\rho b}\} + 2HU\frac{\partial u_b'}{\partial s} + \frac{\partial HUv'}{\partial n} - gI\{\frac{c'U^2b}{g} - \frac{T\beta'(y)\cos\beta_0}{g\rho b}\} + C_fUu_b' + \frac{TH}{\rho}\frac{\partial}{\partial s}\{\frac{\beta'(y)\cos\beta_0}{b}\} = 0 \tag{15}$$

Let ζ be the rate of normal movement of the side of the water rivulet. Then, the geometrical consideration of Fig.7 gives that

$$\zeta = \frac{\partial y}{\partial t}\cos\theta \tag{16}$$

This equation describes north side movement of the water rivulet. If ζ is positive (negative), the water rivulet advances (recedes). ζ is assumed to be a function of tangential and normal flow velocities near the side of the water rivulet, $u|_{n=b} = U + u'(s, b)$ and $v|_{n=b} = v'(s, b)$. Thus,

$$\begin{aligned}\zeta &= \zeta(u, v)\\ &= \zeta(U+u', v')\end{aligned} \tag{17}$$

To evaluate v', let us consider a dimensionless form of Eq.(9c). The following new variables are defined as:

$$u' = U\,u_*';\ v' = U\,v_*';\ h' = H\,h_*';\ s = s_*/c;\ \text{and}\ n = b\,n_* \tag{18}$$

in which "*" represents a dimensionless variable. Substitution of Eq.(18) into Eq.(9c) and consideration of bc << 1 (Assumption (2)) yield

$$\frac{\partial v_*'}{\partial n_*} = 0 \tag{19a}$$

$$\frac{\partial u_*'}{\partial s_*} + \frac{\partial h_*'}{\partial s_*} = 0 \tag{19b}$$

Since the across-stream velocity is zero on the both sides of the water rivulet, v' = 0 is derived by Eq.(19a). Therefore, since u' >> v' consists due to the assumption (2), the rate of normal movement of the side of the water rivulet is obtained by

$$\zeta = \zeta(U, 0) + \frac{\partial\zeta}{\partial u}\bigg|_{u=U} u'(s, b)$$

$$= \zeta(U, 0) + E(U)u'(s, b) \tag{20}$$

in which $E(U) = \frac{\partial\zeta}{\partial u}\Big|_{u=U}$. Accordingly, ζ is considered to be a function of the only tangential flow velocity near the side of the water rivulet. It is the same as the case of river meandering (Ikeda et al.[6]. In Eq.(20), $\zeta(U, 0)$ is assumed to vanish. From Eq.(16),

$$\gamma \frac{\partial y}{\partial t} = E(U)\, u' \tag{21}$$

in which $\gamma = \cos\theta = [\,1 + (\frac{\partial y}{\partial x})^2]^{-1/2}$. The transformations from intrinsic to Cartesian co-ordinate dictate that

$$\frac{\partial}{\partial s} = \gamma \frac{\partial}{\partial x}\ ;\ c = -\gamma^3 \frac{\partial^2 y}{\partial x^2} \tag{22}$$

Eq.(15), (21), and (22) yield the nonlinear bend equation expressed in Cartesian co-ordinate (with $v' = 0$):

$$-(U^2+gH)\gamma \frac{\partial}{\partial x}\{\frac{U^2 b}{g}\gamma^3 \frac{\partial^2 y}{\partial x^2} + \frac{Tk_*\cos\beta_0}{g\rho b}\} + \frac{2HU\gamma^2}{E}\frac{\partial^2 y}{\partial x\partial t}$$

$$+gI\{\frac{U^2 b\gamma^3}{g}\frac{\partial^2 y}{\partial x^2} + \frac{Tk_* y\cos\beta_0}{g\rho b}\} + \frac{C_f U\gamma}{E}\frac{\partial y}{\partial t} + \frac{TH}{\rho}\frac{\partial}{\partial x}(\frac{k_* y\cos\beta_0}{b}) = 0 \tag{23}$$

In the above equation the constant $E = E(U)$ is absorbed by the transformation $E(U)\rightarrow t$. For small-amplitude bends, $\gamma \simeq 1$ and Eq.(23) is linearized to

$$(U^2+gH)\frac{U^2 b}{g}\frac{\partial^3 y}{\partial x^3} + (U^2+gH)\frac{Tk_*\cos\beta_0}{g\rho b}\frac{\partial y}{\partial x} - 2HU\frac{\partial^2 y}{\partial x\partial t}$$

$$-IU^2 b\frac{\partial^2 y}{\partial x^2} - \frac{ITk_*\cos\beta_0}{\rho b}y - C_f U\frac{\partial y}{\partial t} - \frac{THk_*\cos\beta_0}{\rho b}\frac{\partial y}{\partial x} = 0 \tag{24}$$

On the left hand side of the above equation, the second, the fifth, and the seventh terms show the effect of the surface tension.

LINEAR STABILITY ANALYSIS

To study the stability of the straight water stream (rivulet), the following stream geometry fluctuation is considered:

$$y = \varepsilon e^{\alpha t}\cos(kx - \omega t) \tag{25}$$

in which α = growth rate of bend amplitude; ω = migration frequency; $C = \omega/k$ = migration speed of bend; $k = 2\pi/\lambda$; λ = wavelength of meander. The substitution of Eq.(25) into (24) yields the dispersion relations:

$$k^3(U^2+gH)\frac{U^2b}{g} - (U^2+gH)\frac{Tk_*\cos\beta_0}{g\rho b} + 2HUk\,\alpha - C_fU\omega - k\frac{THk_*\cos\beta_0}{\rho b} = 0 \tag{26a}$$

$$-2HUk\omega + IU^2bk^2 - \frac{ITk_*\cos\beta_0}{\rho b} - C_fU\alpha = 0 \tag{26b}$$

Thus, α is obtained as:

$$\alpha = \frac{-k^4\phi + k^2\psi - \Theta}{4H^2k^2 + C_f^2} \tag{27}$$

in which $\phi = (U^2 + gH)\dfrac{2UbH}{g}$;

$$\psi = \frac{2HTk_*\cos\beta_0}{\rho b}\left(\frac{U}{g} + \frac{2H}{U}\right) + C_fIUb; \text{ and}$$

$$\Theta = \frac{C_fITk_*\cos\beta_0}{\rho bU}.$$

The instability occurrs or the meandering is generated if α is positive such as

$$k_d < k < k_u \tag{28}$$

in which $k_d = [\{\psi - \sqrt{\psi^2 - 4\phi\Theta}\}/ 2\phi]^{1/2}$; and $k_u = [\{\psi + \sqrt{\psi^2 - 4\phi\Theta}\}/ 2\phi]^{1/2}$. The wave number k_M at which stability is maximized can be obtained from the condition $\partial\alpha/\partial k = 0$; it is found to be

$$k_M = \left[-\frac{C_f^2}{4H^2} + \frac{1}{2}\sqrt{\frac{C_f^4}{4H^4} + \frac{4\Theta H^2 + C_f^2\psi}{\phi H^2}}\right]^{1/2} \tag{29}$$

RESULT AND DISCUSSION

To compare the theoretical results with the experimental data, the following equation and numerical value are employed:

$$C_f = \frac{8}{Re} \tag{30a}$$

$$T = 0.75\times10^{-4}\ \text{kg/m} \tag{30b}$$

in which Re = the rivulet Reynolds number. For the numerical values of U and H, the cross-sectional means are used. Fig.9 explains the suitability of Eq.(29) for the value of $k_*\cos\beta_0$, 21.93, which is computed by the least square method. With the value of $k_*\cos\beta_0$, the relation $k = k_M$ holds. The validity of Eq.(28) is tested in Fig.10. This indicates that the wave number of meandering k obtained by the experiments are between k_d and k_u. k_d and k_u show the lower and the upper limits which instability occurrs, respectively.

CONCLUDING REMARKS

The meandering properties of water rivulets are studied theoretically and experimentally. Assuming that the restraint force acting on the water rivulet due to the surface tension is proportional to the deviation of the present position from the initial straight trace of the rivulet and the solution of the zeroth order are constant, the linear stability theory gives the relationship between the meander wavelength of water rivulet and flow properties. The result to give the criterion which instability occurrs is in good agreement with the experimental data. This fact shows that the present theory is applicable to analyze the phenomena of the meandering of the water rivulets.

REFERENCES

1. Allen, R.F. & Biggin, C.M. Longtudinal flow of a lenticular liquid filament down an inclined plane. The Physics of Fluids, Vol.17,No.2,pp.287-291, 1974.
2. Culkin, J.B. & Davis, S.H. Meandering of water rivulets. A.I.Ch.E.Jour., Vol.30,No.2,pp.263-267, 1984.
3. Davis, S.H. Moving contact lines and rivulet instabilities. Part 1. The static rivulet. J.Fluid Mech., Vol.98,part 2, pp.225-242, 1980.
4. Davis, T.R.H. & Tinker, C.C. Fundamental characteristics of stream meanders. Geological Society of American Bulletin, Vol.95,pp.505-512, 1984.
5. Gorycki, M.A. Hydraulic drag: a meandering initiating mechanism. Bull. Geol. Soc. Am.,Vol.84,pp.175-186, 1973.
6. Ikeda, S., Hino, M. & Kikkawa, H. Theoretical study on the free meandering of rivers. Proc.Japan Soc. Civ. Eng., No. 225,pp.63-73, 1976 (in Japanese).
7. Ikeda, S., Parker, G. & Sawai, K. Bend theory of river meanders. Part 1. linear development. Jour. Fluid Mech., Vol.112,pp.363-377, 1981.
8. Kern, J. Zur hydrodynamik der rinnsale. Verfahrenstechnik, 3, pp.425-430, 1969.
9. Kern, J. Stabilitätsproblem der rinnsalströmung. Verfahrenstechnik, 5, pp.289-294, 1971.
10. Nakagawa, T. On role of discharge in sinuosity of stream on a small plate. Natuewiss, 69, pp.142-143, 1982.
11. Nakagawa, T. & Scott, J.C. Stream meanders on a smooth hydrophobicsurface. Jour. Fluid Mech., Vol.149, pp.89-99, 1984.
12. Rayleigh, L. Proc. Roy. Soc. A10, 4, 1879.
13. Sieber, M. Experimentelle Untersuchungen zum rinnsalemäander. Max-Plank-Institut Fur Stromungsforschung, Gottingen, Bericht 5.
14. Stepanov, V.G., Volyak, L.D. & Tarlakov, Y.V. Wetting contact angles of certain systems. J. Eng. Phy., Vol.32, pp.646, 1977.
15. Tanner, W.F. Helicoidal flow, a possible cause of meandering. J. Geophys. Res., 65, pp.983, 1960.
16. Towell, G.D. & Rothfeld, L.B. Hydrodynamics of rivulet flow. A.I.Ch.E.Jour., Vol.12, No.5, pp.972-980, 1966.
17. Weiland, R.H. & Davis, S.H. Moving contact lines and rivulet instabilities. Part 2. long waves on flat rivulets. Jour.

Fluid Mech., Vol.107, pp.261-280, 1981.
18. Young, T. An essay on the cohesion of fluids. Philos. Terms. R. Soc. London, Vol.95, pp.65-87, 1805.

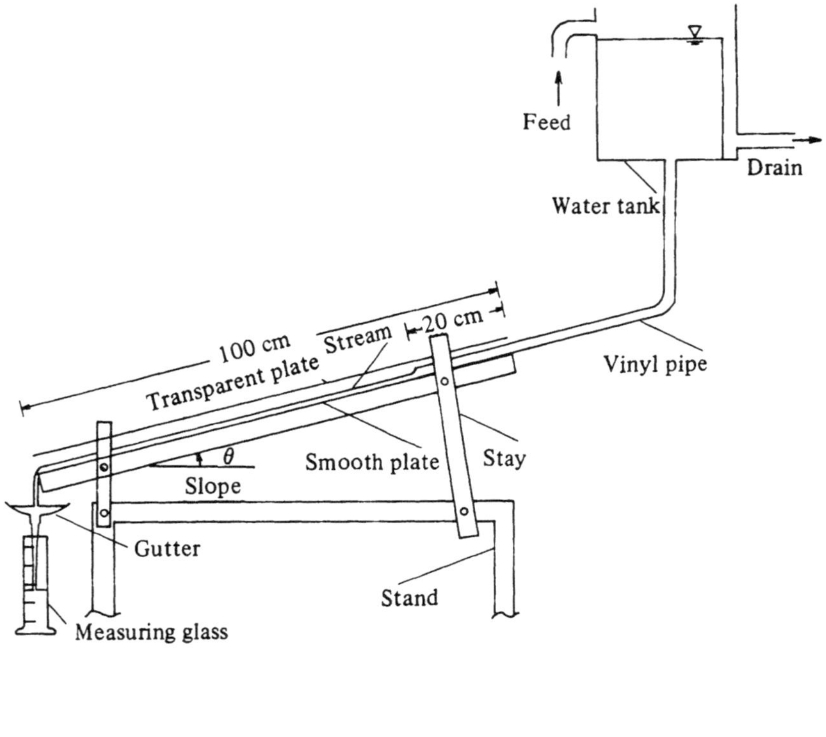

Fig. 1 Schematic diagram of the experiment

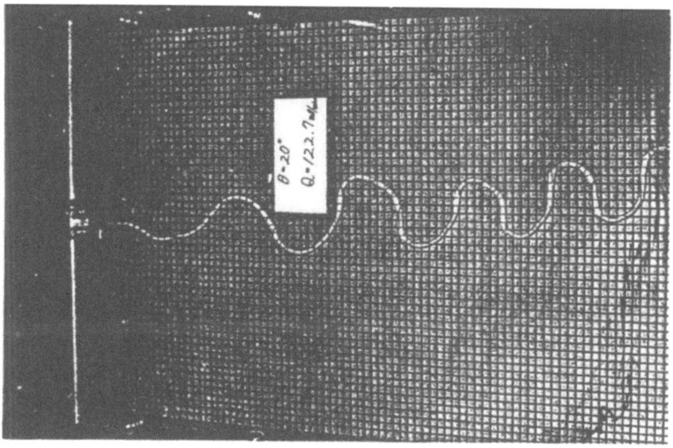

Photo 1. A stable meandering stream on the smooth plate. Discharge = 122 mℓ/sec; surface slope = 20°. The flow direction is from top to bottom of the Photo.

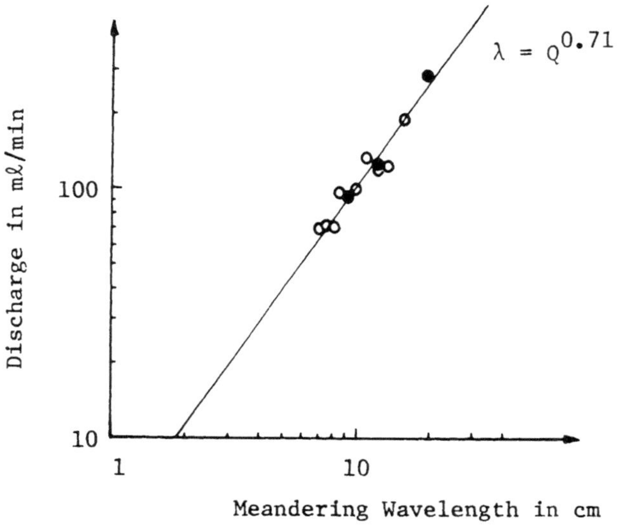

Fig. 2 Relationship between wavelength & discharge. Solid symbols show Sieber's data.

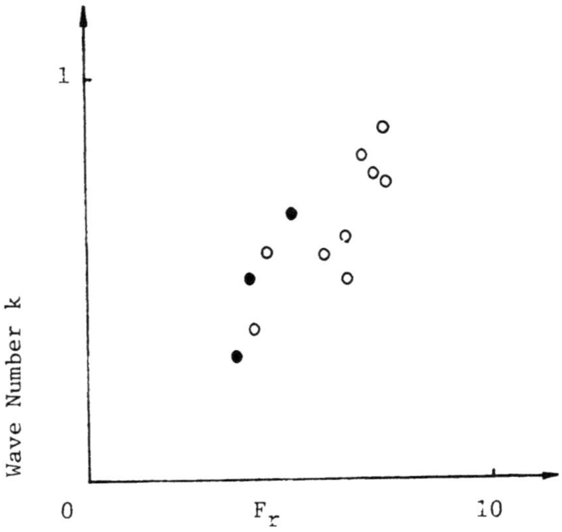

Fig. 3 Relationship between Froude number and Wave number. Solid symbols show Sieber's data.

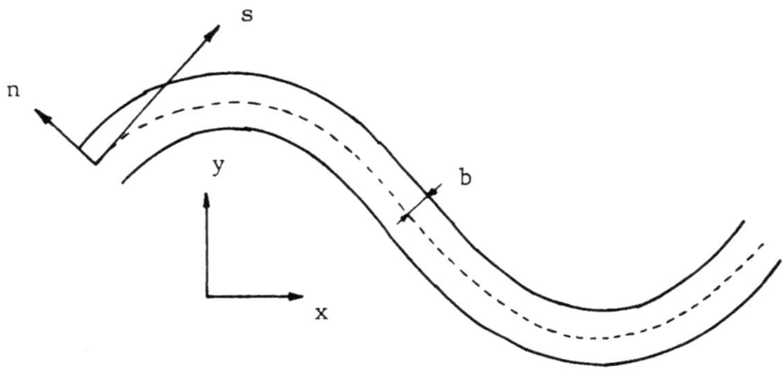

Fig. 4 Definition sketch for a sinuous rivulet

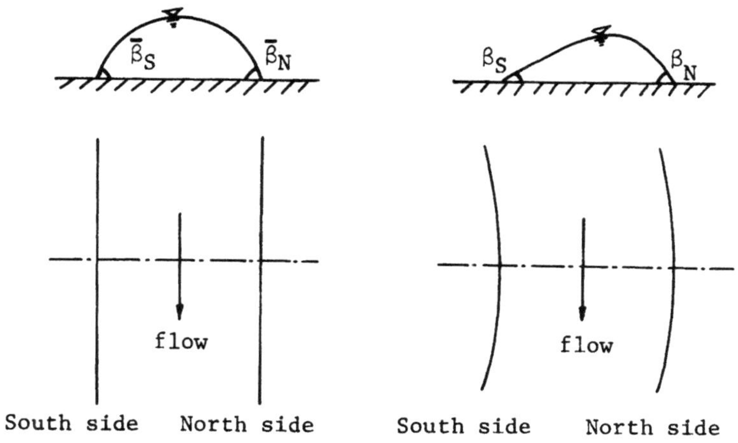

Fig. 5 Definition of contact angle

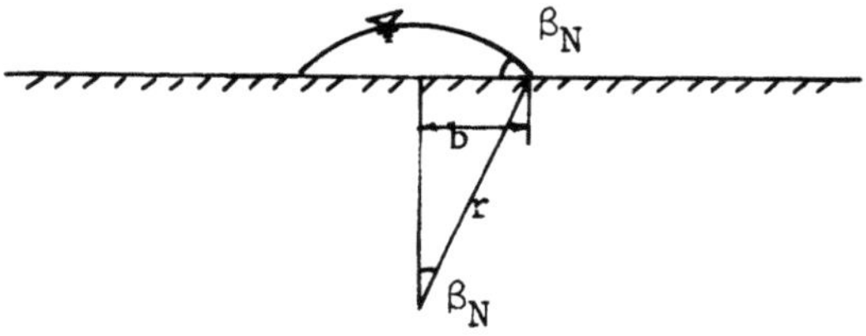

Fig. 6 Definition sketch of cross section of rivulet

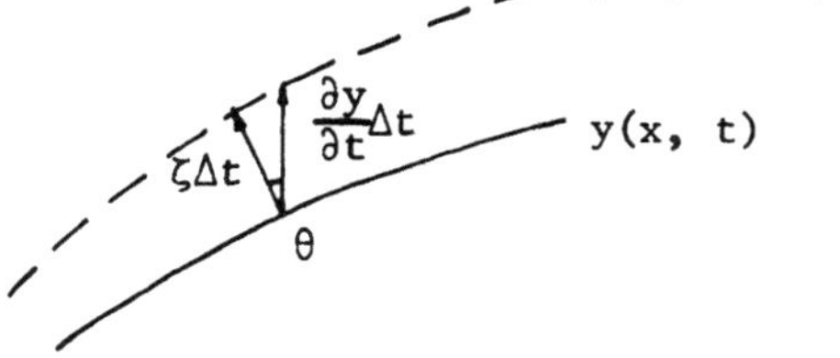

Fig. 7 Diagram for derivation of the equation of rivulet movement

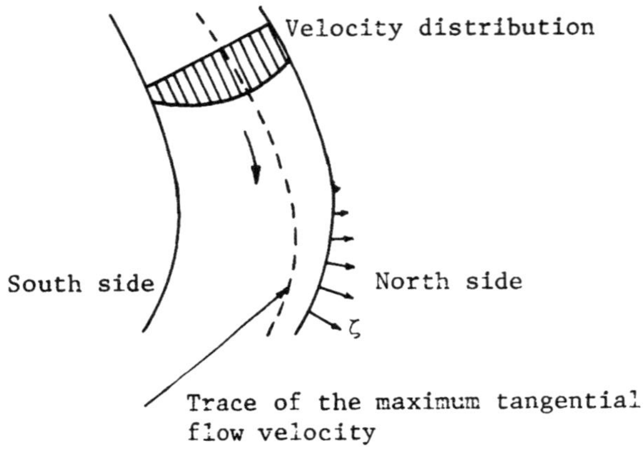

Fig. 8 Explanation for occurrence of water rivulet side

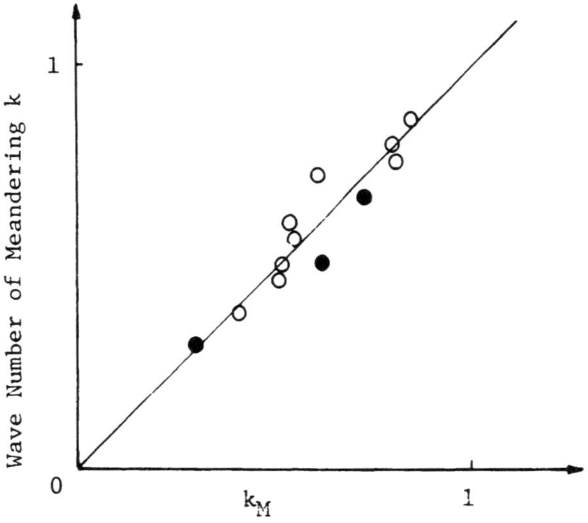

Fig. 9 Test of Eq.(29) for experimental data. Solid symbols show Sieber's data.

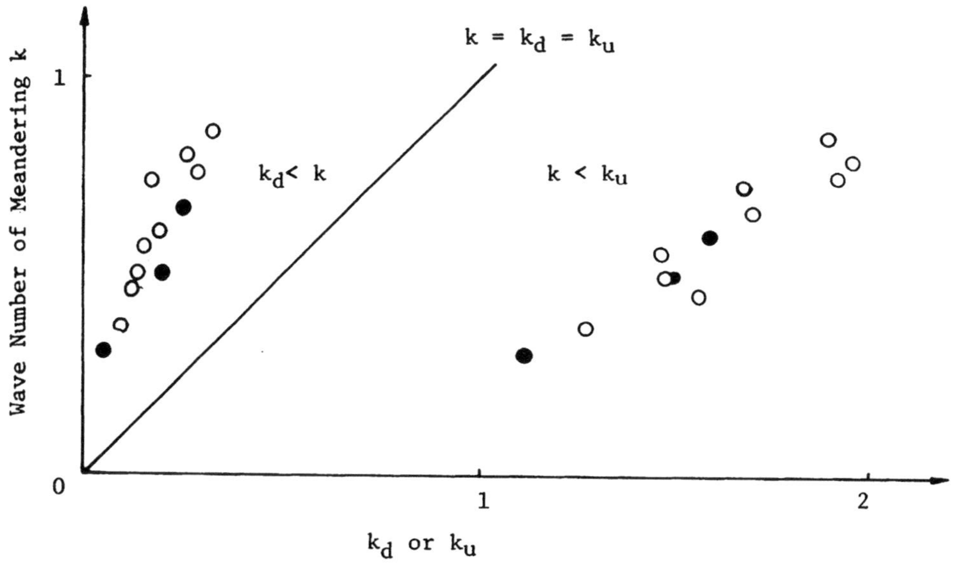

Fig. 10 Test of Eq.(28) for experimental data. Solid symbols show Sieber's data.

Shock Modeling in Variable Bubble Point Problems of Petroleum Engineering

I. Herrera (*), A. Galindo (**), R. Camacho (**)
(*) Instituto de Geofísica, National University (UNAM), Mexico City
(**) Instituto Mexicano del Petróleo and National University, Mexico City, Mexico

ABSTRACT

The treatment of discontinuities is incorporated in the modeling of a gas front which invades undersaturated liquid oil, allowing in this manner a more complete treatment of these kind of problems. In this connection, a methodology of wider applicability is introduced with the following features: Systematic formulation of jump conditions; Finite differences schemes for functions with jumps and Eulerian-Langrangian approach to front tracking. Using this approach quite satisfactory numerical results are obtained.

INTRODUCTON

Several methods have been presented in the literature for representing saturation fronts accurately. A class of such methods is based on the method of characteristics and is usually limited to miscible flow problems (Douglas[1], Douglas and Russell[2], Ewing et al.[3,4], Russell[5], Herrera and Ewing[6]). Front tracking methods (Glimm et al.[7]) have some attractive features, but in its present form they are restricted to cases in which capillary effects can be ignored (Bratvedt et al.[8]). Buckley-Leverett frontal advance theory together with a linear approximation of the fractional flow function has been applied in simple reservoir studies (Kale[9]).

For the treatment of sharp discontinuities, the equations are usually separated into pressure and saturations equations and the diffusion terms associated with capillary pressure are ignored in the saturation. Then, Buckley-Leverett Theory is applied (Buckley-Leverett[10], Kale[9]). Another approach due to Glimm[7] (Bratvedt et al.[8]), is to apply the theory of non-linear conservation laws to the resulting saturation equation.

In this paper, the modeling of discontinuities which occur when a gas front invades a region of undersaturated liquid oil is discussed and in this connection, an alternative approach to sharp fronts is presented. The general formulation is equally applicable to other sharp discontinuities ("shocks"), such as oil displaced by a water front. The main contributions are:

a).- A very systematic formulation of jump conditions, including the velocity of advancing fronts;

b).- Introduction of finite difference schemes for functions with jumps and using them, development of an Eulerian-Lagragian approach to front tracking; and

c).- A rigorous mathematical setting is given for more general problems.

The case of an advancing front of gas into undersaturated liquid oil has been treated previously (Raghavan[11] and Camacho and Raghavan[12,13]), but to our knowledge this is the first time that jumps are incorporated in the model. The modeling procedure followed up to now, which does not incorporate jumps, in what follows will be called the "traditional approach".

JUMP CONDITIONS

To give to our developments a sound and firm mathematical basis, we start from first principles. The basis of the fundamental equations that govern the flow and transport of fluids in a reservoir are the balance equations of Continuous Mechanics. The synthesis of this theory that has been given by Allen, Herrera, and Pinder[14]; and by Herrera and Allen[15], are very convenient for our purposes. In the case of multi-phase systems, each phase α

moves with its own velocity $\underline{v}^{\alpha}$. The balance laws satisfied by any intensive property ψ^{α} belonging to such phase, are:

$$\psi^{\alpha}_{t} + \nabla\cdot(\psi^{\alpha}\underline{v}^{\alpha})-\nabla\cdot\underline{\tau}^{\alpha}= g^{\alpha} \quad (1a)$$

and

$$[\psi^{\alpha}(\underline{v}^{\alpha}-\underline{v}_{\Sigma})- \underline{\tau}^{\alpha}]\cdot\underline{n} = g^{\alpha}_{\Sigma} \quad (1b)$$

Equations (1a) and (1b), are the general "differential balance law" and the general "jump condition", respectively. The vector $\underline{\tau}^{\alpha}$, is the **flux** of ψ^{α} across surfaces in space, while the quantities g^{α} and g^{α}_{Σ} represent **external supply** of ψ^{α} (Allen, Herrera and Pinder[14]; Herrera and Allen[15]), per unit volume and unit time, in the case of g^{α}, while g^{α}_{Σ} represents **external supply** of ψ^{α} through the discontinuity, per unit area and unit time.

The black oil (or beta) model that will be considered is based on the following hypothesis:

a).- There are three phases: water, liquid oil and gas;
b).- Water and oil are immiscible, while gas is soluble only in liquid oil; i.e. the water and gas phases consist of only one component, while the liquid oil is made of two components: dissolved gas and non-volatile oil.
c).- No physical diffusion is present (this includes both, molecular diffusion and that induced by the randomness of the porous media).

With each one of the components there is associated one intensive property, which represents the mass per unit volume of that component. Thus, one can write:

$$\psi^{w}=\phi S_{w}\rho_{w},\ \bar{\psi}^{o}=\phi S_{o}\bar{\rho}_{o},\ \bar{\psi}^{dg}=\phi S_{o}\bar{\rho}_{dg},\ \psi^{g}=\phi S_{g}\rho_{g} \quad (2)$$

for the intensive properties associated with water, non-volatile oil, dissolved gas and gas in the gas phase, respectively. Also, the **flux** $\underline{\tau}$ corresponding to each one of these intensive properties vanishes identically, since no physical diffusion is present. Applying Equation (1a) to each one of these intensive properties on gets

$$(\phi\, S_{w}\rho_{w})_{t} + \nabla\cdot(\phi\, \rho_{w}S_{w}\underline{v}^{w}) = 0 \quad (3)$$

$$(\phi\, S_{o}\bar{\rho}_{o})_{t} + \nabla\cdot(\phi\, \bar{\rho}_{o}S_{o}\underline{v}^{o}) = 0 \quad (4)$$

$$(\phi\, S_o \bar{\rho}_{dg})_t + \nabla\cdot(\phi\, \bar{\rho}_{dg} S_o \underline{v}^o) = g^o_{Ig} \tag{5a}$$

$$(\phi\, S_g \rho_g) + \nabla\cdot(\phi\, \rho_g S_g \underline{v}^g) = g^g_{Io} \tag{5b}$$

where no extraction terms have been included, so that for each one of the components the external supply terms vanish, except for those corresponding to the interchange of mass between gas and dissolved gas. Thus, g^o_{Ig} is the mass of gas that is dissolved in the liquid oil per unit volume per unit time, while g^g_{Io} is the mass of dissolved oil that goes into the gas phase per unit volume per unit time. Clearly

$$g^o_{Ig} + g^g_{Io} = 0 \tag{6}$$

Adding up Equations (5), one gets

$$\{\phi\, (S_o \bar{\rho}_{dg} + S_g \rho_g)\}_t + \nabla\cdot\{\phi(\bar{\rho}_{dg} S_o \underline{v}^o + \rho_g S_g \underline{v}^g)\} = 0 \tag{7}$$

Introducing the formation volume factors, Darcy's Law, as well as the relation

$$\rho = \frac{\rho_{gSTC}}{\rho_{oSTC}} R_s\, \bar{\rho}_o \tag{8}$$

the system of Equations (3), (4) and (7), becomes the familiar system of equations of black oil:

$$\nabla\cdot[\lambda_w(\nabla p_w - \gamma_w \nabla z)] = \frac{\partial}{\partial t}\left(\frac{\phi S_w}{B_w}\right) \tag{9a}$$

$$\nabla\cdot[\lambda_o(\nabla p_o - \gamma_o \nabla z)] = \frac{\partial}{\partial t}\left(\frac{\phi S_o}{B_o}\right) \tag{9b}$$

$$\nabla\cdot[R_s \lambda_o(\nabla p_o - \gamma_o \nabla z) + \lambda_g(\nabla p_g - \gamma_g \nabla z)]$$

$$= \frac{\partial}{\partial t}\left[\phi\left(R_s \frac{S_o}{B_o} + \frac{S_g}{B_g}\right)\right] \tag{9c}$$

In a similar fashion applying Equations (1b), to each one of the four components, it is obtained:

$$[\phi\, \rho_W S_W (\underline{v}^W - \underline{v}_\Sigma)]\cdot\underline{n} = 0 \tag{10a}$$

$$[\phi\, \bar{\rho}_o S_o (\underline{v}^o - \underline{v}_\Sigma)]\cdot\underline{n} = 0 \tag{10b}$$

$$[\phi\, \bar{\rho}_{dg} S_o (\underline{v}^o - \underline{v}_\Sigma)]\cdot\underline{n} = g^o_{\Sigma g} \tag{11a}$$

$$[\phi\, \rho_g S_g (\underline{v}^g - \underline{v}_\Sigma)]\cdot\underline{n} = g^g_{\Sigma o} \tag{11b}$$

In addition, Darcy's Law requires:

$$[p_l] = 0\ ;\ l = w,\ o,\ g \tag{12}$$

Above, the quantities $g^{o}_{\Sigma g}$ and $g^{g}_{\Sigma o}$ stand for the exchange of mass between the gaseous phase and the component of dissolved gas which is contained in the liquid oil phase. Mass conservation requires that:

$$g^{o}_{\Sigma g} + g^{g}_{\Sigma o} = 0 \qquad (13)$$

So that, adding up Equations (13), it is obtained:

$$[\phi\, \bar{\rho}_{dg} S_o \underline{v}^o + \phi\, \rho_g S_g \underline{v}^g]\cdot\underline{n} - [\phi\, \bar{\rho}_{dg} S_o + \phi\, \rho_g S_g]\underline{v}_\Sigma\cdot\underline{n} = 0 \qquad (14)$$

Since the pressures of the different phases are continuous, so is the porosity ϕ and ρ_w. This allows cancelling these factors in Equations (10) and (14). Thus, introducing the volume factors, they can be written as:

$$[S_w]\underline{v}_\Sigma\cdot\underline{n} - [S_w \underline{v}^w]\cdot\underline{n} = 0 \qquad (15a)$$

$$[S_o/B_o]\underline{v}_\Sigma\cdot\underline{n} - [(S_o/B_o)\underline{v}^o]\cdot\underline{n} = 0 \qquad (15b)$$

$$[R_s S_o/B_o + S_g/B_g]\underline{v}_\Sigma\cdot\underline{n} - [(R_s S_o/B_o)\underline{v}^o + (S_g/B_g)\underline{v}^g]\cdot\underline{n} = 0 \qquad (15c)$$

Equations (15), together with (12), constitute the desired system of jump conditions for the three-phase (four component) oil reservoir. They relate the jumps of the physical variables with the velocity $\underline{v}_\Sigma$ of the advancing front. Observe that, in general, the jumps of the volume factor B_o and of R_s may be non-zero since they are functions of bubble point, in addition to pressure.

THE VELOCITY OF THE ADVANCING FRONT

Thus far the discussion has been completely general. Consider now an advancing front of gas into a non-saturated liquid oil. For this case, taking the unit normal $\underline{n}$ to Σ in the opposite sense to $\underline{v}^g$ (i.e. $\underline{v}_g\cdot\underline{n} < 0$), it is seen that

$$[S_g/B_g] = S_g/B_g \qquad \text{and} \qquad [(S_g/B_g)\underline{v}^g]\cdot\underline{n} = (S_g/B_g)\underline{v}^g\cdot\underline{n} \qquad (16a)$$

and

$$[(R_s S_o/B_o)\underline{v}^o]\cdot\underline{n} = \dot{R}_s[(S_o/B_o)\underline{v}^o]\cdot\underline{n} + [R_s]\overline{(S_o/B_o)\underline{v}^o}\cdot\underline{n} \qquad (16b)$$

Writing

$$\underline{v}_\Sigma\cdot\underline{n} = \eta\, \underline{v}_g\cdot\underline{n} \qquad \underline{v}_o\cdot\underline{n} = \zeta\, \underline{v}_g\cdot\underline{n} \qquad \text{and} \qquad \omega = \frac{S_o B_g}{B_o S_g} \qquad (17)$$

Equations (15b) and (15c) can be written as :

$$\eta = 1 - [R_s]\frac{[\zeta]}{[\omega]}\omega_+\omega_- \;; \qquad \dot{\zeta} = \eta - \dot{\omega}\frac{[\zeta]}{[\omega]} \qquad (18)$$

Observe that, in view of Equation (18), η represents the factor by which the advancing gas front is slowed down with respect to the particle velocity of the gas phase. It is similar to the factor F_w' which occurs in the Buckley-Leverett frontal advance theory (Kale[9]). In general, $\eta > \dot{\zeta}$, since otherwise the gas front is not really advancing with respect to the unsaturated oil. Therefore, $\frac{[\zeta]}{[\omega]} > 0$, by virtue of the second of Equations (18).

NUMERICAL FORMULATION OF JUMP CONDITIONS

Equations (18) are informative and permit acquiring insight on the manner in which the advance of the gas front occurs. However, the jump conditions, in the manner in which they are presented in Equations (18), are not in their most convenient form to be used in numerical applications. In this Section they will be transformed into the form they will be used in the numerical formulations that follow.

Introducing the mobilities, Equations (16) can be written as:

$$[\lambda_w(\nabla p_w - \gamma_w \nabla z)]\cdot\underline{n} + \left[\frac{\phi S_w}{B_w}\right]\underline{v}_\Sigma\cdot\underline{n} = 0 \qquad (19a)$$

$$[\lambda_o(\nabla p_o - \gamma_o \nabla z)]\cdot\underline{n} + \left[\frac{\phi S_o}{B_o}\right]\underline{v}_\Sigma\cdot\underline{n} = 0 \qquad (19b)$$

$$[R_s\lambda_o(\nabla p_o - \gamma_o\nabla z) + \lambda_g(\nabla p_g - \gamma_g\nabla z)]\cdot\underline{n} + \left[\phi\left(R_s\frac{S_o}{B_o} + \frac{S_g}{B_g}\right)\right]\underline{v}_\Sigma\cdot\underline{n} = 0 \qquad (19c)$$

For simplicity, in what follows only two phases will be considered: oil and gas. In addition, only a 1-D formulation will be presented, so that gravity effects will be left out and it will be assumed that the gas moves towards the left (i.e., $v_\Sigma < 0$). In this case Equations (19) become:

$$\left[\lambda_o\frac{\partial p_o}{\partial x}\right] + \left[\frac{\phi S_o}{B_o}\right]v_\Sigma = 0 \qquad (20a)$$

$$\left[R_s\lambda_o\frac{\partial p_o}{\partial x} + \lambda_g\frac{\partial p_g}{\partial x}\right] + \left[\phi\left(R_s\frac{S_o}{B_o} + \frac{S_g}{B_g}\right)\right]v_\Sigma = 0 \quad (20b)$$

Since $\lambda_g=0$ on $\Sigma-$, it is clear that

$$\left[\lambda_g\frac{\partial p_g}{\partial x}\right] = \left(\lambda_g\frac{\partial p_g}{\partial x}\right)_{\Sigma+}$$

and Equation (20b) can be written as:

$$\left[R_s\lambda_o\frac{\partial p_o}{\partial x}\right] + \left[\phi\left(R_s\frac{S_o}{B_o} + \frac{S_g}{B_g}\right)\right]v_\Sigma = -\left(\lambda_g\frac{\partial p_g}{\partial x}\right)_{\Sigma+} \quad (21)$$

Using the general relation $[ab]=\dot{a}[b]+[a]\dot{b}$, it is obtained:

$$\dot{\lambda}_o\left[\frac{\partial p_o}{\partial x}\right] + \left[\frac{\phi S_o}{B_o}\right]v_\Sigma = -\left[\lambda_o\right]\dot{\overline{\frac{\partial p_o}{\partial x}}} \quad (22a)$$

$$\dot{\overline{R_s\lambda_o}}\left[\frac{\partial p_o}{\partial x}\right] + \left[\phi\left(R_s\frac{S_o}{B_o} + \frac{S_g}{B_g}\right)\right]v_\Sigma = -\left(\lambda_g\frac{\partial p_g}{\partial x}\right)_{\Sigma+} - \left[R_s\lambda_o\right]\dot{\overline{\frac{\partial p_o}{\partial x}}} \quad (22b)$$

Introducing the notation

$$\Delta = \left[\frac{\partial p_o}{\partial x}\right]; \qquad m = \dot{\overline{\frac{\partial p_o}{\partial x}}} \quad (23a)$$

it is seen that

$$\left(\frac{\partial p_g}{\partial x}\right)_{\Sigma+} = m + \Delta/2 \quad (23b)$$

Observe that capillary pressure has been neglected when writing Equation (23b). Then, the system of Equations (22) can be easily transformed into:

$$\dot{\lambda}_o\Delta + \left[\frac{\phi S_o}{B_o}\right]v_\Sigma + \left[\lambda_o\right]m = 0 \quad (24a)$$

$$\left(2\dot{\overline{R_s\lambda_o}} + \lambda_{g+}\right)\Delta + 2\left[\phi\left(R_s\frac{S_o}{B_o} + \frac{S_g}{B_g}\right)\right]v_\Sigma + 2\left(\lambda_{g+} + [R_s\lambda_o]\right)m = 0 \quad (24b)$$

For the numerical treatment, Equations (24) will be used as system of equations for the unknowns Δ, v_Σ and m, all of them defined on the space-time surface of discontinuity $\Sigma(t)$. However, this system is not determined since it is 2 by 3. The additional equation that is needed, will be obtained weighting the differential equations (9) in a suitable manner, as it is explained in the next Section.

THE WEIGHTED EQUATIONS

The one-dimensional version of Equations (9b and c), is:

$$\frac{\partial}{\partial x}\left(\lambda_o \frac{\partial p_o}{\partial x}\right) - \frac{\partial}{\partial t}\left(\frac{\phi S_o}{B_o}\right) = 0 \qquad (25a)$$

$$\frac{\partial}{\partial x}\left(R_s \lambda_o \frac{\partial p_o}{\partial x} + \lambda_g \frac{\partial p_g}{\partial x}\right) - \frac{\partial}{\partial t}\left\{\phi\left(R_s \frac{S_o}{B_o} + \frac{S_g}{B_g}\right)\right\} = 0 \qquad (25b)$$

In this Section these equations will weighted using convenient space-time weights, deriving in this manner equations suitable for discretization. The weights that will be chosen lead to what is essentially a cells method.

It will be assumed that $Cu = -V_\Sigma \Delta t/\Delta x < 1$. The position of the gas front will be denoted by $x_\Sigma(t)$. The space interval will be divided into a finite number of equally spaced cells. Taking time t_n as starting time, a procedure for contructing the solution at time t_{n+1} will be developed. The index "i" will be reserved to denote the cell containing the front at time t_{n+1}. Since $Cu<1$ and $V_\Sigma<0$, only two cases must be distinguished. Either, during the time interval (t_n, t_{n+1}) the gas front does not cross any inter-cell boundary (Case A); or it crosses one inter-cell boundary (Case B). Observe that in Case B the inter-cell boundary that is crossed is $x_{i+1/2}$.

Case A

The system of space-time weights to be used in cell "i" (i.e.; $[x_{i-1/2}, x_{i+1/2}]$), is:

$$w_1^o(x,t) \equiv 1 \qquad (26A)$$

$$w_2^o(x,t) = \begin{cases} 0\ , & x < x_\Sigma(t) \\ 1\ , & x > x_\Sigma(t) \end{cases} \qquad (26B)$$

for the oil and

$$w^g(x,t) = \begin{cases} 0\ , & x < x_\Sigma(t) \\ 1\ , & x > x_\Sigma(t) \end{cases} \qquad (26C)$$

for the gas. The resulting equations are:

$$\int_{t_n}^{t_{n+1}} \left(\lambda_o \frac{\partial p_o}{\partial x}\right)_{i+1/2} dt + \int_{x_{i-1/2}}^{x_{i+1/2}} \left(\phi \frac{S_o}{B_o}\right)^{n} dx =$$

$$\int_{t_n}^{t_{n+1}} \left(\lambda_o \frac{\partial p_o}{\partial x}\right)_{i-1/2} dt + \int_{x_{i-1/2}}^{x_{i+1/2}} \left(\phi \frac{S_o}{B_o}\right)^{n+1} dx \qquad (27A)$$

$$\int_{t_n}^{t_{n+1}} \left(\lambda_o \frac{\partial p_o}{\partial x}\right)_{i+1/2} dt + \int_{x_{\Sigma(t_n)}}^{x_{i+1/2}} \left(\phi \frac{S_o}{B_o}\right)^{n} dx =$$

$$\int_{t_n}^{t_{n+1}} \left(\lambda_o \frac{\partial p_o}{\partial x} + \phi \frac{S_o}{B_o} V_\Sigma\right)_{\Sigma(t)} dt + \int_{x_{\Sigma(t_{n+1})}}^{x_{i+1/2}} \left(\phi \frac{S_o}{B_o}\right)^{n+1} dx \qquad (27B)$$

$$\int_{t_n}^{t_{n+1}} \left\{R_s \lambda_o \frac{\partial p_o}{\partial x} + \lambda_g \frac{\partial p_g}{\partial x}\right\}_{i+1/2} dt + \int_{x_{\Sigma(t_n)}}^{x_{i+1/2}} \left\{\phi\left(R_s \frac{S_o}{B_o} + \frac{S_g}{B_g}\right)\right\}^{n} dx =$$

$$\int_{t_n}^{t_{n+1}} \left\{R_s \lambda_o \frac{\partial p_o}{\partial x} + \lambda_g \frac{\partial p_g}{\partial x} + \phi\left(R_s \frac{S_o}{B_o} + \frac{S_g}{B_g}\right) V_\Sigma\right\}_{\Sigma(t)} dt + \int_{x_{\Sigma(t)}}^{x_{i+1/2}} \left\{\phi\left(R_s \frac{S_o}{B_o} + \frac{S_g}{B_g}\right)\right\}^{n+1} dx$$

(27c)

In these equations the usual notation for line integrals has been used. In particular $dt=(1+V_\Sigma^2)^{-1/2}d\sigma$, where $d\sigma$ is the length in space-time.

Equations (27) constitute a system of three equations for cell "i", which must be coupled with the system of jump conditions (24) and the equations of the remaining cells (two for each cell, when a two-phase system is considered), to obtain a determined system. Assume, for the sake of definiteness, that we are solving for pressure and saturation of the gas.Cells, other than cell "i", can be treated in a standard manner, solving for example, for oil pressure and gas saturation at the center of the cell. However, cell "i" must be treated in a special manner. We have available the system of five equations constituted by Equations (24) and (27). A possible choice (and this one was used in the present study) of the corresponding five unknowns is: the jump and average of the pressure gradient (Δ and m, respectively), the velocity v_Σ of the gas front (these three unknown functions defined on $\Sigma(t)$), the saturation and the pressure. Then, one can solve the value p_{oi}^{n+1} at the center of cell "i" and the value $S_{g\Sigma}^{n+1}$ of gas saturation at the gas front, together with Δ^{n+1}, m^{n+1} and $\underline{v}_\Sigma^{n+1}$. A special feature of this procedure is that one does not solve for saturation at the center of cell "i".

Case B

The system of space-time weights to be used in cell "i" (i.e.; $[x_{i-1/2}, x_{i+1/2}]$), is:

$$w_1^o(x,t) \equiv 1 \tag{28A}$$

for the oil and

$$w_1^g(x,t) = \begin{cases} 0, & x < x_\Sigma(t) \\ 1, & x > x_\Sigma(t) \end{cases} \tag{28B}$$

for the gas. The system of space-time weights to be used in cell "i+1" (i.e.; $[x_{i+1/2}, x_{i+3/2}]$), is:

$$w_2^o(x,t) \equiv 1 \tag{28C}$$

for the oil and

$$w_2^g(x,t) = \begin{cases} 0, & x < x_\Sigma(t) \\ 1, & x > x_\Sigma(t) \end{cases} \tag{28D}$$

for the gas. In addition, a space-time weight with support in the union of cells "i" and "i+1", is applied to the oil equation. It is defined by

$$w_3^o(x,t) = \begin{cases} 0, & x < x_\Sigma(t) \\ 1, & x > x_\Sigma(t) \end{cases} \tag{28E}$$

Corresponding to these five weighting functions, five equations are derived for the union of cells "i" and "i+1". Putting them together with the jump conditions (Equations 24) a system of seven equations is obtained. The corresponding seven unknowns selected for the applications of the present paper were:

the jump and average of the pressure gradient at the gas front at time t_{n+1}, the velocity of the advancing front $\underline{v}_{\Sigma}$, the values p^{n+1}_{oi} and p^{n+1}_{oi+1}, of the oil pressure at cells "i" and "i+1", respectively, and the saturations of the gas S^{n+1}_{gi+1} and $S^{n+1}_{g\Sigma}$ at the center of cell "i+1" and at the gas front, respectively. Again, as in Case A, a special feature of the procedure is that one does not solve for the saturation at the center of cell "i".

NUMERICAL RESULTS

As a test for the formulation presented in this paper, the numerical results that were obtained using it in a black oil simulator, neglecting capillary pressure, are presented.

A linear reservoir is considered, producing at a constant oil rate of 800 m^3/day at the left boundary, and a no flow boundary at the right end (L = 550 meters) . This reservoir is composed of two zones, a left undersaturated oil zone and a right saturated zone, where the oil and gas phases coexist. Initially, the pressure is the same for all positions, S_g = 0.7 in the saturated zone, and the front is located at x= 275 meters.

The PVT properties for the oil phase consist of constant viscosity and two values of B_o and R_s, one at the left side of the front and another at the right side, respectively. For the gas phase the PVT properties used, are typical in the oil industry. Straight line relationships are used for the relative permeabilities.

In order to check the validity of the simulator results, we have considered two cases: firstly, the undersaturated zone was taken as incompressible, and secondly, a non-zero value was given to the rock compressibility.

The analytical solution for the first case in the undersaturated zone dictates a constant pressure gradient. Figure 1, shows a comparison of pressure profiles at different times for this case. Two kinds of profiles are presented. The solid lines correspond to the solution obtained when the proposed formulation is included, and the dashed lines correspond to the traditional formulation in which jumps are not considered. As already mentioned, the pressure gradient is constant in the incompressible zone.

To correlate the transient (one phase) solution for a slightly compressible liquid with a multiphase flow solution, Raghavan[11] and, Camacho and Raghavan[12,13], have proposed the use of a function of pressure and saturation, called pseudopressure, which is defined as follows:

$$m(p) = \frac{k\ h\ W}{q} \int_{x=0}^{x=L} \left[\frac{k_{ro}}{\mu_o\ B_o}\ \frac{\partial P}{\partial x} \right] dx \tag{29}$$

For a linear system containing a slightly compressible liquid, Nabor and Barham[16] have shown that the pressure drop behaves linearly with time in a log-log plot, with a slope of one-half during the transient period. Once the outer boundary is manifested in the response, the solution deviates from the straight line, moving up for a closed outer boundary and achieving a constant value, whena constant pressure outer boundary condition is prescribed. When, as in the traditional approach, the jumps are not considered in the formulation, the pseudopressure defined by Equation (29) will deviate below the straight line for a closed outer boundary, after the transient period has ended. To over come this limitation, Camacho and Raghavan[12,13] have suggested to include an additional integral in the right hand side of Equation (29). In this manner, variations in average properties are taken into account.

For a constant pressure (constant saturation) outer boundary, Equation (29) yields a constant value, equal to the position of this boundary, after the transient period has ended. Figure 2 shows the behavior of the pseudopressure for the compressible case. The solid line corresponds to the solution obtained when our formulation for the jumps is used, and the dots correspond to the traditional formulation. It can be observed the presence a straight line with one-half slope during the transient period, in the solution for both formulations. Observe that when the jumps are considered in the formulation, the pseudo pressure is equal to the position of the front, since this has a similar effect to a constant pressure outer boundary. For the traditional formulation, the pseudopressure falls below because of the reason explained above.

The behavior of pressure at x=0 is shown in Figure 3. In Figure 4 it is presented a comparison of pressure profiles at different times. The results of Figures 3 and 4 correspond to those of Figure 2. The solid lines represent the proposed formulation and the dots the traditional formulation.

Figures 5 to 8 show the behavior of v_Σ ,m , Δ, y x_Σ versus time, respectively, for both compressible and incompressible inner zone cases. The solid lines represent the compressible case and the dots the incompressible one.

REFERENCES

1. Douglas, Jr. J. "Simulation of Miscible Displacement in Porous Media by a Modified Method of Characteristic Procedure". Numerical Analysis, Dundee 1981, Lecture Notes in Mathematics, 912, Springer-Verlag, Berlin 1982

2. Douglas, Jr., J. and Russell, T.F.: "Numerical Methods for Convection Dominated Diffussion Problems Based on Combining the Method of Characteristic with Finite Difference or Finite Difference or Finite Element Procedures". SIAM J. Numer. Anal., 19(5), (1982).

3. Ewing, R.E., Russell, T.F. and Wheeler, M.F.: "Simulation of Miscible Displacement Using Mixed Methods and a Modified Method of Characteristics", paper SPE 12241 presented at the 1983 Reservoir Simulation Symposium, San Francisco, November 1983.

4. Ewing, R.E., Russell, T.F. and Wheeler, M.F.: "Convergence Analysis of an Approximation of Miscible Displacement in Porous Media by Mixed Finite Elements and a Modified Method of Characteristics," Computer Methods in Applied Mechanics and Engineering 47, 1984, 73-92

5. Russell, T.F.: "Finite Elements with Characteristics for Two-Component Incompressible Miscible Displacement", paper SPE 10500 presented at the Sixth SPE Symposium on Reservoir Simulation, New Orleans, January-February 1982.

6. Herrera, I. and Ewing, R.E. "Localized Adjoint Methods: Applications to Multiphase Flow Problems". Proceedings Fifth Wyoming Enhanced Oil Recovery Symposium, May 10-11, 1989, Casper Wyoming, pp 155-173, 1990

7. Glimm, J., Lindquist, B., McBryan, O.M., Plohr, B, and Yaniv, S: "Front Tracking for Petroleum Reservoir Simulation", paper

SPE 12238 presented at the SPE 1983 Reservoir Simulation Symposium, San Francisco, November 15-18.

8. Bradvedt, F. Bradvedt, and K. Buchholz, C.F., "A New Front Tracking Method for Reservoir Simulation", Paper SPE 19805 presented at the 64th SPE Annual Technical Conference and Exhibition at San Antonio Texas, October 8-11, 1989, 343-355.

9. Bluckley-Levert., "Dynamics of Moving Interfaces in Porous Media: III Extension of Buckley-Levertt Theory", paper SPE 11249 presented at the 57th Annual Fall Technical Conference and Exhibition of the Society of Petrolum Engineers of AIME, Dallas Texas September 26-29, 1982.

10. Buckley, S.E. and Leverett, M.C.: "Mechanics of Fluid displacement in sands". Trans., AIME (1942) 146.

11. Raghavan, R., "Pressure Transient Analysis of a Vertically Fractured Well Produced by Solution Gas Drive", Soc. Pet. Eng. J. (1977), 369-376.

12. Camacho-V., R.G. and Raghavan, R., "Some Theoretical Results Useful in Analyzing Well Performance Under Solution Gas Drive", Paper SPE 16580.

13. Camacho-V., R.G. and Raghavan, R., "Boundary-Dominated Flow in Solution Gas-Drive Reservoirs", SPE Reservoir Engineering, Dec. 1989.

14. Allen, B.M, Herrera I. and Pinder. G.F., "Numerical Modeling in Science and Engineering", John Wiley & Sons, New York, 1988.

15. Herrera, I. and Allen, M.B., "Modelación Computacional de Sistemas en Ciencias e Ingeniería", Comunicación Técnica No 9, Serie Docencia e Investigación,Instituto de Geofísica, UNAM 1986.

16. Nabor, G.W. and Barham, R.H., "Linear Aquifer Behavior", JPT, May, 1964, 561-563.

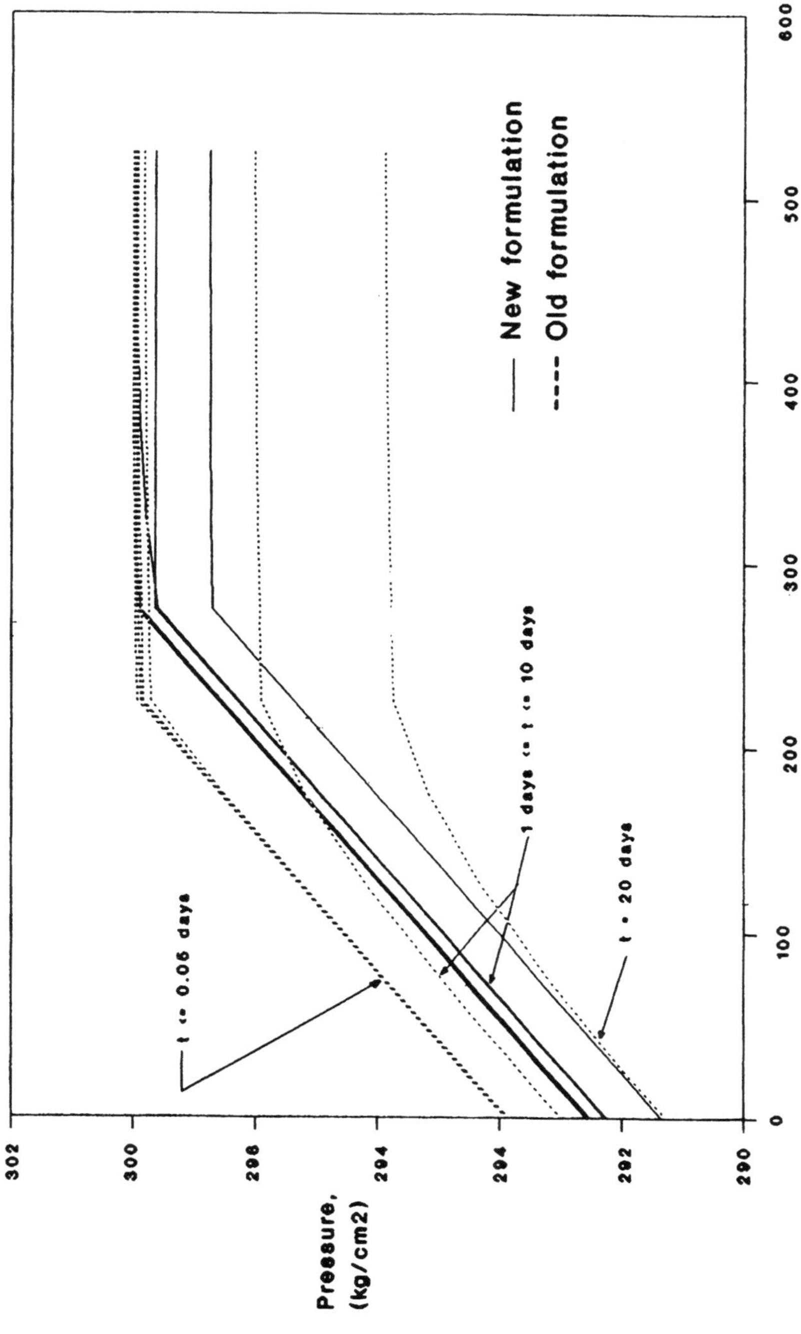

Fig. 1 Comparison of pressure profiles for new and old formulation, incompressible case.

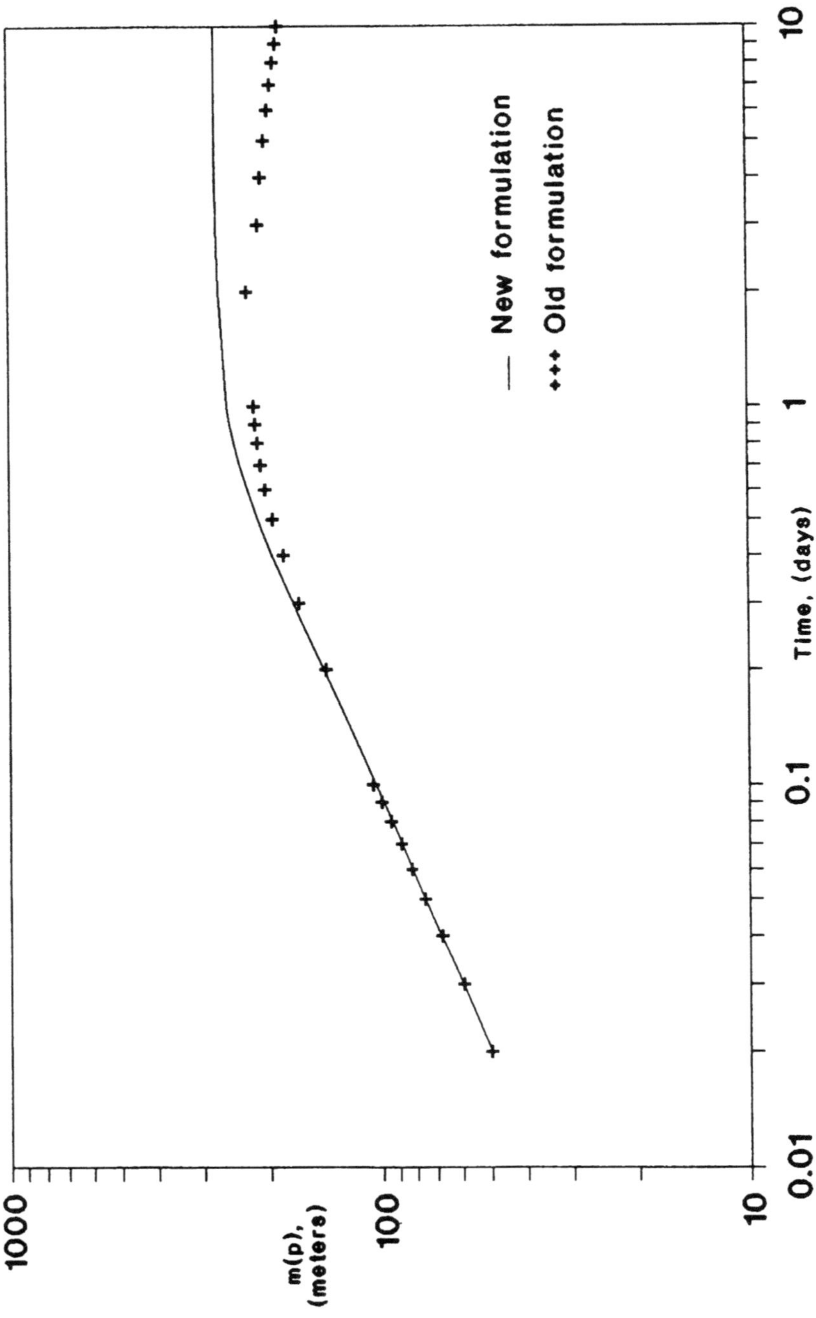

Fig. 2 Comparison of pseudopressure behavior for new and old formulations.

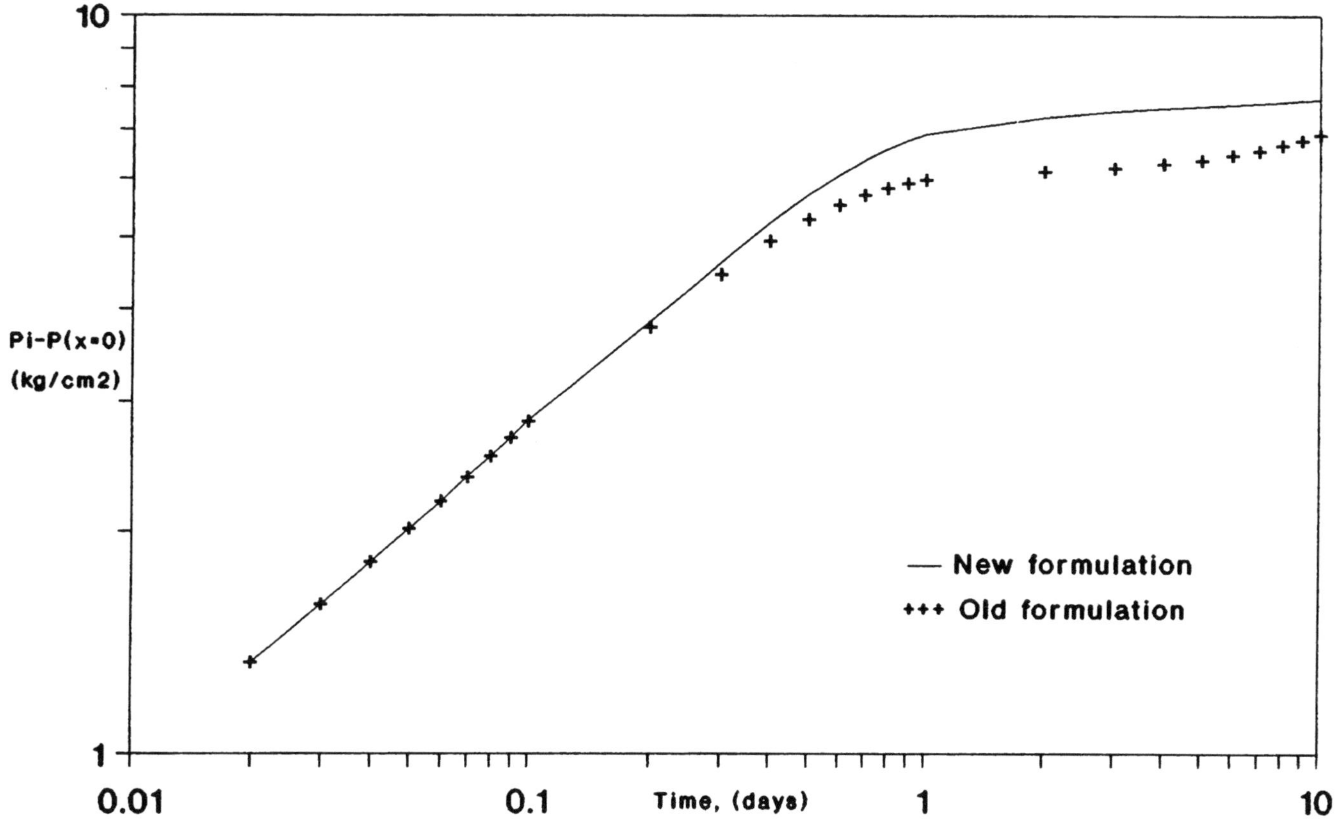

Fig. 3 Comparison on pressure drop behavior for new and old formulations.

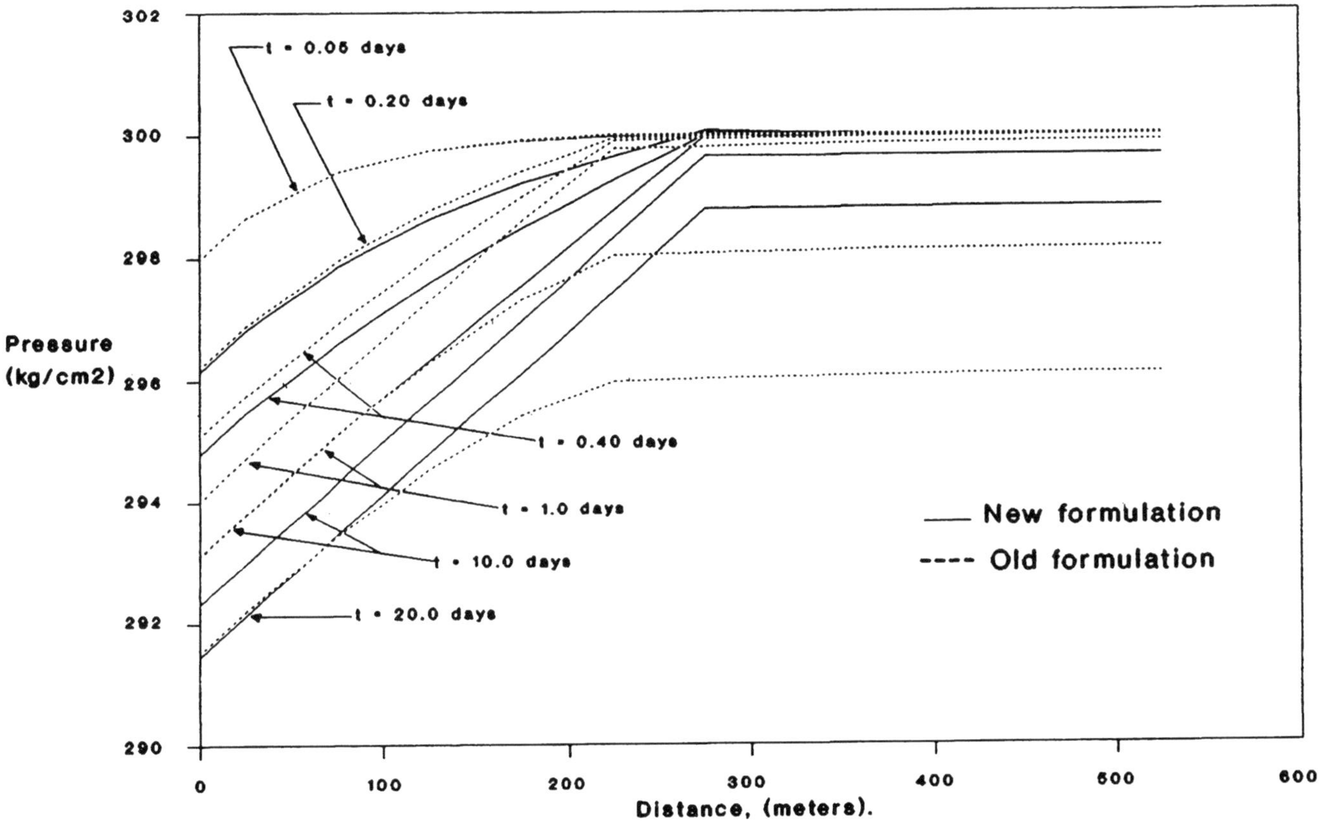

Fig. 4 Comparison of pressure profiles for new and old formulations compressible case.

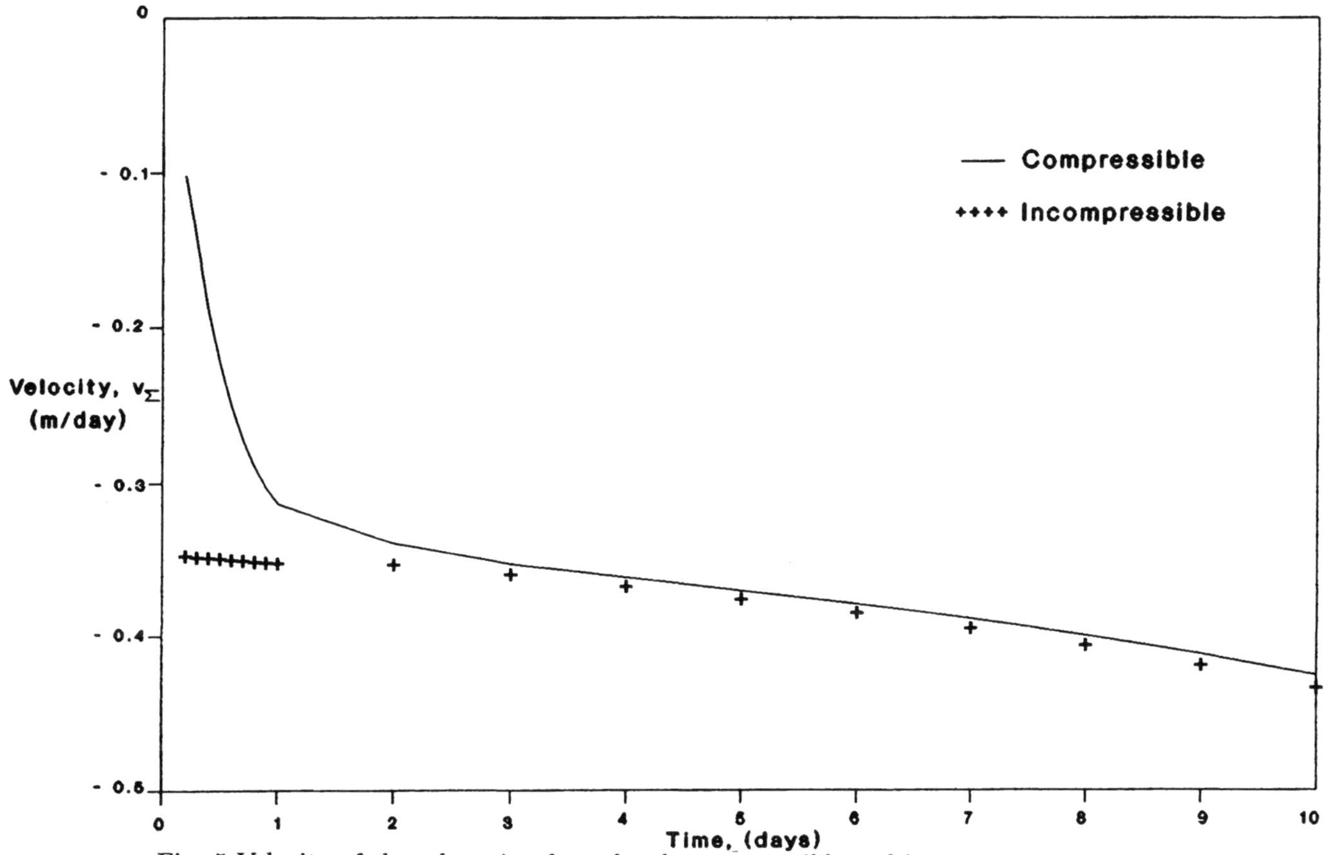

Fig. 5 Velocity of the advancing front for the compressible and incompressible cases.

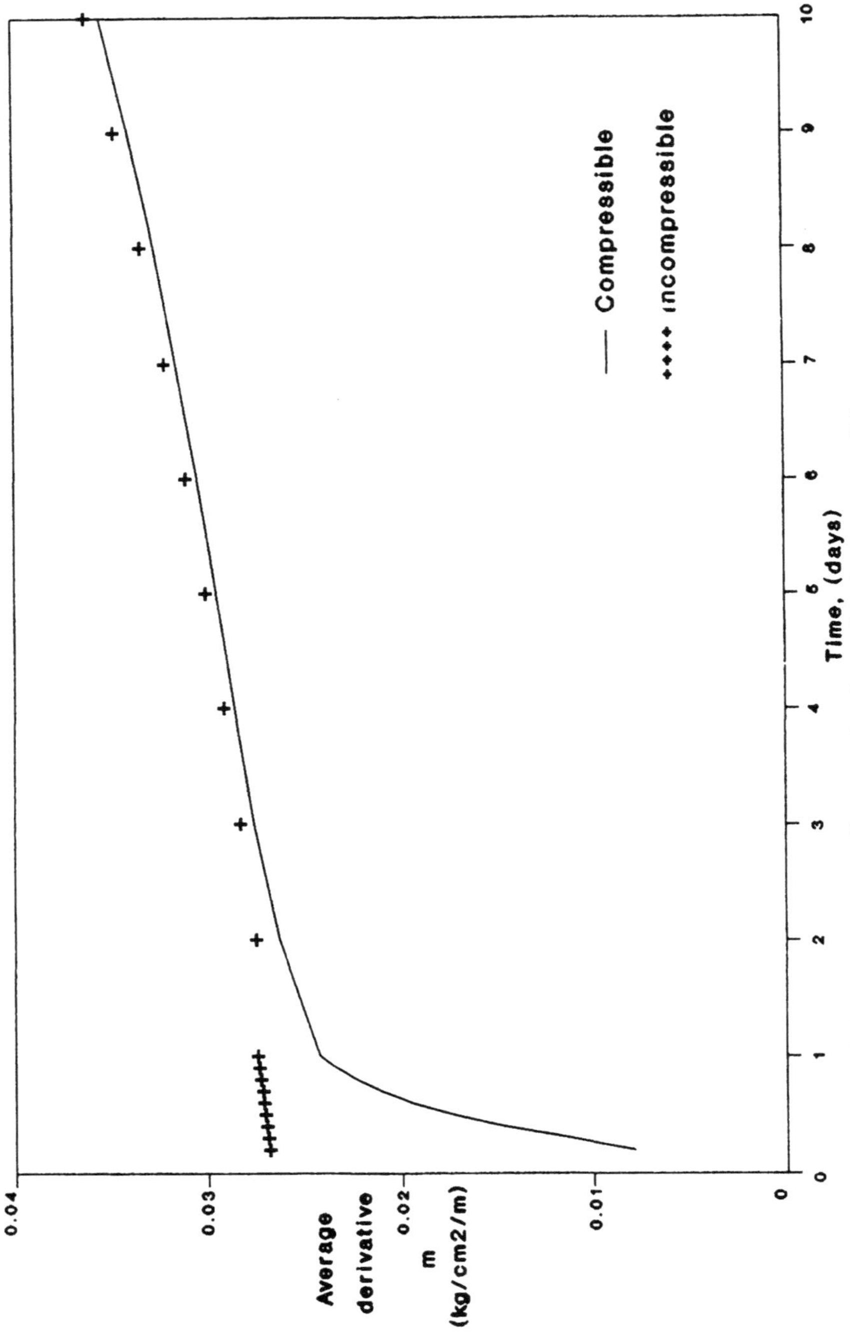

Fig. 6 Average of the pressure gradient at the front for the compressible and incompressible cases.

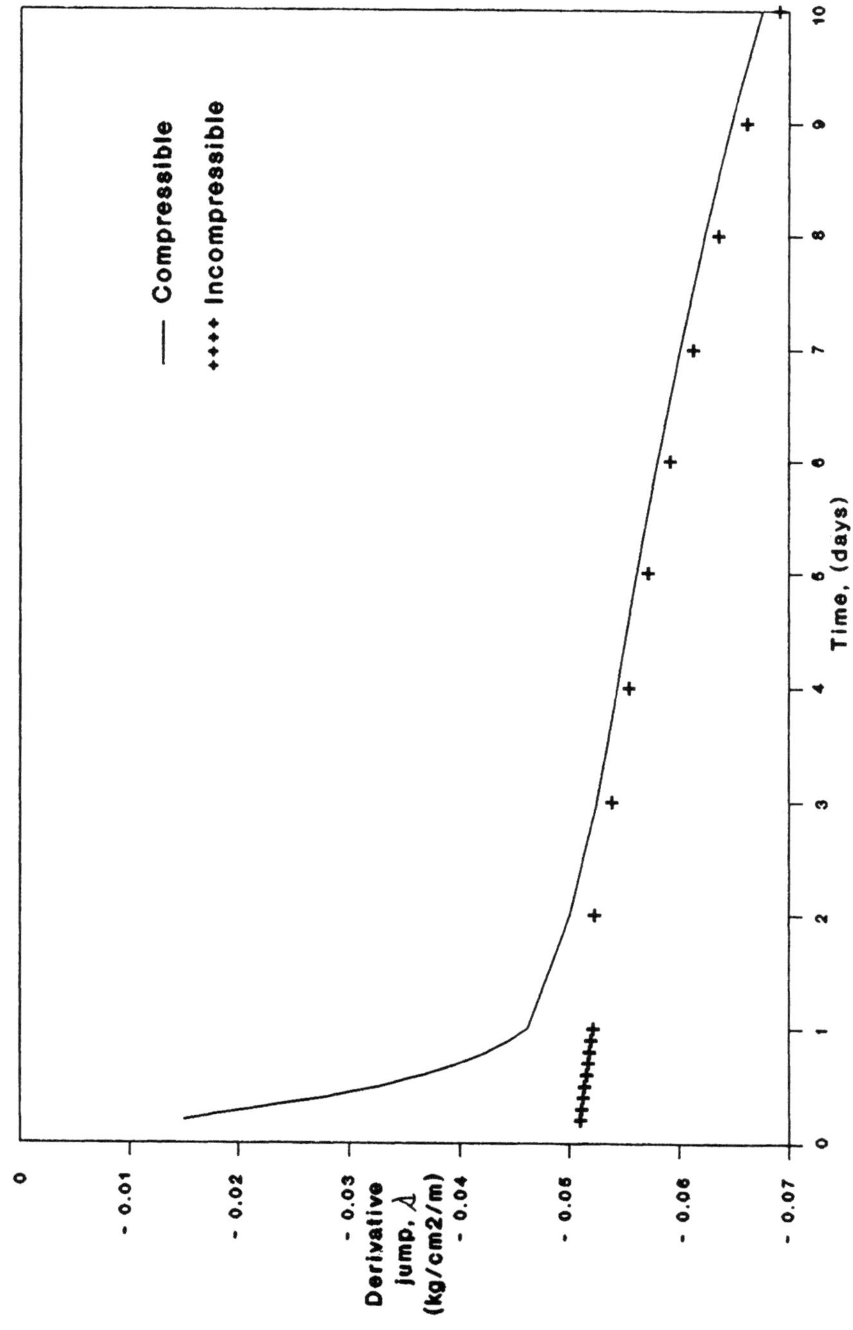

Fig. 7 Jump in pressure gradient at the front for the compressible and incompressible case.

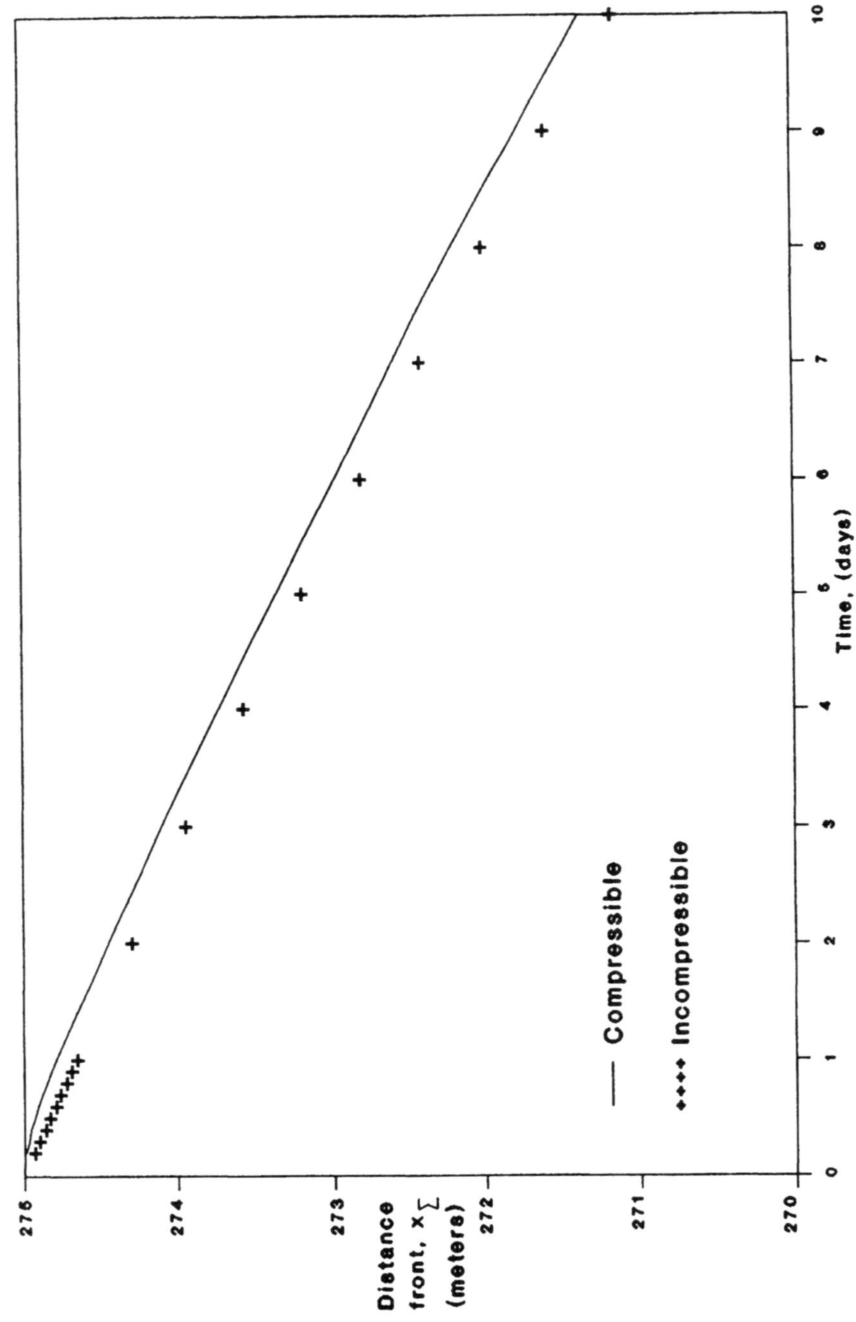

Fig. 8 Position of the front for the compressible and incompressible cases.

Multidimensional Modelling of Uniflow Scavenging in Two-stroke Reciprocating Piston Engines

M.R. Ravi, A.G. Marathe

Department of Mechanical Engineering, Indian Institute of Science, Bangalore, India

Introduction

Design of internal combustion cylinders has always been on the basis of qualitative simplified theoretical analysis and a trial and error process of fabrication and testing. Realistic numerical predictions of actual processes taking place inside the cylinder of an internal combustion engine are extremely complex to understand, comprehend, model and predict. The developments in the areas of turbulent flow and combustion modelling and the advent of powerful computers has rendered the goal of predicting the in-cylinder processes in an internal combustion engine a realizable one.

The flow processes in an internal combustion engine are complex because of their inherent three dimensionality. Besides, no unified models exist that can consistently represent realistic turbulent flows and combustion. Many research groups all over the world have worked for over two decades on the multidimensional modelling and simulation of flow and combustion in internal combustion engine cylinders. The research group in Los Alamos National Laboratories of the U.S.A. has developed a series of computer programs to simulate the internal combustion engine flows. The most versatile of these are the CONCHAS - SPRAY [1] and the KIVA [2,3]. The former solves multicomponent, chemically reactive flows with fuel sprays in arbitrary shaped two-dimensional (axisymmetric) computational domains. The latter is the three-dimensional version of the former, and has a few improvements in the treatment of convective and acoustic terms in the momentum equations. Gosman et al [4,5,6] of the Imperial College of Science and Technology, U.K., have developed an implicit solution methodology called RPM (Reciprocating Piston Motion) for two- and three- dimensional turbulent flows in arbitrary geometries with chemical reactions and fuel sprays. Errera et al [7,8] have developed a computer program called MATHILDA to solve multidimensional fluid flow problems in complicated geometries like cylinder – inlet manifold – valve combinations.

Two-stroke engines have an additional feature of simultaneous intake and exhaust process, i.e., the *scavenging* process. Diwakar [9,10] has studied the effect of inlet swirl on the scavenging characteristics of the General Motors EMD 710 turbocharged two-stroke diesel engine [11], using the two- and three dimensional approaches. Uzkan [12,13,14] has studied the effects of inlet swirl, engine stroke and engine speed respectively, on the scavenging process in the same engine geometry, using the CONCHAS – SPRAY computer program [1]. In all the above works, the initial thermodynamic state and the initial velocity distributions inside the cylinder are assumed at the exhaust valve opening (EVO) timing, and the simulation is carried out for the scavenging period alone. Ravi et al [15,16] have pointed out that the initial thermodynamic state of the cylinder contents at EVO affects the scavenging characteristics considerably, and hence, the choice of the initial thermodynamic state cannot be arbitrary for realistic prediction of scavenging characteristics.

This paper presents the method used in the computer program *CARE* (Cyclic Analysis of Reciprocating Engines) developed by the authors [15] and its salient features. A cylinder geometry as shown in figure 1 is used for the present computations. This geometry is the same as that used by Diwakar [9] and Uzkan [12,13,14], except that the piston bowl is absent. The inlet and exhaust manifold pressures are also the same as those used in [9,10,12,13,14]. The engine dimensions and operating conditions are as follows:

Bore	:	0.2302m
Stroke	:	0.2794m
Connecting rod length	:	0.5842m
Exhaust valve opening (EVO)	:	105° ATDC
Inlet port opening (IPO)	:	135° ATDC
Inlet port closing (IPC)	:	225° ATDC
Exhaust valve closing(EVC)	:	239° ATDC
Compression ratio	:	16
Inlet pressure	:	0.26104 MPa
Exhaust pressure	:	0.23904 MPa
Inlet temperature	:	379 K

The inlet gases are assumed to be a homogeneous mixture of air and octane at stoichiometric proportions, and for simplicity, their diffusion coefficients are assumed to be equal.

Method of Analysis

Global and Local Features

The fluid flow and gas exchange processes in an internal combustion engine cylinder are primarily governed by the flow through the inlet and the exhaust ports and the reciprocating motion of the piston. The flow rates across the inlet and exhaust ports are functions solely of the pressure histories inside the cylinder and at the inlet and exhaust manifolds. These pressure histories are functions mainly of time, and are relatively insensitive to the changes in details of the flow field inside the cylinder.

This gives rise to a distinct possibility of dividing the problem of engine cycle simulation into two parts:

- the global part, which computes flow caused at the inlet and exhaust ports by the pressure differences at the inlet and exhaust, and
- the local part, which computes the distribution of the flow field variables inside the cylinder.

The global part is solely time-dependent, while the local part solves partial differential equations governing the conservation of mass, momentum, energy and species masses to determine the spatial distribution of flow field variables at various levels of time.

This kind of division of the problem is strictly valid only if the cylinder can be assumed to be at uniform pressure. This assumption is valid only when the flow inside the cylinder is incompressible. The fact that the density of the gases inside the cylinder is a function of space and time owing to the effects of piston action and spatial gradients of mass fractions and temperature makes the flow appear compressible. But compressibility of the flow is determined only by the magnitude of the velocities inside the cylinder and their ratio with the velocity of pressure waves (acoustic velocity) at the local temperature and composition. The velocities inside the cylinder during gas exchange seldom exceed 100 m/s except at the exhaust port, while the acoustic velocity at the local temperature is usually higher than 800 m/s. This means the largest Mach number (ratio of fluid velocity to acoustic velocity) encountered inside the cylinder is less than 0.125, and the flow can well be treated as incompressible at such low Mach numbers.

Once the pressure phenomena have been isolated, we have an incompressible fluid flow problem to solve. The governing conservation equations for mass, momentum, energy and species masses are solved using the inflow and outflow boundary conditions derived from the global part of the problem. A discussion on the global and local parts of the algorithm used in the *CARE* computer program are given in detail in the succeeding sections.

Global Part of the Problem

This part of the solution algorithm is so named because it governs the total mass and energy flow into and out of the cylinder, without much dependence on the detailed flow field information. It accounts for the following processes:

- Flow across the inlet and exhaust ports by virtue of the inlet-to-cylinder and cylinder-to-exhaust pressure differences.
- Work done by or on the cylinder contents due to piston motion and
- Heat added and composition change caused by combustion.

The flow across the inlet or exhaust ports is computed using the compressible orifice flow equations which provide a relation between the pressure difference, area of the port and the mass flow rate. Even though the flow inside the cylinder is incompressible at large, the velocity at the inlet and exhaust ports could be high for certain time intervals during which the pressure difference is high. If the compressibility caused by such high velocities is not accounted for, one would slightly overpredict the mass flow across the ports. In the present analysis, the inlet and exhaust manifold pressures are assumed to be constant, and only the cylinder pressure varies with time, giving rise to time-variation of the inlet and exhaust mass flows.

The time step value needed for a compressible orifice flow equation to be applied to the intake and exhaust flows is governed by the CFL criterion, based on the acoustic velocity at mean cylinder temperature and the cylinder dimensions. But the fluid flow computations are subsonic, and at very low Mach numbers. Thus, the time step used for fluid flow computations can be much larger than that used for inflow/outflow conditions. This is incorporated by allowing for many sub-time steps to compute the mass flow and pressure history within one fluid flow time step [15,16].

At every sub-time step, the cylinder pressure is held at its value at the end of the previous sub-time step, and mass flow into and out of the cylinder are computed using the orifice flow equations, using a constant discharge coefficient of 0.8 for the inlet port and 0.6 for the exhaust port. This would give the new value of the mass in the cylinder. The new mean density can be calculated knowing the new volume of the cylinder.

The energy balance on the cylinder needs to account for piston work and heat release. The cylinder is assumed to be adiabatic, and the piston work is computed as a thermodynamic compression or expansion work. Heat release due to combustion is modelled by assuming instantaneous conversion of a stoichiometric proportion of fuel and air in any grid control volume in the cylinder into a stoichiometric quantity of the product, with proportionate heat release, when the cell temperature exceeds 650°C. The change in energy content of each control volume is converted into equivalent temperature rise and added to the corresponding temperature values. Energy change due to piston work is also modelled in a similar fashion. Mass fractions of species are altered on control volumes where the "Combustion" has taken place. The pressure inside the cylinder is then computed using the equation of state.

Local Part of the Problem

This part of the algorithm solves the governing equations of conservation of mass, momentum, energy and species masses. It uses the boundary conditions for mass and energy influx and efflux from the output of the global part to obtain the flow field distribution of velocities, pressure, temperature and mass fractions of species. After delinking the global part, the local part is just an incompressible transient fluid flow problem in a cavity with one solid boundary (the piston face) moving back and forth periodically and the inflow and outflow boundaries of various sizes opening and closing at various time intervals.

The inlet port is modelled as a sleeve opening all around the cylinder liner, while the exhaust valve is located centrally on the cylinder head. The opening of the exhaust valve is assumed to be in the form of an annular slot letting only axial outflow, and valve lift into the cylinder is ignored. All cylinder walls, piston face, cylinder head and the exhaust valve are assumed to be thermally insulated.

Treatment of Moving Boundary

The movement of the piston into and out of the cylinder makes the computational domain expand and contract as a function of time. Since the piston position and its velocity are solely functions of time, the axial coordinate can be transformed to make the computational domain have stationary boundaries, i.e., boundaries with fixed coordinates. If $\delta(t)$ is the instantaneous distance from piston face to cylinder head, then, $\eta = \frac{z}{\delta(t)}$ normalizes the axial coordinate to values between 0.0 and 1.0, at all points of time. This means that any grid line represented by $\eta = \eta_1$ moves with a velocity proportionate to the instantaneous piston velocity. If $\eta = 0$ represents cylinder head, which is stationary, and $\eta = 1$ represents piston face, which moves with a velocity $w_p(t)$ at any given time t, then any grid line in between the two, parallel to them, moves with a velocity $\eta w_p(t)$. Thus, for advective fluxes in the axial direction, one has to consider the flow velocity relative to the grid line velocity. Defining a relative velocity $w_1 = w - \eta w_p(t)$, we must use w_1 in place of w in advection flux terms in all conservation equations.

Since the physical domain is expanding and contracting with time, it is not possible to ensure one grid line to coincide with the interface between solid and open boundaries at the inlet port, at all time steps. In other words, when the piston moves past the inlet port, the grid lines also move along with it, and so, the control volumes near the $r = r_0$ boundary (r_0 = cylinder bore radius) would have part of their boundary surface as open boundary and part of it as solid wall boundary. This has been taken care of in calculating the fluxes through these boundary control surfaces. Every boundary control surface is checked at every time step for such an occurrence and the computation of boundary fluxes is made accordingly. For exhaust ports, whose locations are fixed, and no grid lines move past them, no such treatment is necessary. However, to avoid the limitation on the smallest grid spacing due to the size of the exhaust port opening, a similar treatment is used for exhaust port as for the inlet port. Thus, the *CARE* computer program gives qualitatively correct trend and quantitatively reasonable accuracy even with a grid as coarse as 5 x 5 or 10 x 10.

Governing Equations

The flow inside the cylinder of an engine is essentially turbulent, especially due to the inlet flow and its mixing with the contents of the cylinder. Though the Navier-Stokes equations can represent the in-cylinder turbulence fully, the grid and time step resolutions needed by such a full simulation would be enormous. Hence it was chosen to use the Density Weighted Ensemble Averaged (DWEA) Navier-Stokes equations, with an eddy viscosity model to represent turbulent mixing.

The equations are averaged and the velocities, density, temperature and other variables are split into the mean and the fluctuating components. All the governing equations would be written in terms of the mean component, and the effects of fluctuating components are modelled for. We can, without ambiguity, choose to drop the notation of *mean* component, and denote the mean components by the same names that would be used for the unsplit variables, for convenience. For example, the notation u in the following equations means the density weighted ensemble average of the radial velocity, and the instantaneous radial velocity does not appear anywhere in the equations.

With the above background, we can write the equations of conservation of mass and momentum as follows, in the conservation law form. These equations are derived for the axisymmetric $(r - \eta)$ plane.

(i) Conservation of mass:

$$\frac{1}{\delta}\frac{\partial(\rho\delta)}{\partial t} + \frac{1}{r}\frac{\partial(\rho r u)}{\partial r} + \frac{1}{\delta}\frac{\partial(\rho w_1)}{\partial \eta} = 0 \tag{1}$$

(ii) Conservation of radial momentum:

$$\frac{1}{\delta}\frac{\partial(\rho u\delta)}{\partial t} + \frac{1}{r}\frac{\partial(\rho r u u)}{\partial r} + \frac{1}{\delta}\frac{\partial(\rho w_1 u)}{\partial \eta} - \frac{\rho v^2}{r} =$$

$$-\frac{\partial p}{\partial r}+\frac{1}{r}\frac{\partial}{\partial r}\left[2r\mu_{eff}\frac{\partial u}{\partial r}-\frac{2}{3}r\mu_{eff}\,\nabla\cdot\vec{\mathbf{V}}\right]$$
$$-\frac{\mu_{eff}}{r}\left[\frac{2u}{r}-\frac{2}{3}\nabla\cdot\vec{\mathbf{V}}\right]+\frac{1}{\delta}\frac{\partial}{\partial\eta}\left[\mu_{eff}\left(\frac{\partial w}{\partial r}+\frac{1}{\delta}\frac{\partial u}{\partial\eta}\right)\right] \qquad (2)$$

(iii) Conservation of circumferential momentum:

$$\frac{1}{\delta}\frac{\partial(\rho v\delta)}{\partial t}+\frac{1}{r}\frac{\partial(\rho ruv)}{\partial r}+\frac{1}{\delta}\frac{\partial(\rho w_1 v)}{\partial\eta}+\frac{\rho uv}{r}=$$
$$\frac{1}{r^2}\frac{\partial}{\partial r}\left[r^3\mu_{eff}\frac{\partial}{\partial r}\frac{v}{r}\right]+\frac{1}{\delta}\frac{\partial}{\partial\eta}\left[\frac{\mu_{eff}}{\delta}\frac{\partial v}{\partial\eta}\right] \qquad (3)$$

(iv) Conservation of axial momentum:

$$\frac{1}{\delta}\frac{\partial(\rho w\delta)}{\partial t}+\frac{1}{r}\frac{\partial(\rho ruw)}{\partial r}+\frac{1}{\delta}\frac{\partial(\rho w_1 w)}{\partial\eta}=$$
$$-\frac{1}{\delta}\frac{\partial p}{\partial\eta}+\frac{1}{r}\frac{\partial}{\partial r}\left[r\mu_{eff}\left(\frac{\partial w}{\partial r}+\frac{1}{\delta}\frac{\partial u}{\partial\eta}\right)\right]$$
$$+\frac{1}{\delta}\frac{\partial}{\partial\eta}\left[\frac{2\mu_{eff}}{\delta}\frac{\partial w}{\partial\eta}-\frac{2}{3}\mu_{eff}\,\nabla\cdot\vec{\mathbf{V}}\right] \qquad (4)$$

(v) Conservation of Energy:

$$\frac{1}{\delta}\frac{\partial(\rho C_pT\delta)}{\partial t}+\frac{1}{r}\frac{\partial(\rho ruC_pT)}{\partial r}+\frac{1}{\delta}\frac{\partial(\rho w_1 C_pT)}{\partial\eta}=$$
$$\frac{1}{r}\frac{\partial}{\partial r}\left[rk_{eff}\frac{\partial T}{\partial r}\right]+\frac{1}{\delta}\frac{\partial}{\partial\eta}\left[\frac{k_{eff}}{\delta}\frac{\partial T}{\partial\eta}\right] \qquad (5)$$

(vi) Conservation of Species:

$$\frac{1}{\delta}\frac{\partial(\rho\omega_k\delta)}{\partial t}+\frac{1}{r}\frac{\partial(\rho ru\omega_k)}{\partial r}+\frac{1}{\delta}\frac{\partial(\rho w_1\omega_k)}{\partial\eta}=$$
$$\frac{1}{r}\frac{\partial}{\partial r}\left[r\rho D_k\frac{\partial\omega_k}{\partial r}\right]+\frac{1}{\delta}\frac{\partial}{\partial\eta}\left[\frac{\rho D_k}{\delta}\frac{\partial\omega_k}{\partial\eta}\right] \qquad (6)$$

where

$$\nabla\cdot\vec{\mathbf{V}}=\left[\frac{1}{r}\frac{\partial(ur)}{\partial r}+\frac{1}{\delta}\frac{\partial w}{\partial\eta}\right] \qquad (7)$$

In the above equations, ρ, u, v, w, T and p are the density weighted ensemble averaged density, radial velocity, circumferential velocity, axial velocity, temperature and pressure respectively; C_p is the specific heat and k_{eff} is the effective turbulent thermal conductivity; ω_k is the mass fraction of a species k, defined as $\omega_k = \frac{m_k}{m}$ where m_k is the mass of the species k and m

is the total mass in the region of interest; D_k is the diffusivity of the species k into the remaining mixture; δ is the instantaneous distance between the cylinder head and piston face, t, r and η are the time, radial and axial coordinates respectively; w_1 is the axial velocity of the fluid relative to grid lines, and μ_{eff} is the effective turbulent viscosity, equal to the sum of the laminar viscosity μ_l and turbulent viscosity μ_t. $\vec{\mathbf{V}}$ is the velocity vector, defined as

$$\vec{\mathbf{V}} = u\vec{i} + v\vec{j} + w\vec{k}$$

$\vec{i}$, $\vec{j}$ and $\vec{k}$ being the unit vectors along radial, circumferential and axial directions respectively. The model used for determining the turbulent viscosity is described in a later section.

The sum of the mass fractions of all species at any given region is unity, and so, it is sufficient that if N species are present, we solve for $(N-1)$ species. The mass fraction of the Nth species can directly be obtained by subtracting the remaining $(N-1)$ species mass fraction from unity. In the present work, we consider only three species, namely, fuel, air and the products. The mass fractions of fuel and air are solved for, and that of the product is obtained by simple deduction.

The density of individual species is computed using the ideal gas equation of state for the respective species using the local temperature and cylinder pressure, and the density of the mixture is calculated using the mass fractions of the species. This can be done based on the assumption that the gases are ideal and they form an ideal mixture.

Boundary Conditions

The boundaries for the present problem can be divided into four broad categories – the stationary solid walls, the moving solid wall, the inlet boundary and the outlet boundary. On the stationary solid walls, the velocities are zero, i.e., no-slip condition is applied. The moving solid wall, i.e., piston face, has only the axial velocity equal to the instantaneous piston velocity. All the walls are assumed to be adiabatic, and so, zero-flux condition is used for the energy equation, on the solid walls. They are impervious to mass, and so, zero-flux conditions are applied to species conservation equations too, on the solid walls.

The inlet velocities are determined from the global part as described earlier. The inlet velocity so obtained would only have a radial component. This is converted to the actual velocity vector by using the input information available on the inlet direction, and its inclination with respect to the radius vector in the $r-\theta$ and $r-z$ planes. In other words, the input inclination angles are used to compute the inlet circumferential velocity or swirl velocity, and the inlet axial velocity. The species mass fractions, temperature and density boundary conditions are determined from the global part [15].

The outlet velocity is assumed to have only its axial component. This velocity is computed in the global part from the mass flow rates through the exhaust port. The details of how the boundary conditions for density, temperature and species mass fractions are computed have been described in detail elsewhere [15].

Turbulence Model

Turbulence has been modelled as an additional diffusivity added to the flow by eddy action. This kind of a model is a simple case of eddy viscosity models, in which the additional viscosity due to the turbulent eddies is assumed to be uniform throughout the computational domain at any given instant, and varies with time accordingly as the length and time scales deciding the viscosity change.

The model used in the present work is a minor modification over the one proposed by Mansour [17] and used by Ikegami et al. [18]. The models used by the above workers assumes the eddy viscosity to be a constant with space and time, for a given engine size and speed. Ikegami [18] takes the mean piston velocity $\overline{w}_p$ as the velocity scale and the bore diameter D as the length scale to determine the eddy viscosity μ_t. Schwarz et al [19] assume a constant eddy viscosity of 0.01 $kgm^{-1}s^{-1}$ for their computations. In the present work, it is proposed to use the instantaneous distance between the cylinder head and piston crown ($\delta(t)$) to be length scale and the instantaneous piston velocity ($w_p(t)$) to be the velocity scale defining the instantaneous eddy viscosity ($\mu_t(t)$). This is more logical since the turbulent eddies are likely to be broken up with compression, and turbulence is likely to die down to smaller values when the piston is near the top dead centre. The velocity of the piston contributes to the generation of further turbulence during expansion. The expressions for computing the eddy viscosity can be written as follows:

$$\begin{aligned} k(t) &= 0.74 w_p(t)^2 \\ \ell(t) &= 0.14\delta(t) \end{aligned} \qquad (8)$$

$$\mu_t(t) = \overline{\rho}(t) k(t)^{\frac{1}{2}} \ell(t) \qquad (9)$$

where $k(t)$ is the instantaneous turbulence kinetic energy and $\ell(t)$ is the instantaneous mixing length, $\overline{\rho}(t)$ being the global mean density.

Discretization and Solution Procedure

The governing equations written in the general conservation law form are integrated over the control volumes corresponding to the different flow field variables to obtain the algebraic equations for the values of the variables at the grid points. The continuity equation gives rise to the equations for flow field pressure and corrections to the velocity field obtained by solving the momentum equations. The solutions are iterated to convergence at every time step. The method of discretization, correcting velocities and updating for iterations follow the SIMPLER algorithm [20].

The algebraic equations derived by using the control volume approach give rise to a set of simultaneous equations for each flow field variable. The coefficient matrix has only five non-zero diagonals, and has been solved by the tridiagonal matrix algorithm for each row of grid points and line-iteration. Less than 20 iterations were sufficient for convergence of the system of equations at every time step. The convergence was decided based on the mass balance residue at the grid control volumes. The root mean square of these residuals going to less than 10^{-4} was the criterion for convergence. A time step value of 5° crank angle is used for the solution. The grid is a 24 x 24 non-uniform mesh, with very fine mesh near the solid walls and likely port locations.

Interfacing and Cross-checking Global and Local Parts

The global part is connected loosely to the local part, but uses information from local part results from the previous time level to derive the boundary conditions for the local part equations at the current time level. The local part, on the other hand, totally depends on the global part, and uses all boundary conditions provided by the global part of the problem. The cross-checking of the results obtained from the global and local parts can be made by comparing the masses of individual species and the total energy content of the cylinder computed at every time step from global and local parts. While the energy content computed from both parts agree excellently at all time steps, the masses of individual species and total charge mass differ by 3-4% during some parts of the cycle. This discrepancy decreases with finer grids, but is present even with the finest grid examined.

Sample Results

To illustrate the capabilities of the computer program *CARE*, a few sample results are presented here, showing the effect of engine speed on scavenging. Figure 2 shows flow field results for 400rpm, at 162° ATDC. Figure 3 shows the same results for 1200rpm. One can clearly see from the fresh charge mass fraction contours in the above two figures that the penetration of fresh charge into the cylinder is much deeper for the given instant at a low speed compared to that at a high speed. This is because the real time elapsed between inlet port opening (IPO, 135° ATDC) and 162° ATDC is thrice for the 400rpm case compared to the 1200rpm case.

The velocity vectors show a small recirculation zone close to the axis of the cylinder in the 400rpm case, where fresh gases have penetrated the cylinder contents, impinged on the central portion of the cylinder head and deflected inwards, i.e., towards the axis of the cylinder. The penetration is far from complete in the 1200rpm case, and so, no such recirculation zone is seen, except for the small pockets in the vicinity of the exhaust port. However, both cases show the presence of the "wall vortex" at the inlet port, which is much more elongated in the 400rpm case owing to the strong through flow compared to 1200rpm case.

Figure 4 illustrates the variation of gas exchange efficiencies with engine speed. At low speeds, longer time is available for fluid flow and so, residual gases are replaced with fresh gases to a greater extent at low speeds. Hence, scavenging efficiency is higher at low speeds. Moreover, a larger quantity of mass of fresh gases is delivered at lower speeds, and a larger proportion of it flows through the exhaust, resulting in a high scavenge ratio (delivery ratio) and low trapping efficiency at lower speeds. Filling of the cylinder or its charging with fresh gases is also better at lower speeds. Figure 4 illustrates all these facts well.

Conclusion

A computer program, *CARE*, has been developed for engine flow applications. The procedure involves splitting the problem into its global and local parts, and computing the flow field with the boundary conditions derived from the global part. The code has been successfully applied for the multidimensional modelling of scavenging in an axisymmetric uniflow scavenged two stroke engine. Sample results have been presented to demonstrate the capability of the computer program, and the results are found to be in accordance with realistic trends.

References

[1] Cloutman, L.D., Dukowicz, J.K., Ramshaw, J.D. and Amsden, A.A., 'CONCHAS – SPRAY : A computer code for reactive flows with fuel sprays', Los Alamos Scientific Laboratory report LA –9294 – MS, May 1982.

[2] Amsden, A.A., Ramshaw, J.D., O'Rourke, P.J. and Dukowicz, J.K., 'KIVA : A computer program for two- and three- dimensional fluid flows with chemical reactions and sprays', Los Alamos Scientific Laboratory report LA –10245 – MS, February 1985.

[3] Amsden, A.A., Butler, T.D., O'Rourke, P.J. and Ramshaw, J.D., 'KIVA – A comprehensive model for 2-D and 3-D engine simulations', SAE paper no.850554, 1985.

[4] Gosman, A.D., Johns, R.J.R. and Watkins, A.P., 'Development of prediction methods for in-cylinder processes in reciprocating engines' in *Combustion Modelling in Reciprocating Engines* ed. Mattavi, J.N. and Amann, C.A., Plenum Press, 1980.

[5] Gosman, A.D. and Harvey, P.S., 'Computer analysis of fuel - air mixing and combustion in an axisymmetric direct injection Diesel', SAE paper no.820036, 1982.

[6] Gosman, A.D., Tsui, Y.Y. and Watkins, A.P., 'Calculation of three-dimensional air motion in model engines', SAE paper no.840229, 1984.

[7] Errera, M.P., 'Numerical prediction of fluid motion in the induction system and the cylinder in reciprocating engines', SAE paper no.870594, 1987.

[8] Errera, M.P., Labbe, J. and Jerot, A., 'Three-dimensional numerical and experimental analysis of in-cylinder flow in an internal combustion engine', SAE paper no.880106, 1988.

[9] Diwakar, R., 'Multidimensional modelling of the gas exchange processes in a uniflow scavenged two-stroke diesel engine', *International Symposium on Flows in I.C. Engines III*, FED **28**, ed. Uzkan, T., Tiederman, W.G. and Novak, J.M., 1985.

[10] Diwakar, R., 'Three-dimensional modelling of in-cylinder gas exchange processes in a uniflow scavenged two-stroke engine', SAE paper no.870596, 1987.

[11] Kotlin, J.J., Dunteman, N.R., Chen, J. and Heilenback, J.W., 'The general motors EMD model 710G series of turbocharged two-stroke diesel engine', *ASME Diesel and Gas Engines Symposium*, Dallas, Texas, Feb. 17-20, 1985.

[12] Uzkan, T. and Hazelton, J.R., 'The influence of swirl on the fresh charge stratification in an I.C. engine combustion chamber', SAE paper no.860466, 1986.

[13] Uzkan, T., 'Analytically predicted effects of engine stroke on the scavenging characteristics of a uniflow two-stroke engine', in *On Flows in Internal Combustion Engines – IV*, ed. Uzkan, T., ASME – FED **65**, pp 67-73, 1988.

[14] Uzkan, T., The effects of engine speed on the scavenging characteristics of a two-cycle engine', Trans. ASME., *J. Engg. Gas Turbines and Power,* **110**, No. 3, pp 523 – 530, July 1988.

[15] Ravi, M.R., 'Multidimensional Modelling of Uniflow Scavenging in Two-stroke Reciprocating Piston Engines', Ph.D. Thesis, Indian Institute of Science, 1991 (in preparation).

[16] Ravi, M.R. and Marathe, A.G., 'On the importance of periodically repeating pressure histories to the gas exchange characteristics of a uniflow scavenged two-stroke engine', Trans. ASME., *J. Engg. Gas Turbines and Power,*(in review).

[17] Mansour, N.N., in *Flows in Internal Combustion Engines*, ed. Uzkan, T., p39, 1982.

[18] Ikegami, M., Komatsu, G. and Nishiwaki, K., 'Numerical simulation of flow in an engine cylinder (1st report, formulation and algorithm)', Paper No.248-27, *Bull. JSME*, **29**, No. 248, pp 508 - 515, February 1986.

[19] Schwarz, M.P., Thornton, G.J. and Gilbert, W., 'In-cylinder flow simulation for a motored two-stroke engine', *Computational Fluid Dynamics*, ed. de Vahl Davis, G. and Fletcher, C., pp 631 - 640, North Holland, 1988.

[20] Patankar, S.V., *Numerical Heat Transfer and Fluid Flow*, Hemisphere Publishing Co., 1980.

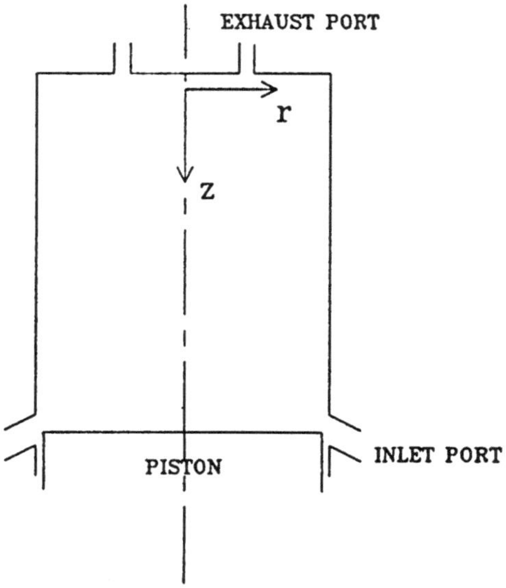

Figure 1. Geometry of the engine cylinder

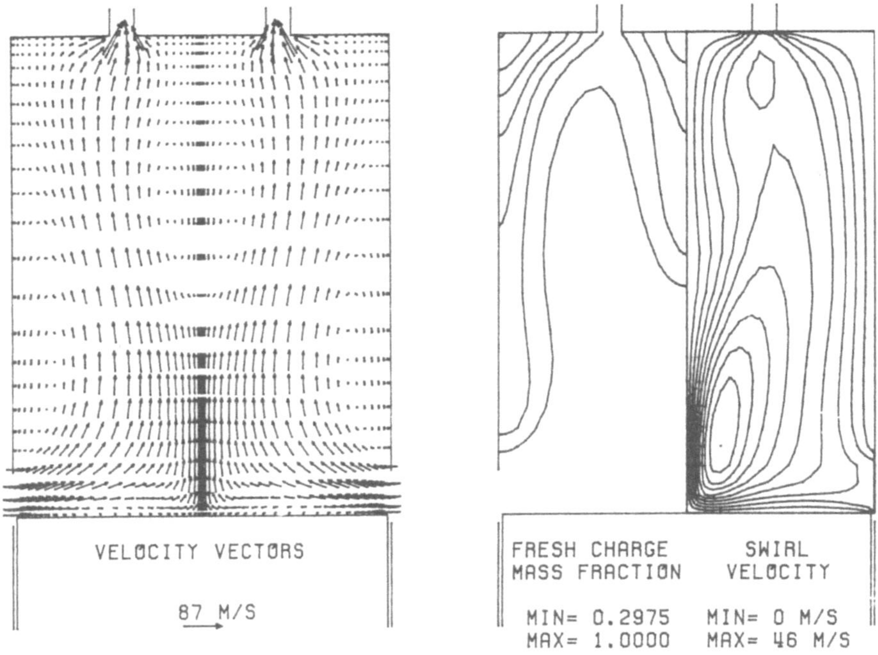

FIG. 2 FLOW FIELD RESULTS AT 162 DEG ATDC FOR 400 RPM

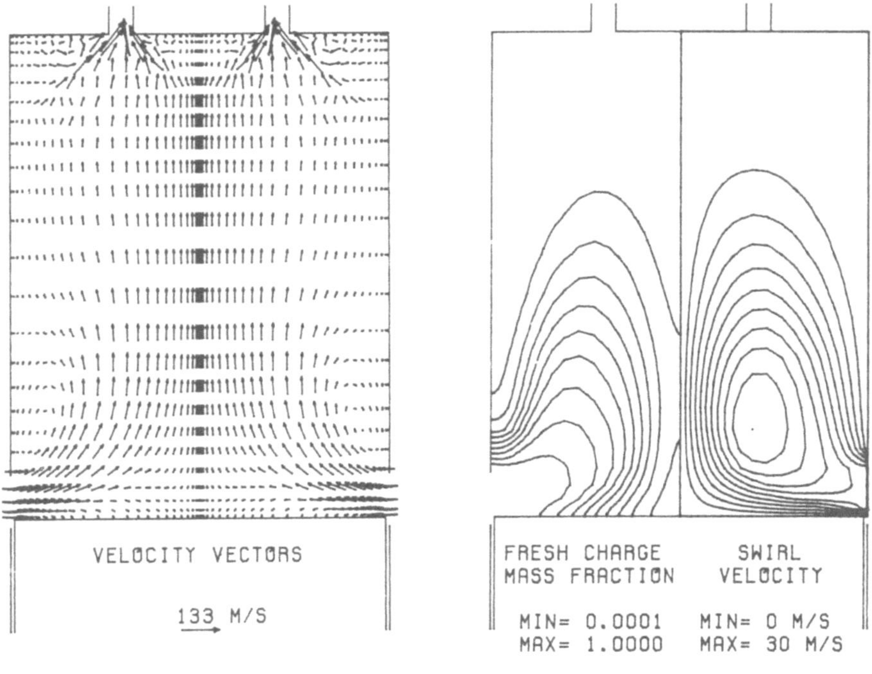

FIG. 3 FLOW FIELD RESULTS AT 162 DEG ATDC FOR 1200 RPM

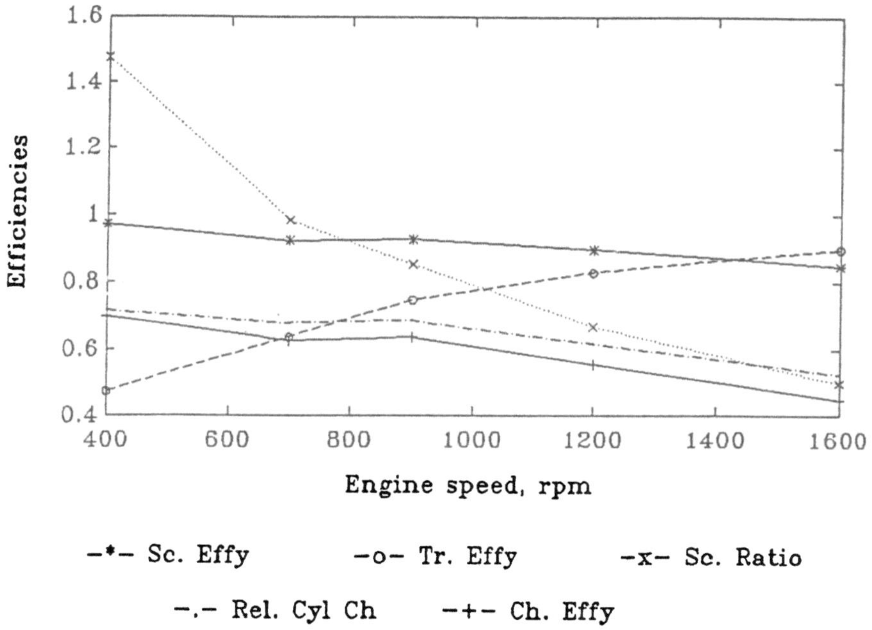

Figure 4. Effect of engine speed on efficiencies

Cubic Spline Technique for the Solution of Quasi-Linear Hyperbolic System in Non-Conservative Form

R. Shankar, P.C. Jain (*), T.V. Singh

Department of Mathematics, Indian Institute of Technology, Hauz Khas, New Delhi -110016, India

(*) Present Address: 3389 D/3, Vasant Kunj, New Delhi -110030, India

ABSTRACT

By using splitting up technique and cubic spline method, a numerical method for quasilinear-hyberbolic system of partial differential equations in non-conservative form has been developed. The method has been applied to study the unsteady motion of converging cylindrical shock problem in non-equilibrium gas dynamics. The effects of vibrational relaxation on such shock waves have been obtained by comparing these results with the results in ideal gas dynamics. It has been observed that effect of relaxation variable is to decrease the growth rate of shock strength when it is propagating towards the axis. Further, it has been observed that the non-equilibrium character of the gas is to delay the shock convergence with axis and thus to increase the convergence time.

INTRODUCTION

In general, the partial differential equations governing the hydrodynamic-time-dependent, invisid, compressible flow can be written as a hyperbolic system of equations in conservation form [1]. In several cases of physical interest one may not be able to get the system of equations in conservation form [2]. In the present investigation, we have developed a numerical method by using a combined approach of the spliting up and cubic spline technique to study the unsteady motion of converging cylindrical shock waves in non-equilibrium gas dynamics. The method for Burgers' equation has already been developed by us [3], which gives consistent and stable results for all Reynold's numbers. Infact, first we have here extended our analysis for system of nonlinear partial differential equations whichisin non-conservative form. Then, we have applied this method to study the unsteady motion of converging cylindrical shock problem in non-equilibrium gas dynamics. More specifically, by explaining the physical problem and the equations of motion together with the initial and boundary conditions and applying aforementioned method we have studied the effects of vibrational relaxation on such shock waves.

DIFFERENCE SCHEME

We consider the following system of equations

$$U_t + [F(U)]_x = S(U,x,t), \tag{1}$$

where the independent variables are x and t; U is a vector having p-components $(u_1,u_2,...,u_p)$; F(U) and S(U,x,t) are also p-dimensional vectors.

By using two-time-step splitting technique,eq. (1) can be written as

$$\frac{1}{2} U_t = - F_x , \tag{2}$$

$$\frac{1}{2} U_t = S . \tag{3}$$

Approximating the time derivative by forward differences and the space derivative by the first order derivative of the cubic spline function $S_n^*(x)$, interpolating F_i^n's (i = 1,2,...,K), equation (2) is written as

$$\frac{1}{k} (U_i^{n+1/2} - U_i^n) = -\theta_1 \underline{m}_i^{n+1/2} - (1-\theta_1)\underline{m}_i^n , \tag{4}$$

where U_i^n denote the value of U at ith and nth level in space and time respectively; k is the increment in time and $\theta_1 \in [0,1]$ is a cubic spline parameter. We have denoted the first derivative of $S_n^*(x_i)$ by $\underline{m}_i^n$.

Now, using the relation between $\underline{m}_i^n$ [4], we have

$$\underline{m}_{i+1}^n + 4\underline{m}_i^n + \underline{m}_{i-1}^n = \frac{3}{h}(F_{i+1}^n - F_{i-1}^n), \tag{5}$$

where h is the space mesh size.
Next, eliminating $\underline{m}$ from (4) and (5), we get

$$(1+\frac{1}{6}\delta_x^2)I(U_i^{n+1/2} - U_i^n) + \frac{1}{2}r\theta_1\mu\delta_x(F_i^{n+1/2} - F_i^n)$$

$$= -\frac{1}{2}r\mu\delta_x F_i^n, \tag{6}$$

where $\delta_x^2 U_i^n = U_{i+1}^n - 2U_i^n + U_{i-1}^n$,

$$\mu\delta_x U_i^n = U_{i+1}^n - U_{i-1}^n,$$

$r = \frac{k}{h}$ and I is the identity matrix of order p.
Further, approximating of eq.(3) yields

$$U_i^{n+1} = U^{n+1/2} + K\theta_2 S_i^{n+1} + K(1-\theta_2)S_i^{n+1/2}, \tag{7}$$

where $\theta_2 \in [0,1]$.
Now, we remove the nonlinearities in the difference schemes (6) and (7) by setting

$$F_i^{n+1/2} = F_i^n + A_i^n \,\delta U^n \tag{8}$$

$$S_i^{n+1} = S_i^{n+1/2} + B_i^n \,\delta U^{n+1/2}, \tag{9}$$

where $A = \frac{\partial F}{\partial U}$, $B = \frac{\partial S}{\partial U}$,

$\delta U^n = U_i^{n+1/2} - U_i^n$, and neglect terms of higher order differences.
Then, eqs. (6) and (7) take the form

$$[(1+\frac{\delta_x^2}{6})I + \frac{r\theta_1}{2}\mu\delta_x A_i^n]U_i^{n+1/2}$$

$$= (1+\frac{\delta_x^2}{6})I\,U_i^n - \frac{(1-\theta_1)}{2}r\mu\delta_x F_i^n \tag{10}$$

and

$$(I-k\theta_2 B_i^{n+1/2})U_i^{n+1} = U_i^{n+1/2} + k(1-\theta_2)S_i^{n+1/2} . \quad (11)$$

Eqs. (10) and (11) are the basic equations of the algorithm with appropriate initial and boundary conditions. It can be checked by expanding (10) and (11) about the point (in) that the scheme is of order $0(k^2 + h^4)$ for $\theta_1 = \theta_2 = \frac{1}{2}$; for other values of θ_1 and θ_2, the scheme is of order $0(k+h^4)$. To check the stability of the scheme we apply (10) and (11) to the linear form of (1) and by using Von-Naumann method of stability we see that this technique provides an unconditionally stable method for all $\theta_1, \theta_2 \in [1/2,1]$.

For computational work we have used this scheme augumented by a smoothing process (Vorozhtsov [5]) as given by

$$\bar{U}_i^{n+1} = \frac{1}{(2+\phi)} [\phi U_i^{n+1} + U_{i+1}^{n+1} + U_{i-1}^{n+1}] , \quad (12)$$

where $\bar{U}_i^{n+1}$ is a vector of smoothed solution and ϕ is empirical chosen parameter. An appropriate choice for ϕ would be some sufficiently large number which we have taken to be 200 in our computation.

THE EQUATION OF MOTION AND PHYSICAL PROBLEM

As pointed out in the introduction, we are here concerned to apply the aforementioned method to study the entitled shock problem in non-equilibrium gas dynamics.

The system of partial differential equations governing the one dimensional unsteady vibrationally relaxing gas in either slab, cylindrical or spherical symmetry is given by (Abarbanel and Goldberg [2], Vincenti [6]).

$$\frac{\partial U}{\partial t} + \frac{\partial[F(U)]}{\partial r} = S(U,r) , \quad (13)$$

where

$$U = \begin{bmatrix} R \\ M \\ E \end{bmatrix} \qquad F(U) = \begin{bmatrix} M \\ (\gamma_f - 1)E + \left(\frac{3-\gamma_f}{2}\right)\frac{M^2}{2} \\ \frac{M}{R}\left(\gamma_f E - (\gamma_f - 1)\frac{M^2}{2R}\right) \end{bmatrix}$$

$$S = \begin{bmatrix} 0 \\ p r^{\alpha-2} \\ -\psi^* \frac{r^{\alpha-1}}{\gamma_f - 1} \end{bmatrix}.$$

In the above equation

$$\psi^* = \frac{q-q^*}{\tau}\left(\frac{1-\gamma_f}{\gamma_f}\right)\frac{\rho^2}{p},$$

$$R = r^{\alpha-1}\rho, \quad M = r^{\alpha-1}\rho u, \quad E = r^{\alpha-1}(e+q\rho)$$

$$= r^{\alpha-1}\rho(\varepsilon + \frac{1}{2}u^2 + q)$$

and

$$p = \frac{\gamma_f - 1}{r^{\alpha-1}}\left(E - \frac{M^2}{2R}\right),$$

where $\rho, u, \varepsilon, p, t$ and r are respectively, the fluid density, particle velocity, internal energy per unit mass, pressure, time and space coordinate of symmetry. Thus when $\alpha = 1$, we have slab symmetry and $r = x$, for $\alpha = 2$, 3 we have respectively, cylindrical or spherical symmetry and r is the radius; γ_f is frozen gamma; q is the vibrational energy per unit mass; τ is the vibrational relaxation time and q^* is the vibrational energy per unit mass at equilibrium.

Here we consider the case of $\alpha = 2$, that is, the following physical problem, which we shall examine and solve numerically. A cylindrical diaphragm of radius r_0 separates the internal flow from the external one and also the flows on both the sides are at rest. The conditions in the two regions are homogeneous and the outer pressure and density are higher than the inner one. Initially the ratio between the outer and inner pressure as well as density are greater than one. Further, at $t = 0$ the diaphragm is removed and the flow begins inward, creating a shock wave in the gas under consideration. The motion of collapsing cylindrical shock is not uniform and the axis is a singular point of the solution.

In a shock tube, if the temperature (which is uniform) on the two sides is the same, then a shock propagates into the low pressure region followed by a contact surface and an expansion fan travels into the high pressure region. Infact, with increasing time the temperature field becomes non-uniform. Our main academic interest here is to study the behaviour of such shock waves.

EQUATIONS IN NON DIMENSIONAL FORM

In order to transform the system (13) into a non-dimensional form, we define the following dimensionless quantities

$$r' = \frac{r}{r_o}\ ,\ \rho' = \frac{\rho}{\rho_o}\ ,\ p' = \frac{p}{p_o}\ ,\ u' = \frac{u}{u_o}$$

$$t' = \frac{t}{t_o}\ ,\ u_o^2 = \gamma_f\ c_{fo}^2 = \frac{p_o}{\rho_o}\ ,\ t_o = \frac{r_o}{u_o}$$

$$e = p_o e',\ \beta = \frac{t_o}{p_o}\ (\ \frac{q-q^*}{\tau}\ )\ \frac{\rho^2}{p}\ u_o^2\ ,$$

where c_{f_o} is the initial speed of sound in the inner-region, r_o is the diaphragm radius, p_o and ρ_o are respectively the pressure and density in the inner region.

Now replacing the system (13) in non-dimensional form after dropping the primes for convenience sake, the new system is identical to the system (13).

INITIAL AND BOUNDARY CONDITIONS

Next, we consider the dimensionless scheme in which the diaphragm is 1; the internal density and pressure are also 1; i.e. $p_o = \rho_o = 1$; the radius is denoted by Kh and the following conditions are satisfied

$$p_j^o = \rho_j^o = 1 \quad \forall\ j < K$$

$$p_j^o = p^*,\ \rho_j^o = \rho^* \quad \forall\ j > K$$

$$p_k^o = \frac{p^*+1}{2} + \frac{h}{2}(p^*-1) \qquad (14)$$

$$\rho_k^o = \frac{\rho^*+1}{2} + \frac{h}{2}(\rho^*-1)$$

and $u_j^o = 0 \quad \forall\ j$.

We have also the boundary conditions

$$R_o^n = M_o^n = E_o^n = 0. \qquad (15)$$

In order to compute p, ρ, u and e at $r = 0$, we use the following [2] relations to calculate density and energy at $r = 0$

$$\rho_o^{n+1} = \rho_o^n + \rho_1^n - \rho_1^{n+1} - \frac{k}{h}(\rho_1^n + \rho_1^{n+1})(u_1^n + u_1^{n+1}), \qquad (16)$$

$$e_o^{n+1} = e_o^n + e_1^n - e_1^{n+1} - \frac{k}{h}(e_1^n + e_1^{n+1} + p_1^n + p_1^{n+1}) \times (u_1^n + u_1^{n+1}). \qquad (17)$$

The problem defined above is actually in the semi-infinite domain $0 \leq r < \infty$. In practice, however, one must specify a finite domain. It turns out that if the outer boundary is taken twice the diaphragm radius (i.e. $0 \leq r \leq 2$) then at $r = 2$ there is very small effect due to the shock collapsing and rebounding from the axis upto the time the rebounding shock returns to the diaphragm location. We calculate right hand boundary by using the expression

$$U_{2K}^{n+1} = U_{2K}^n - \frac{K}{2h}(3F_{2K}^n - 4F_{2K-1}^n + F_{2K-2}^n) + \frac{K^2}{2h^2}[3A_{2K}^n(F_{2K}^n - F_{2K-1}^n) + 4A_{2K-1}^n (F_{2K-1}^n - F_{2K-2}^n) + A_{2K-2}^n(F_{2K-2}^n - F_{2K-3}^n)]. \qquad (18)$$

Until the shock converges upon the axis $u(h,t) \leq 0$, then the convergence time t_c can be expressed as

$$t_c = \sum_{m=1}^{n+1} \Delta t_m , \tag{19}$$

where (n+1) is the least natural number for which $u_1^{n+1} > 0$.

NUMERICAL RESULTS

By using the initial conditions and space mesh size h = 1/50, we have studied the propagation of converging cylindrical shock in a vibrationally relaxing gas, the initial strength of which is 1.93. Fig. 1-3, show the plotting of numerical values obtained by applying our scheme for $\gamma_f = 9.0/7.0$. These figures show the distribution of pressure, density and velocity in a relaxing gas at time interval 0.2 respectively. In order to compare our results, with those which are previously available, we have also solved the equations (13) in ideal gas dynamics by aforementioned scheme and found our results in agreement with the existing results (Abarbanel and Goldberg [2]).

It is clear from Fig. 1 that the strength of the shock is seen to decrease from 1.93, while it is propagating towards the axis. We have also observed that the rise in pressure is very high but finite value when it touches the axis and reflected diverging shock appears. After shock converges with the axis, the pressure behind the reflected shock decreases as time increases. In Fig. 2, the distribution of density is similar to Fig. 1, except the appearance of the contact surface. Unlike the shock, the contact surface is far from being a discontinuity in pressure and velocity. The change in density appeared gradually and spreading over an increasing number of mesh points. The contact surface also propagates toward the axis behind the converging shock and is traversed by the diverging shock. From the distribution of velocity in Fig. 3, it is observed that while the converging shock decreases flow velocity from zero to a negative value, the diverging shock increases the velocity to a small positive value. The flow velocity at a fixed point behind the converging shock increases with time and decreases when diverging shock has passed. The effect of vibrational relaxation is felt on the strength of the shock. and also on the time taken by the shock to reach the axis. Figs. 4 and 5 show that the vibrational energy - causes to increase the growth rate of pressure and thus leading to a decrease in shock strength. It accelerates the shock convergence with axis and thus causing to decrease the convergence time.

From the above study we have concluded that a cylindrical shock wave existing in non-equilibrium gas dynamic flow decreases in strength while propagating towards the axis and when it converges with the axis, the diverging shock appears. By comparing our results with previous workers (Abarbanel and Goldberg [2]), we observed that the vibrational relaxation is to increase the growth rate of shock strength, when it is propagating towards the axis. Further, it is concluded that the effect of vibrational energy is to accelerate the shock convergence with axis and thus decreases the convergence time.

REFERENCES

1. Lax,P and Wendroff, B. Systems of Conservation laws Commun. Pure Appl. Math., Vol. 13, pp.217-237,1960.

2. Abarbanal, S. and Goldberg, M. Numerical Solution of Quasi-conservative Hyperbolic Systems - The Cylindrical Shock Problem, Journal of Comput. Phys., Vol. 10, pp. 1-21, 1972.

3. Jain, P.C., Shankar, R. and Singh, T.V. Numerical Technique for Solving Burgers' and Convective Diffusion Equations at Low and High Reynolds Numbers (Communicated for Publication).

4. Ahlberg, J.H., Nilson, E.N., Walsh,J.L. The Theory of Splines and their Applications,Academic Press Inc. (London) Ltd., 1967.

5. Vorozhtsov, E.V., On the Classification of Discontinuities by the Pattern Recognition Methods, Computers & Fluids, Vol. 18, No. 1, pp. 35-74, 1990.

6. Vincenti, W.G. and Charles, H. Kruger,Jr., Introduction to Physical Gas Dynamics, John Wiley & Sons, Inc. (USA), 1965.

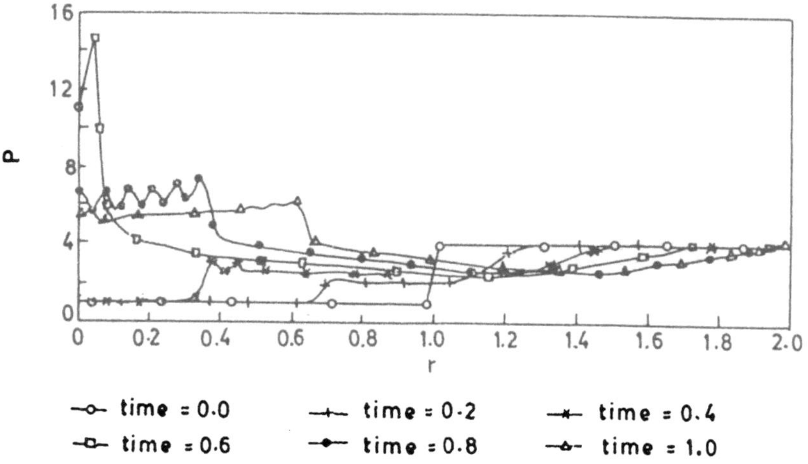

Fig.1 Distribution of pressure in a vibrationally relaxing gas flow at time intervals 0.2 for $\beta = 1.0$

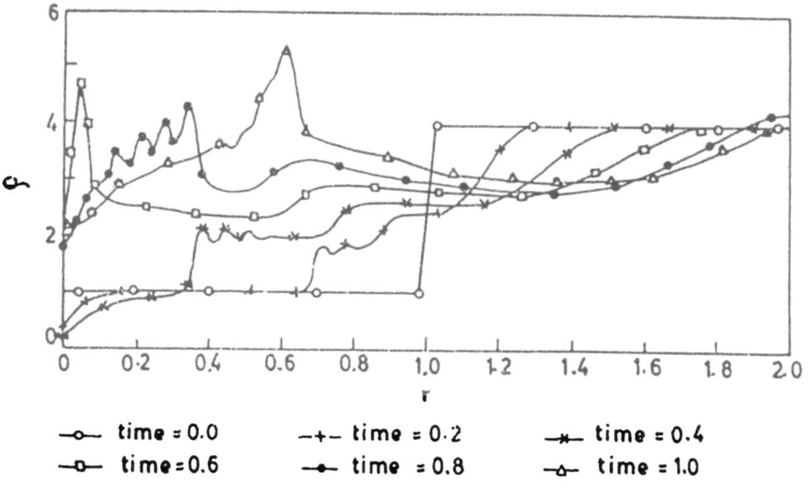

Fig.2 Distribution of density in a vibrationally relaxing gas flow at time intervals of 0.2 for $\beta = 1.0$

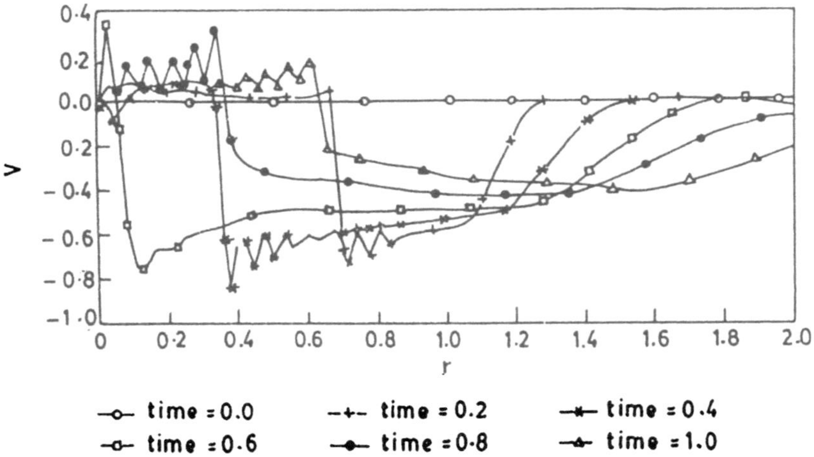

Fig·3 Distribution of velocity in a vibrationally relaxing gas flow at time intervals 0·2 for $\beta = 1.0$

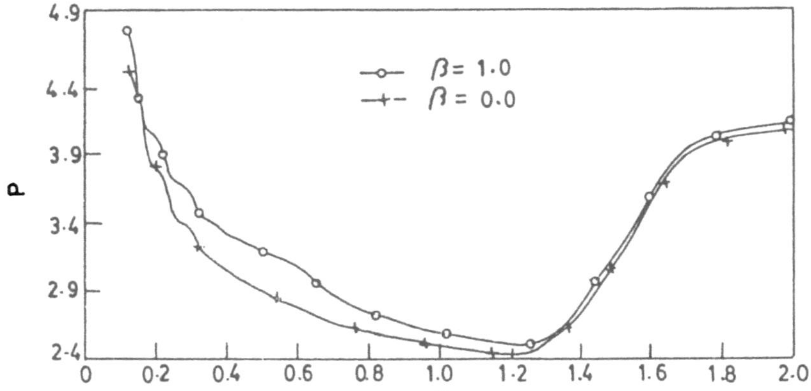

Fig·4 Effect of vibrational relaxation on the distribution of pressure at $t = 0.6$

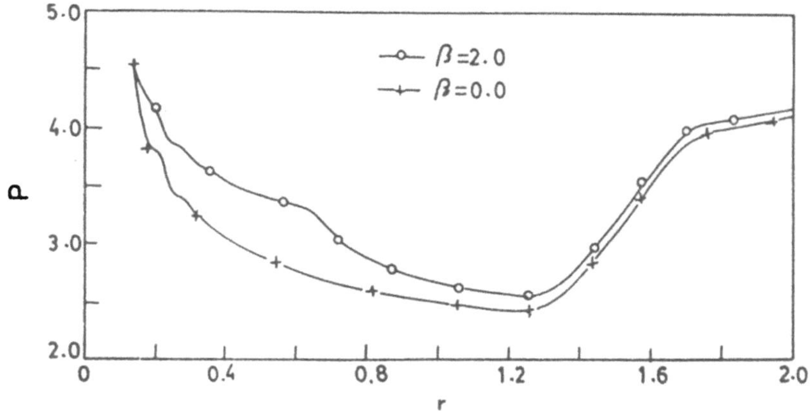

Fig. 5 Effect of vibrational relaxation on the distribution of pressure at t = 0.6

Anvarbek M. Meirmanov (Nowosibirsk, USSR)

The Stefan Problem

Translated from the Russian by Marek Niezgódka and Anna Crowley

1991. Approx. 300 pages. 17 x 24 cm. Cloth ISBN 3-11-011479-8

de Gruyter Expositions in Mathematics, Volume 3

Editors: *O. H. Kegel, V. P. Maslov, W. D. Neumann* and *R. O. Wells, Jr.*

The Stefan problem is one of the most classical free boundary problems of parabolic type. It arises from modelling phase-change phenomena, such as phase transitions between, for instance, liquid and solid states of a material. Since the appearance of Rubinstein's important monograph in 1967 this book provides the first systematic analysis of Stefan-type problems. The existence of classical solutions for the multidimensional Stefan problem was a long-standing problem. The author's approach to the solution of this problem forms the central part of the book. Together with a complete constructive proof of the classical solvability (local in time), examples of critical developments showing the lack of global-in-time solutions in the general setting are given. A careful analysis of the intrinsic structure of the free boundaries, that can have the form of mushy zones is provided. For one-dimensional Stefan problems, qualitative properties of global classical solutions are studies, including an analysis of their asymptotic behaviour and periodicity. The role of compatibility conditions is discussed.

This book is addressed to advanced students and research mathematicians, in particular applied mathematicians and engineers.

Contents:

Preliminaries · Classical solution of the multidimensional Stefan problem · Existence of the classical solution to the multidimensional Stefan problem on an arbitrary time interval · Lagrange variables in the multidimensional one-phase Stefan problem · Classical solution of the one-dimensional Stefan problem for the homogeneous heat equation · Structure of the generalized solution to the one-phase Stefan problem. Existence of a mushy region · Time-periodic solutions of the one-dimensional Stefan problem · Approximate approaches to the two-phase Stefan problem · Appendix: I. G. Götz, A. M. Meirmanov, Modelling of binary alloy crystallization.

Bei Fragen zur Produktsicherheit wenden Sie sich bitte an:
If you have any questions regarding product safety,
please contact:

Walter de Gruyter GmbH
Genthiner Straße 13
10785 Berlin
productsafety@degruyterbrill.com